Spark
零基础实战

王家林　孔祥瑞　等编著

化学工业出版社
·北京·

Spark 是业界公认的近几年发展最快、最受关注度的一体化多元化的大数据计算技术，可以同时满足不同业务场景和不同数据规模的大数据计算的需要。

本书首先通过代码实战的方式对学习 Spark 前必须掌握的 Scala 内容进行讲解并结合 Spark 源码的阅读来帮助读者快速学习 Scala 函数式编程与面向对象完美结合的编程艺术，接着对 Hadoop 和 Spark 集群安装部署以及 Spark 在不同集成开发环境的开发实战作出了详细的讲解，然后基于大量的实战案例来讲解 Spark 核心 RDD 编程并深度解密 RDD 的密码，并且通过实战的方式详解了 TopN 在 Spark RDD 中的实现，为了让读者彻底了解 Spark，本书用了大量的篇幅详细解密了 Spark 的高可用性、内核架构、运行机制等内容。

Spark 零基础实战这本书定位于零基础的学员，也可以作为有一定大数据 Hadoop 经验的从业者以及对大数据非常感兴趣的学生的第一本 Spark 入门书籍。

图书在版编目（CIP）数据

Spark 零基础实战/王家林等编著. —北京：化学工业出版社，2016.10（2018.7重印）
ISBN 978-7-122-28017-6

Ⅰ.①S… Ⅱ.①王… Ⅲ.①数据处理软件 Ⅳ.①TP274

中国版本图书馆 CIP 数据核字（2016）第 215244 号

责任编辑：王淑燕　宋湘玲　　　　装帧设计：关　飞
责任校对：宋　玮

出版发行：化学工业出版社（北京市东城区青年湖南街 13 号　邮政编码 100011）
印　　装：大厂聚鑫印刷有限责任公司
787mm×1092mm　1/16　印张 20　字数 503 千字　　2018 年 7 月北京第 1 版第 2 次印刷

购书咨询：010-64518888（传真：010-64519686）　售后服务：010-64518899
网　　址：http://www.cip.com.cn
凡购买本书，如有缺损质量问题，本社销售中心负责调换。

定　　价：68.00 元

前言

大数据已经成为公众流行词多年，不管在业界还是在其他领域都紧随时代发展的潮流，人类社会的发展已经进入到大数据时代。我们生活的今天大到互联网公司，小到每一个个体或者每一台移动设备其每天都会产生海量的新数据，那么对于这些海量数据的处理就面临着巨大的考验，而在此过程中为了满足业务需要，各类技术如雨后春笋般出现并得到 IT 企业的实践应用和发展，就应对海量数据的处理框架而言，于 2006 年诞生的 Hadoop，使业界掀起一股热潮，它改变了企业对数据的存储、处理和分析的过程，加速了大数据的发展，形成了自己的极其火爆的技术生态圈，并受到非常广泛的应用。而 Spark 在 2009 年最初来源于伯克利大学的研究性项目，于美国加州大学伯克利分校的 AMPLab 实验室诞生，2010 年实现开源并在 2013 年成为 Apache 的基金孵化器项目并在不到一年的时间成为其的顶级项目，在短短几年的时间内获得极速发展并被各大互联网公司应用于实际项目中以实现海量数据的处理，可以毫不夸张地讲 Spark 是大数据时代发展的必然产物，势必会成为最好的大数据处理框架之一。

根据 Stackoverflow 调查显示 Spark 是 2016 年 IT 从业者获得薪水最高的技术之一，从事 Spark 开发的 IT 人员年薪达到 125000 美元，从事 Scala 开发的 IT 人员年薪同从事 Spark 的 IT 人员保持一致的水平，可见 Spark 已经成为开发人员在大数据领域收入最好的技术之一。了解 Spark 或者读过 Spark 源码的人都知道 Spark 主要是 Scala 语言开发的，而 Scala 语言是一门面向对象与函数式编程完美结合的语言。因此本书主要以零基础实战掌握 Spark 运行机制为导向详细对 Scala 的语法和重要知识点进行实战讲解，通过源码对 Spark 的内核架构进行剖析并赋予实战案例来引导读者能够在掌握 Scala 的同时快速进行 Spark 的深入学习。

Spark 基于 RDD（弹性分布式数据集）实现了一体化、多元化的大数据处理体系，是目前最热门最高效的大数据领域的计算平台。Spark 框架完美融合了 Spark SQL、Spark Streaming、MLLib、GraphX 子框架，使得各子框架之间实现数据共享和操作，强大的计算能力和集成化使得 Spark 在大数据计算领域具有得天独厚的优势，因此国际上很多大型互联网公司均使用 Spark 实现海量数据的处理，如国内的 BAT 等，有超过千台节点组成的集群高效快速地处理每日生成的海量数据。

Spark 在大数据处理领域的迅猛发展，给了很多互联网公司高效处理海量数据的方案，但是 Spark 人才的稀缺使得很多公司心有余而力不足，以至于不能将企业的生产力量化提高成了很多企业面临的主要问题，大数据 Spark 工程师的缺少直接制约了很多公司的转型和发展，在此情况下本书以零基础实战为主导，由基础部分细致地带领初学者从零基础入门直到深入学习 Spark。本书主要面向的对象是预从事大数据领域的初学者、高校学生以及有一定大数据从事经验的工作人员等。

本书以零基础实战Spark为主导，首先实战讲解Scala基础语法与定义、Scala面向对象编程、Scala函数式编程、Scala类型系统模式匹配、Scala因式转换以及Scala并发编程等，基本包含了Scala所有重要内容并且每一部分在实战的同时配合Scala在Spark源码中的应用带领读者彻底理解Scala语言的艺术。其次对Spark源码在不同方式下的编译进行演示，对Hadoop不同模式的集群搭建、Spark集群的搭建以及Spark在IDE、IntelliJ IDEA不同工具下的实战和源码导入均作了细致讲解，相信通过源码的学习和不同工具下对Spark程序的开发实战可以帮助读者对Spark有一个全面的理解和认识，并能快速投入到实际开发中。然后对Spark中最为重要的核心组件之一RDD（弹性分布式数据集）进行了详细地解析，并介绍Spark Master HA的4种策略，解密如何通过ZOOKEEPER这种企业经常使用的策略来保证Spark Master HA。本书最后一部分综合讲解了Spark内核架构以及实战解析Spark在不同模式下的运行原理。希望本书可以引领读者细致高效地学习Spark框架，并成为企业渴求的Spark高端人才。

参与本书编写的有王家林、孔祥瑞等。本书能顺利出版，离不开化学工业出版社的大力支持与帮助，包括进度把控、技术服务、排版等各个方面，在此表示诚挚地感谢。

在本书阅读过程中，如发现任何问题或有任何疑问，可以加入本书的阅读群（QQ：302306504）提出讨论，会有专人帮忙答疑。同时，该群中也会提供本书所用案例代码。

如果读者想要了解或者学习更多大数据的相关技术，可以关注DT大数据梦工厂微信公众号DT _ Spark及QQ群437123764，或者扫描下方二维码咨询，也可以通过YY客户端登录68917580永久频道直接体验。王家林老师的新浪微博是http://weibo.com/ilovepains/欢迎大家在微博上进行互动。

由于时间仓促，书中难免存在不妥之处，请读者谅解，并提出宝贵意见。

王家林 2016.8.13 于北京

目 录

第 1 章 Scala光速入门 1

第 2 章 Scala面向对象彻底精通及Spark源码阅读 19

第 3 章 Scala函数式编程彻底精通及Spark源码阅读 30

第 4 章 Scala模式匹配、类型系统彻底精通与Spark源码阅读 58

第 5 章 Scala隐式转换等彻底精通及Spark源码阅读 66

第 16 章 Spark内核架构解密 248

第 17 章 Spark运行原理实战解析 260

第1章 Scala光速入门

Spark 以其极快的发展速度和大规模数据集优秀的处理计算能力而被业界所推崇，很多人对 Spark 优秀的处理能力产生了极强的兴趣，很多公司为了更好地处理大规模数据的业务选择了 Spark 进行开发。要学习 Spark 就需要学习 Scala，因为 Spark 主要是由 Scala 语言编写的，如果想更好地学习 Spark 的内核架构和底层实现，就需要熟练掌握 Scala 编程语言。本章通过 Scala 基础与语法以及 Array、Map 等数据结构的丰富的代码实战带领大家快速掌握 Scala 入门需要的知识。

1.1 Scala 基础与语法入门实战

1.1.1 Scala 基本数据类型

Scala 包括 8 种常用数据类型：Byte、Char、Short、Int、Long、Float、Double 和 Boolean。基本数据类型取值范围及示例如表 1-1 所示。

表 1-1 基本数据类型

基本类型	取值范围或示例
Byte	范围在－128～127
Char	范围 U＋0000～U＋FFFF
Short	范围在－32768～32767
Int	范围－2147483648～2147483647
Long	范围－9223372036854775808～9223372036854775807
Float	单精度浮点数例如：0.0f 或者 0.0F 若没有 f 或者 F 后缀则是 Double 类型
Double	双精度浮点数例如：0.11
Boolean	布尔类型表示为 true 和 false

String 也是 Scala 基本数据类，String 属于 java.lang 包，其余的所有基本类型都是 Scala 的包成员，例如 Int 是 scala.Int。Scala 中基本类型包和 java.lang 包是默认导入的，

可以直接使用。

1.1.2 Scala 变量声明

Scala 通过 var 和 val 来声明变量。

【例 1-1】 定义 val 类型变量

val 类型变量定义后不可以重新赋值。val 类型变量相当于 Java 中的 final 修饰的变量。

```
scala> val i = 10
i: Int = 10                    //Scala根据类型推断,推断出i的类型为Int并把10赋值给i
scala> println(i)
10

scala> i = 11                  //i定义为val类型,是不可变的。重新给i赋值会报如下错误
<console>:8: error: reassignment to val
       i = 11
       ^
scala> val i,j = 10            //Scala允许一次定义多个变量
i: Int = 10
j: Int = 10

scala> val a = 1f              //定义Float类型的数据，同样适用了类型推断
a: Float = 1.0

scala> val i :Int = 10         //显示指定数据类型
i: Int = 10

scala> val name:String = "zhangsan"  //显示指定定义的字符串类型
name: String = zhangsan
```

【例 1-2】 定义 var 类型变量

var 类型变量确定以后值可以修改。

```
scala> var x = 5               //定义可变变量x，类型推断出是整数
x: Int = 5

scala> x = 10                  //重新赋值变量x
x: Int = 10

scala> var  sex = "male"       //定义字符串类型变量
sex: String = male

scala> sex = "famale"          //给sex重新赋值
sex: String = famale

scala> val name:String = "zhangsan"    //var类型变量定义时显示指定变量类型
name: String = zhangsan
```

1.1.3 算术操作符介绍

Scala 有丰富的内置运算符，Scala 的操作符也是函数，可以通过“对象．运算符（）”

方式使用。下面分别介绍不同类型的运算符，如表 1-2 所示。

表 1-2　算术运算符

运算符	描述	运算符	描述
+	两个数相加	/	两个数相除
−	两个数相减	%	模运算,整数除法后的余数
*	两个数相乘		

【例 1-3】 算术运算符实战

Scala 是一种面向函数和面向对象相结合的语言。其算术操作符是函数的另一种展现。如表 1-3 所示。

```
scala> 1 + 2                    //+ 在scala中也是对象的函数，这里可以写成1.+(2),下面- * / 等都和+ 一样

res0: Int = 3
scala> 5- 1                     //两个数相减
res2: Int = 4
scala> 3 * 5                    //两个数相乘
res3: Int = 15
scala> 6 / 3                    // 两个数相除
res4: Int = 2
scala> 5 % 2                    //求余
res5: Int = 1
```

表 1-3　关系运算符

运算符	描　述
==	判断两个数的值是否相等,如果是的话那么条件为真,否则为假
! =	判断两个数的值是否相等,如果值不相等,则条件为真,否则为假
>	判断左边数的值是否大于右边数的值,如果是的话那么条件为真,否则为假
<	判断左边数的值是否小于右边数的值,如果是的话那么条件为真,否则为假
>=	判断左边数的值是否大于或等于右边数的值,如果是的话那么条件为真,否则为假
<=	判断左边数的值是否小于或等于右边数的值,如果是的话那么条件为真,否则为假

【例 1-4】 关系运算符实战

所有的关系判断返回值都为 Boolean 类型。如表 1-4 所示。

```
scala> 3 == 2                   //判断3是否等于2，返回false

res6: Boolean = false
scala> 3 != 2                   //判断3是否不等于2
res7: Boolean = true
scala> 3 > 1                    //判断3是否大于1
res8: Boolean = true
scala> 3 < 1                    //判断3是否小于1
res9: Boolean = false
```

表 1-4 逻辑运算符

运算符	描　述	示　例
&&	所谓逻辑与操作。如果两个操作数为非零则条件为真	(A && B)为 false
||	所谓的逻辑或操作。如果任何两个操作数是非零则条件变为真	(A || B)为 true
!	所谓逻辑非运算符。使用反转操作数的逻辑状态。如果条件为真，那么逻辑非操作符作出结果为假	！(A && B)为 true

【例 1-5】 逻辑运算符实战

逻辑运算符返回结果为 Boolean 类型。

```
scala> true && false                //逻辑运算符的两侧是结果为Boolean类型的表达式
res11: Boolean = false
scala> true && false
res11: Boolean = false
scala> 3 > 1 && 2 != 1              //如果左侧表达式为假，则不再判断右侧表达式
res13: Boolean = true
scala> 3 > 1 || 2 != 1              //如果左侧表达式为真，则不再判断右侧表达式
res14: Boolean = true
scala> !(3 > 1)                     //！否定表达式的结果
res15: Boolean = false
```

位运算符适用于位和位操作，主要包括三种：&，｜，和^，其真值表如表 1-5 所示。

表 1-5 位运算符

p	q	p & q	p | q	p ^ q
0	0	0	0	0
0	1	0	1	1
1	1	1	1	0
1	0	0	1	1

【例 1-6】 位运算符实战

位运算是指按照计算数据在计算机中的二进制进行 0、1 的按位“与”、“或”等的操作。如表 1-6 所示。

```
scala> 0 & 0                //按位与运算
res2: Int = 0
scala> 0 | 0                //按位或运算
res3: Int = 0
scala> 0 ^ 0                //按位异或运算
res4: Int = 0
scala> 0 & 1
res5: Int = 0
scala> 0 | 1
res6: Int = 1
scala> 0 ^ 1
res7: Int = 1
scala> 1 & 1
res8: Int = 1
scala> 1 | 1
res9: Int = 1
scala> 1 ^ 1
res10: Int = 0
```

表 1-6　赋值运算符

运算符	描　述	示　例
=	简单地将右侧的值赋于左侧	C=A+B 将分配 A+B 的值到 C
+=	先执行加法再赋值	C+=A 相当于 C=C+A
-=	先执行减法再赋值	C-=A 相当于 C=C-A
*=	先执行乘法再赋值	C *=A 相当于 C=C *A
/=	先执行除法再赋值	C/=A 相当于 C=C/A
%=	先执行模量去余再赋值	C%=A 相当于 C=C%A
<<=	左移位并赋值	C<<=2 等同于 C=C<<2
>>=	向右移位并赋值	C>>=2 等同于 C=C>>2
&=	按位与赋值	C&=2 等同于 C=C&2
^=	按位异或赋值	C^=2 等同于 C=C^2
|=	按位或赋值	C|=2 等同于 C=C|2

【例 1-7】 赋值运算符实战

```
scala> var a = 3 + 8                //3 + 8的结果值赋值给a
a: Int = 11
scala> a += 3                       //a加3，Scala中没有a++表达式，该表达式可以写成 a = a+ 3
scala> println(a)                   //打印出当前a的值
14
scala> a-= 5                        //a减5，Scala中没有a- -表达式,该表达式可以写成 a = a - 5
scala> println(a)
9
scala> a *= 2                       //a乘以2，该表达式可以写成a = a * 2
scala> println(a)
22
```

1.1.4　条件语句

条件语句语法结构的两种方式如下。

方式一：写在同一行。

if（布尔表达式）x else y

方式二：写在不同行并且有多条语句组成的结构体。

```
if（布尔表达式）{
    ……
} else {
    ……
}
```

Scala 中条件语句是有返回值的。可以将其赋值给某个变量，类似于 Java 中的三目运算符。

【例 1-8】 if 语句实战

if 语句通过判断条件是否为真来执行后面的结构体。

```
scala> val i =  if (3 > 1) 3 else 1
i: Int = 3                          //3赋值给了变量i
```

【例 1-9】 if 语句中变量的应用

定义一个变量 x，并将这个变量应用到 if 语句的条件判断中。

```
scala> var x = 10                          //定义一个中间变量
x: Int = 10
scala> if (x < 20 ){                       //如果x<20为true，则执行大括号中的语句
     |     println("This is if test")
     | }
This is if test
```

【例 1-10】 将 if 语句赋值给一个变量

Scala 中 if 语句是有返回值的，可以将 if 语句的返回结果赋值给一个变量。

```
scala> val result = if(x == 10 ){ 10 }       //将if语句的结果赋值给result
result: AnyVal = 10
scala> println(result)
10
```

【例 1-11】 if 与 else 结合使用

将 if-else 的返回结果赋值给 result。

```
scala> val x = 10
x: Int = 10
scala> val result = if(x > 5) x else 5      //if-else类似于java中的三目运算符
result: Int = 10
scala> println(result)
10
scala>
```

Scala 中的条件语句在其他的一些结构中被称为守卫。例如：for 循环中、match 模式匹配中等。

1.1.5 循环

Scala 中循环结构有 for、while、do-while 循环等。

（1）for 循环

for 循环语句可以重复执行某条语句，直到某个条件得到满足，才退出循环。

for 语法：

```
for（变量<－集合）
    {
     循环体
    }
```

【例 1-12】 利用 for 循环打印出 1 到 10

当满足后面的循环条件时执行 for 循环体，示例代码如下：

```
scala> for(i <- 1 to 10) print(i + " ")      // 1 to 10 生成Range(1, 2, 3, 4, 5, 6, 7, 8, 9, 10)
1 2 3 4 5 6 7 8 9 10
```

【例 1-13】 利用 for 循环打印出 1 到 9，to 包含尾部，until 不包含尾部

```
//1 until 10 生成的是Range(1, 2, 3, 4, 5, 6, 7, 8, 9)
scala> 1 until 10
res0: scala.collection.immutable.Range = Range(1, 2, 3, 4, 5, 6, 7, 8, 9)

scala> for(i <- 1 until 10) print(i + " ")
1 2 3 4 5 6 7 8 9
```

【例 1-14】 for 循环中使用守卫

for 循环中添加过滤条件 if 语句，示例代码如下：

```
scala> for(i <- 1 to 10 if i % 2 == 0) print(i + " ")              //在循环中添加守卫
2 4 6 8 10
scala> for(i <- 1 to 10 if i % 2 == 0;if i != 2) print(i + " ")     //多个过滤条件之间用分号隔开
4 6 8 10
```

【例 1-15】 嵌套枚举

嵌套枚举是指由多个循环条件的 for 结构体，第二个循环条件依次执行第一个循环条件的每一个变量值，循环次数为 n ∗ m 次。

```
scala> for(i <- 1 to 5 ; j <- 1 to 5 if i % 2 == 0) print(i * j + " ")
2 4 6 8 10 4 8 12 16 20
```

【例 1-16】 利用 yield 关键字返回一个新集合

```
scala> val v1 = for(i <- 1 to 5) yield i
v1: Scala.collection.immutable.IndexedSeq[Int] = Vector(1, 2, 3, 4, 5)

```

（2）while 循环

while 语句为条件判断语句，利用一个条件来控制是否要继续反复执行循环体中的语句。语法如下：

```
while  （条件表达式）
  {
     循环体
}
```

【例 1-17】 while 循环实战

```
scala> while( i <= 5)
    | {
    | i += 1
    |   print(i + " ")
   |
    | }
1 2 3 4 5 6
```

在 for 和 while 中都没有用到 break 和 continue，这是因为在 scala 中没有 break 和 continue 两个关键字，continue 可以通过 if 条件语句来控制是否要向下执行，而 break 语句在 scala 中有特殊的实现，如例 1-18。

【例 1-18】 break 语句的特殊实现

```
import scala.util.control.Breaks._
import scala.util.Random
object ControlStatementBreak {
```

```
    def main(args:Array[String]) {
      breakable {
        while (true) {
          val r = new Random()
          val i = r.nextInt(10)
          println("i==" + i)
          if (i == 5) {
            break
          }
        }
      }
    }
}
```

在这段代码中首先 import 了 scala. util. control. Breaks. _ ，这样就可以使用 breakable，和 break 语句了。在 while 语句的外面用 breakable 语句块包围了，然后在需要 break 的地方调用 break 方法，这里的 break 不是关键字，而是一个 scala 的方法，这个方法会抛出异常，从而中断循环。

(3）do-while 循环

do-while 循环语句和 while 语句类似，两者的区别是 while 语句先判断条件是否成立再执行循环体；而 do-while 循环语句先执行一次循环语句，再判断条件是否成立，这样 do-while 循环中的循环体至少要执行一次；

语法如下：

do ｛

循环体

｝ while（条件表达式）

【例 1-19】 do-while 循环

```
var count = 4
scala> do{
     |   println("count value is:"+count)
     |   count- =  1
     | }while(count > 0)
count value is:4
count value is:3
count value is:2
count value is:1
```

例 1-19 中，println（" count value is：" ＋count）至少会执行一次，然后判断 count 是否大于 0，因为 count 的初始值为 4，经过每次循环减 1 后，count 的大小不再满足 count＞0 这个条件，循环结束。

注意：严格意义上讲，Scala 中 for 被称为表达式，因为它又返回结果，可以赋值给某个变量；而 while 和 do-while 就是循环结构，它们的返回结果为 Unit 类型。

1.1.6 异常控制

Scala 的异常处理有两种方式：捕获异常和抛出异常。

(1) 抛出异常使用关键字 throw

例如：

throw new IllegalArgumentException

(2) 捕获异常

例如：

```
try {
  ……
} catch {
  case ex：NullPointException=>
  case…
}
```

Scala 中异常处理语法结构：

```
try {
  ……
} catch {
  case…
  case…
}
//finally 语句不一定会有，如果有一定会执行
finally {
  println（" 通常用来释放资源!!!" ）
}
```

【例 1-20】 文件读取

当文件不存在时输出文件不存在异常。

```
object ReadFileTest {
  def main(args: Array[String]) {
    try{
      val file = Source.fromFile("F://2.txt")      //当文件不存在时抛出FileNotFoundException异常
      val lines = file.getLines()
      for(content <- lines){
        println(content)
      }
    }catch {
      case ex:FileNotFoundException => println("输入的文件不存在" + ex)
      case ex:Exception => println(ex)
    }finally {
      println("通常用来释放资源！！！")
    }
  }
}
```

（1）当文件存在时运行结果

hi 你好
通常用来释放资源!!!（finally 模块执行结果）

（2）当文件不存在时运行结果

输入的文件不存在 java. io. FileNotFoundException：F：\ 2. txt（系统找不到指定的文件。）

通常用来释放资源!!!（finally 模块执行结果）

1.2 Scala 中 Array、Map 等数据结构实战

1.2.1 定长数组和可变数组

Scala 中数组分为可变和不可变数组。默认情况下定义的是不可变（Immutable）数组；若定义可变数组，需要显示导入包 import Scala. collection. mutable. ArrayBuffer。

【例 1-21】 定义一个长度为 2 的字符串类型数组，默认为不可变数组

```
scala> val arrStr = Array("Scala","Spark")
arrStr: Array[String] = Array(Scala, Spark)
```

【例 1-22】 定长数组

定义长度为 10 的整数类型的数组，初始值为 0，示例代码如下：

```
scala> val arrInt = new Array[Int](10)
arrInt: Array[Int] = Array(0, 0, 0, 0, 0, 0, 0, 0, 0, 0)
```

【例 1-23】 定义可变数组 ArrayBuffer，需要显示导入包

```
scala> import Scala.collection.mutable.ArrayBuffer
import Scala.collection.mutable.ArrayBuffer
scala> val arrBufInt = ArrayBuffer[Int]()
arrBufInt: Scala.collection.mutable.ArrayBuffer[Int] = ArrayBuffer()
```

1.2.2 数组常用算法

【例 1-24】 数组常用算法

数组是非常重要的数据结构，常用算法示例代码如下：

```
arrStr(0) = "Storm"                    //返回结果Array("Storm","Spark")
scala> arrStr.mkString(",")            //指定分隔符
res23: String = Storm,Spark
scala> arrStr.toBuffer                 //将不可变数组转换成可变数组
res27: Scala.collection.mutable.Buffer[String] = ArrayBuffer(Storm, Spark)

遍历数组：
for(i <-0 until arrStr.length) println(arrStr(i))
或者
for(elem <- arrStr) println(elem)      //类似于java的增强for
Array(1,2,3,4,5).sum                   //求和
```

以下是可变数组特有的操作：

```
scala>  arrBufInt += 1                          // 用+=在尾端添加元素
res25: arrBufInt.type = ArrayBuffer(1)
scala>  arrBufInt += (2,3,4,5)                  //  同时在尾端添加多个元素
res26: arrBufInt.type = ArrayBuffer(1, 2, 3, 4, 5)
 arrbuff1++= arrBuff2                           // 可以用 ++=操作符追加任何集合
scala> arrBufInt ++= Array(6,7,8)
res28: arrBufInt.type = ArrayBuffer(1, 2, 3, 4, 5, 6, 7, 8)
arrBufInt .trimEnd(2)                           // 移除最后的2个元素
arrBufInt .remove(2)                            //  移除arr(2+1)个元素
arrBufInt .remove(2,4)                          //从第三个元素开始移除4个元素
val arrBufInt =arrbuff1.toArray                 //将数组缓冲转换为Array
```

1.2.3 Map 映射

Scala 映射就是键值对的集合 Map。默认情况下，Scala 中使用不可变的映射。如果想使用可变集，必须明确地导入 scala. collection. mutable. Map 类。

定义映射：val mapCase=Map（" CHINA" ->" BEIJING"，" FRANCE" ->" PARIS" ）

【例 1-25】 构造不可变映射

Scala 定义的默认 Map 类型为不可变类型，示例代码如下。

```
scala> val bigData = Map("Scala"- > 35,"Hadoop"->30,"Spark"- >50)
bigData: Scala.collection.immutable.Map[String,Int] = Map(Scala-> 35, Hadoop -> 30, Spark-> 50)
scala> bigData("Scala")                        //获取key为Scala的value为35
res29: Int = 35
scala> bigData.contains("Hadoop")              //判断映射中是否包含key为Hadoop的键值对,返回值为Boolean类型
res30: Boolean = true
scala> bigData.getOrElse("Spark",70)           //若映射中存在key为Spark的键值对，则返回对应的 value，若不存在，则返回默认值70
res31: Int = 50
scala> bigData.mkString("{",",","}")           //添加分隔符
res32: String = {Scala -> 35,Hadoop -> 30,Spark -> 50}
scala> bigData.drop(2)                         //以角标0开始，返回角标为2的元素
Map(Spark -> 50)
res33: Scala.collection.immutable.Map[String,Int] = Map(Spark -> 50)
```

【例 1-26】 构造可变映射

构造可变映射时，需要显示导入 Map 包或者在 Map 前面添加全路径，示例代码如下。

```
scala> val bigDataVar = Scala.collection.mutable.Map("Scala"-> 35,"Hadoop"->30,"Spark"->50)
bigDataVar: Scala.collection.mutable.Map[String,Int] = Map(Hadoop -> 30, Spark -> 50, Scala -> 35)

bigDataVar("Spark") = 100        //更新key为Spark的键值对，若对应key不存在则添加对应键值对

bigDataVar += ("Kafka"->69)      //添加键值对
bigDataVar -= ("Kafka"->69)      //删除键值对
//遍历映射
scala> for((k,v) <- bigData) println(k+" " + v)          //打印key和value
Scala 35
Hadoop 30
Spark 50

scala> for(k <- bigData.keySet) println(k)               //只打印key
Scala
Hadoop
Spark
```

1.2.4 Tuple 元组

元组是由两个小括号包住的数据集合，里面可以存放不同的类型。定义元组如下。

```
val tuple1 = (1,"Scala","Hadoop",3)
tuple1对应的结构是tuple1: (Int, String, String, Int) = (1,Scala,Hadoop,3)
```

【例 1-27】 访问元组中的元素

```
scala> val tuple1 = (1,2,3,4,"hello")
tuple1: (Int, Int, Int, Int, String) = (1,2,3,4,hello)
println(tuple1._1)          //打印出第一个元素
println(tuple1._2)          //打印出第二个元素
                            //访问更多元素通过._*
```

1.2.5 List 列表

Scala 中列表是非常类似于数组，这意味着，一个列表的所有元素都具有相同的类型，但有两个重要的区别。首先，列表是不可变的，这意味着一个列表的元素可以不被分配来改变。第二，列表表示一个链表，而数组是平坦的。

【例 1-28】 List 中存放 T 类型数据

具有 T 类型的元素的列表的类型被写为 List［T］。

```
scala> //定义存放String类型的List
scala> val fruit: List[String] = List("apples", "oranges", "pears")
fruit: List[String] = List(apples, oranges, pears)
scala>
scala> // 定义Int类型的List
scala> val nums: List[Int] = List(1, 2, 3, 4)
nums: List[Int] = List(1, 2, 3, 4)
scala>
scala> //定义空List
scala> val empty: List[Nothing] = List()
empty: List[Nothing] = List()
```

```
scala>
scala> // 二维List列表
scala> val dim: List[List[Int]] =
     |   List(
     |     List(1, 0, 0),
     |     List(0, 1, 0),
     |     List(0, 0, 1)
     |   )
dim: List[List[Int]] = List(List(1, 0, 0), List(0, 1, 0), List(0, 0, 1))
scala>
```

【例 1-29】 列表定义的另一种方式

所有的列表可以使用两种基本的构建模块来定义，一个无尾“Nil”和“::”，这有明显的缺点。Nil 也代表了空列表。例 1-28 中的列表可以定义如下。

```
scala> // 定义String类型列表
scala> val fruit = "apples" :: ("oranges" :: ("pears" :: Nil))
fruit: List[String] = List(apples, oranges, pears)
scala>
scala> // 定义Int类型列表
scala> val nums = 1 :: (2 :: (3 :: (4 :: Nil)))
nums: List[Int] = List(1, 2, 3, 4)
scala>
scala> // 定义空列表
scala> val empty = Nil
empty: scala.collection.immutable.Nil.type = List()
scala>
scala> // 二维列表
scala> val dim = (1 :: (0 :: (0 :: Nil))) :: (0 :: (1 :: (0 :: Nil))) :: (0 :: (0 :: (1 :: Nil))) :: Nil
dim: List[List[Int]] = List(List(1, 0, 0), List(0, 1, 0), List(0, 0, 1))
scala>
```

【例 1-30】 列表基本操作

列表中的 head、tail、isEmpty 等方法操作实战。

```
scala> class Test {
     |   def ops {
     |     val fruit = "apples" :: ("oranges" :: ("pears" :: Nil))
     |     val nums = Nil
     |     println( "Head of fruit : " + fruit.head )                 // head方法去除List中的第一个值
     |     println( "Tail of fruit : " + fruit.tail )                 //tail去除List中的除了第一个值外的其他所有值，返回一个List列表

     |     println( "Check if fruit is empty : " + fruit.isEmpty )    //判断List是否为空
     |     println( "Check if nums is empty : " + nums.isEmpty )      //判断List是否为空
     |   }
     | }
defined class Test
```

```
warning: previously defined object Test is not a companion to class Test.
Companions must be defined together; you may wish to use :paste mode for this.

scala> val t = new Test                                    //实例化定义的Test对象
t: Test = Test@68df8c6

scala> t.ops                                               //调用ops方法
Head of fruit : apples
Tail of fruit : List(oranges, pears)
Check if fruit is empty : false
Check if nums is empty : true
scala>
```

【例 1-31】 串联列表

用“:::”运算符或列表“List. ::: ()”方法或“List. concat ()”方法来添加两个或多个列表。

```
scala> class ListConcatTest {
     |   def concatTest = {
     |     val fruit1 = "apples" :: ("oranges" :: ("pears" :: Nil))
     |     val fruit2 = "mangoes" :: ("banana" :: Nil)
     |     //用::: 将两个或更多的List表串联起来
     |     var fruit = fruit1 ::: fruit2
     |      println( "fruit1 ::: fruit2 : " + fruit )
     |     //通过调用.:::()方法串联列表
     |     fruit = fruit1.:::(fruit2)
     |     println( "fruit1.:::(fruit2) : " + fruit )
     |     //通过调用.concat()方法调用列表
     |     fruit = List.concat(fruit1, fruit2)
     |     println( "List.concat(fruit1, fruit2) : " + fruit  )
     |   }
     | }
defined class ListConcatTest

scala>  val lct = new ListConcatTest
lct: ListConcatTest = ListConcatTest@2e19b30

scala>  lct.concatTest
fruit1 ::: fruit2 : List(apples, oranges, pears, mangoes, banana)
fruit1.:::(fruit2) : List(mangoes, banana, apples, oranges, pears)
List.concat(fruit1, fruit2) : List(apples, oranges, pears, mangoes, banana)
scala>
```

1.2.6 Set 集合

Scala 集合是不包含重复元素的集合。Set 集合分为不可改变的和可变的。可变和不可变的对象之间的区别在于，当一个对象是不可变的，对象本身不能被改变。默认情况下，Scala 中使用不可变的集。如果想使用可变集，必须明确地导入 scala. collection. mutable. Set 类。如果想同时使用可变和不可变的集合，那么可以继续参考不变的集合，但可以参考可变设为 mutable. Set。

【例 1-32】 定义不可变 Set 集合

Set 集合中元素不会重复，下面是初始化空集 Set 和有初始值的 Set，示例代码如下。

```
scala> // 定义空Set结合
scala> var s : Set[Int] = Set()
s: Set[Int] = Set()
scala>
scala> // 定义Int类型Set集合
scala> var s : Set[Int] = Set(1,3,5,7)
s: Set[Int] = Set(1, 3, 5, 7)
scala> // 或者
scala> var s = Set(1,3,5,7)
s: scala.collection.immutable.Set[Int] = Set(1, 3, 5, 7)
scala> // 定义String类型Set集合
scala> val str:Set[String] = Set("Scala","Spark","Hadoop")
str: Set[String] = Set(Scala, Spark, Hadoop)
scala>
```

【例 1-33】 Set 基本操作

Set 集合的 head、tail、isEmpty 基本操作，Set 集合的操作和 List 基本一致。

```
scala> class SetTest {
     |   def setMethodTest {
     |    val book = Set("Scala", "Spark", "Hadoop")
     |    val nums: Set[Int] = Set()
     |
     |    println( "Head of book : " + book.head )
     |    println( "Tail of book : " + book.tail )
     |    println( "Check if book is empty : " + book.isEmpty )
     |    println( "Check if nums is empty : " + nums.isEmpty )
     | }
     |}
defined class SetTest

scala> val st = new SetTest
st: SetTest = SetTest@1e27bb89

scala> st.setMethodTest
Head of book : Scala
Tail of book : Set(Spark, Hadoop)
Check if book is empty : false
Check if nums is empty : true
scala>
```

1.2.7 Scala 集合方法大全

Scala 集合的方法大全如表 1-7 所示。

表 1-7　Scala 集合方法

SN	描　述
def +(elem:A):Set[A]	往集合中添加类型为 A 的新元素
def -(elem:A):Set[A]	删除集合中类型为 A 的元素
def contains(elem:A):Boolean	判断集合中是否包含 A 元素
def &(that:Set[A]):Set[A]	返回两个集合的交集
def &~(that:Set[A]):Set[A]	返回两个集合的差集
def +(elem1:A,elem2:A,elems:A *):Set[A]	通过添加传入指定集合的元素创建一个新的不可变集合
def ++(elems:A):Set[A]	连接此不可变的集合使用另一个集合到这个不可变的集合的元素
def -(elem1:A,elem2:A,elems:A *):Set[A]	通过移除传入指定集合的元素创建一个新的不可变集合
def addString(b:StringBuilder):StringBuilder	将不可变集合的所有元素添加到字符串缓冲区
def addString(b:StringBuilder,sep:String):StringBuilder	将不可变集合的所有元素添加到字符串缓冲区，并使用指定的分隔符
def apply(elem:A)	测试如果一些元素被包含在这个集合
def count(p:(A)=>Boolean):Int	计算满足指定条件的集合元素个数
def copyToArray(xs:Array[A],start:Int,len:Int):Unit	复制不可变集合元素到数组
def diff(that:Set[A]):Set[A]	计算这组和另一组的差异
def drop(n:Int):Set[A]	返回除了第 n 个的所有元素
def dropRight(n:Int):Set[A]	返回除了最后 n 个元素的新集合
def dropWhile(p:(A)=>Boolean):Set[A]	丢弃满足谓词的元素最长前缀
def equals(that:Any):Boolean	equals 方法的任意序列，比较该序列到某些其他对象
def exists(p:(A)=>Boolean):Boolean	判断不可变集合中指定条件的元素是否存在
def filter(p:(A)=>Boolean):Set[A]	返回此不可变的集合满足谓词的所有元素
def find(p:(A)=>Boolean):Option[A]	找到不可变的集合满足条件的第一个元素
def forall(p:(A)=>Boolean):Boolean	查找不可变集合中满足指定条件的所有元素
def foreach(f:(A)=>Unit):Unit	应用一个函数 f 在这个不可变的集合中的所有元素
def head:A	返回此不可变的集合的第一个元素
def init:Set[A]	返回除了最后的所有元素
def intersect(that:Set[A]):Set[A]	计算此 set 和另一组 set 之间的交叉点
def isEmpty:Boolean	测试此集合是否为空
def iterator:Iterator[A]	创建一个新的可迭代对象中的所有元素迭代器
def last:A	返回最后一个元素
def map[B](f:(A)=>B):immutable. Set[B]	通过给定的方法将所有元素重新计算
def max:A	查找最大的元素
def min:A	查找最小元素
def mkString:String	显示此不可变的集合字符串中的所有元素
def mkString(sep:String):String	使用分隔符将集合中所有元素作为字符串显示
def product:A	返回此不可变的集合中所有数字元素的积
def size:Int	返回此不可变的集合元素的数量

续表

SN	描　述
def splitAt(n:Int):(Set[A],Set[A])	返回一对不可变的集合组成这个不可变的集合的前 n 个元素，以及其他元素
def subsetOf(that:Set[A]):Boolean	如果集合中含有子集返回 true，否则返回 false
def sum:A	返回此不可变的集合的所有元素的总和
def tail:Set[A]	返回一个不可变集合中除了第一元素之外的其他元素
def take(n:Int):Set[A]	返回前 n 个元素
def takeRight(n:Int):Set[A]	返回最后 n 个元素
def toArray:Array[A]	将集合转换为数字
def toBuffer[B>:A]:Buffer[B]	返回缓冲区，包含了不可变集合的所有元素
def toList:List[A]	返回一个包含此不可变的集合中的所有元素的列表
def toMap[T,U]:Map[T,U]	这种不可变的集合转换为映射
def toSeq:Seq[A]	返回一个包含此不可变的集合的所有元素的序列
def toString():String	返回对象的字符串表示

1.2.8 综合案例及 Spark 源码解析

【例 1-34】 Spark 源码

```
//这个函数是Spark的资源分配最重要的函数
 private def schedule(): Unit = {
//首先要判断是否为ALIVE状态，否则直接返回
  if (state != RecoveryState.ALIVE)  { return }
//把所有等待的driver均匀打乱
   val shuffledWorkers = Random.shuffle(workers)
//利用for循环遍历shuffledWorkers，并判断WorkerState状态是否为ALIVE
    for (worker <- shuffledWorkers if worker.state == WorkerState.ALIVE) {
     //遍历waitingDrivers
      for (driver <- waitingDrivers) {
     //判断可用的内存是否大于driver所需要的内存并判断可用CPU个数是否大于driver需要
        CPU个数
 if (worker.memoryFree >= driver.desc.mem && worker.coresFree >= driver.desc.cores) {
       //把worker和driver信息发出去
        launchDriver(worker, driver)
      //从waitingDrivers 移除分配资源的driver
        waitingDrivers- = driver
      }
    }
  }
  startExecutorsOnWorkers()
}
```

1.3 小结

本章主要介绍了 Scala 基础知识如基本数据类型、变量声明、算术操作符、条件语句、循环和异常控制，以及常用数据结构类型如 Array、Map 等。然后针对每个知识点都给出了示例代码。最后通过结合 Spark 源码分析 Scala 的基础应用。

第2章

Scala面向对象彻底精通及Spark源码阅读

在Scala看来，一切皆是对象，对象是Scala的核心。对象、类、特质（Trait）用来定义Scala的API和库，它们是Scala为面向对象编程提供的基础机制。Scala面向对象涉及的重要概念有class、object、构造器、字段、方法、抽象类、接口等，这些概念很多跟Java面向对象中的概念相似，但也有一些Scala独特之处。下面开始Scala面向对象的详细解析以及最后通过Spark源码中Scala代码的解析加深对大家对Scala面向对象的理解。

2.1 Scala面向对象详解

Scala一切皆对象，对象是Scala的核心。

对象、类、特质（Trait）用来定义Scala的API和库，它们是Scala为面向对象编程提供的基础机制。

2.1.1 Scala中的class、object初介绍

Scala中class定义和Java类似，区别是在Scala中，类名不是必须和文件名相同且一个文件中可以定义多个class。同时Scala的class中不存在静态成员，Java中的静态成员由Scala中的object对象替代，当object对象名和class名相同时则称object为class的伴生对象。

Scala类定义语法：

```
class ClassName {
  ……
}
```

关键字class是Scala中所有类定义的必须修饰符，包含普通类、抽象类、样例类等。定义class时，类体（就是大括号以及其里面的语句）可有可无。

【例 2-1】 Scala 类的定义实战

定义普通类 HiScala，使用关键字 class，示例代码如下。

```
//定义普通类HiScala，一般类名的第一个字母都大写
class HiScala{
    private val name:String = "zhangsan"
    def sayName(){
        println(name)
    }
    def getName  = name
}
```

定义了类名为 HiScala 的一个对象。他有一个私有的 name 属性和两个方法（函数定义在类中称之为方法）。

```
val hiScala = new HiScala
hiScala.sayName                    //打印出"zhangsan"
hiScala.getName                    //返回res0: String = zhangsan
```

在这里无法直接访问 name 属性，因为它是私有字段。

singleton 对象定义语法：

```
object ObjectName {
    ……
}
```

object 定义的对象为单例对象，就是在整个程序中只有这么一个实例。对于同名的 object 和 class 对象互称为伴生对象和伴生类，伴生对象和伴生类必须在同一个文件中。

【例 2-2】 object 定义实战

object 本身有默认的 constructor，这个 constructor 只会被调用一次。

```
object Person {
    println("Scala")
    val age = 10
    def getAge = age
}
```

调用一：

```
scala> Person.getAge
Scala
res4: Int = 10
```

调用二：

```
scala> Person.getAge
res5: Int = 10
```

显然第二次没有再调用 println（" Scala" ）。

注意：①伴生类和伴生对象一定要在同一个文件中；②伴生对象可以访问伴生类的私有成员。

【例 2-3】 继承

继承的关键字是 extends，Scala 是单继承多实现的（用 with 关键字混入多个特质）。

```
//定义一个Animal的抽象类（下面章节介绍），Animal抽象类中定义了抽象方法run和已实现的eat方法
abstract class Animal {
```

```
  def eat = {
    println("Eat food!!!")
  }
  def run
}
//Cat继承Animal类，实现了run方法并重写了eat方法
class Cat extends Animal{
  override def eat: Unit = {
    println("Eat mouse!!!")
  }
  override def run: Unit = {
    println("Running...")
  }
}
class Dog extends Animal{
  override def run: Unit = {
    println("Dog is running...")
  }
}

object AnimalTest{
  def main(args: Array[String]) {
    v al c = new Cat
    c.eat
    c.run
    val d = new Dog
    d.eat
    d.run
  }
}
//运行结果
Eat mouse!!!          //因为在Cat类中重写了eat方法，所以当Cat被实例化并调用的时候，输出重写
                        后的信息
Running...
Eat food!!!           //因为继承了Animal类，所以也继承了其方法run
Dog is running
```

【例 2-4】 子类不能重写父类中被 final 修饰的方法和属性

对于 final 修饰的方法和属性，子类不能进行重写，示例代码如下。

```
scala> class AttributeOverride {
     | val name:String = "Jack"
     | final val sex:String = "male"          //  定义final类型的sex属性
     | final def sayHi = {                    //  定义final类型的方法sayHi
     |   println("Hi Scala...")
     | }
     |}
defined class AttributeOverride

```

```
scala> class SubAttribute1 extends  AttributeOverride{
     |  override val name:String = "Michael"          //重写父类中的final属性并重新赋值为Michael
     | }
defined class SubAttribute1

scala> class SubAttribute1 extends  AttributeOverride{
     |  override val name:String = "Michael"
     |  override val sex:String = "female"
     | }
//当重写父类的final类型的属性时报如下错误
<console>:10: error: overriding value sex in class AttributeOverride of type String;
 value sex cannot override final member
         override val sex:String = "female"
                      ^
//当重写父类的final类型的方法时报如下错误
scala> class SubAttribute1 extends  AttributeOverride{
     |  override val name:String = "Michael"
     |  override def sayHi = {
     |    println("Hi SubAttribute1...")
     |  }
     | }
<console>:11: error: overriding method sayHi in class AttributeOverride of type => Unit;
 method sayHi cannot override final member
         override def sayHi = {
                      ^

```

2.1.2 主构造器与辅助构造器

【例 2-5】 Scala 默认构造器在类上

```
class Person{
   ......
}                    //默认无参构造器
class Person(name:String){
  println(this.name)
}                    //默认带一个字符串参数的构造器，创建对象实例时会打印出name的值

```

【例 2-6】 构造器重载（辅助构造器）

```
class Person(name:String){
    println(this.name)
    def this(name:String,age:Int) {
      this(name)
      println(name+" : "+age)
    }
    def this(name:String,age:Int,sex:String) {
      this(name,age)
      println(name+" : "+age)
    }
}
```

重载构造器，第一行必须是默认构造器或者其他构造器。

【例 2-7】 构造器重载（多参数默认构造器）

```
class Student1(name:String,age:Int,sex:String) {      //定义多参数默认构造器类
  println(name + " : " + age + " : " + sex)
}
object Student1{
  def main(args: Array[String]) {
    val s = new Student1("zhangsan",25,"male")       //创建实例时，必须传入指定的参数
  }
}
//运行结果
zhangsan : 25 : male
```

2.1.3 类的字段和方法彻底精通

【例 2-8】 定义一个 Person 类

```
class Person{
  var name :String = "zhangsan"
  var age = 30
  val sex = "famale"
  def sayHi = println("Hi!!!")
  def increase(age:Int):Int = this.age + age
}
```

在访问对象属性时可以如下定义。

```
val p = new Person
p.name                    //返回zhagnsan
p.name = "lisi"           //将姓名改为lisi
p.sex                     //返回famale
```

在 Person 类中定义了 var 和 val 属性，var 类型的属性 Scala 会帮助生成 public 的 getter 和 setter 方法；val 类型属性因为不可变，所以只有 getter 方法。如果显示定义成 private 类型属性，Scala 会生成私有的对应的 getter 和 setter 方法，外界无法访问。

```
p.sayHi                   //返回Hi!!!
```

Scala 默认的方法是 public 的，若在方法的前面加上 private，则对象实例无法访问该方法。若定义无参方法时不带括号，则调用时也不能写；若定义无参方法带括号，则调用时可写可不写，推荐不写。

【例 2-9】 复写 setter

```
class Person{
   private var myName = "Flink"
   def name = this.myName
   def name_=(newName:String) {
       myName = newName
       println("Hi : "+myName)
   }
}
val person = new Person
person.name = "Scala"          //实际上调用的是name_
```

注意：def name _ = （newName：String）等号两边不能有空格

【例 2-10】 自定义 setter

```
class Person{
  private var myName = "Flink"
  def name = this.myName
  def update(newName:String) {
     myName = newName
     println("Hi : "+myName)
  }
}
```

这时候要修改 name 就不能写成：

```
val person = new Person
person.name = "Scala"          //这样会报找不到name_方法
<console>:9: error: value name_= is not a member of Person
      person.name = "Scala"
           ^
person.update("Scala" )        //修改name的值
```

定义的属性或方法更高级别的访问控制：privatep［this］

如果字段或者方法被 privatep［this］修饰，则该字段或方法只能被当前对象访问，当前对象的其他实例无法访问。类似这样的访问权限控制还有 package 级别控制。

【例 2-11】 private［this］实战

```
object BaseTest {
  def main(args: Array[String]): Unit = {
    val bt = new BaseTest           //
    bt.pri
    println(bt.concat(bt))
  }
}
private[base] class BaseTest{
  private[this] val a =  "NiHao"     //a属性只能被当前对象访问，其他新创建的对象实例无法
                                      访问
  private[base] val b = "Hi"         //同属于base包的对象实例可以访问
  def pri = {
    println(this.a)
  }
  def concat(bt:BaseTest)={          //传入的bt无法访问a属性
    bt.b + " Scala"
  }
}
```

2.1.4 抽象类、接口的实战详解

将有 abstract 关键字的类称为抽象类。抽象类的作用是把相同类型的事物的共同特征提取到同一个对象中。

【例 2-12】 抽象类实战

```
abstract class Person(name:String){
  println(this.name)
```

```
    private var age = 20
    val sex : String = "male"  //抽象类中的字段不能显示用abstract，否则会报
                     "abstract" modifier can be used only for
                      classes; it should be omitted for
                   abstract members
  }
  class Student extends Person("lisi")     //定义Student类继承Person
  val student = new Student             //创建Student对象实例。抽象类不能创建实例
  student.sex                 //Student可以访问父类的非私有属性
```

【例 2-13】 抽象变量 val 和 var 定义，若一个类中定义了抽象成员（变量和方法），则该类必须定义为抽象类。

```
abstract class AbstractTest{
  val str:String                                  //定义抽象字段
  val strImpl:String = "zhangsan"
  val integer:Int
  var bool:Boolean
  defadd(a:Int,b:Int):Int
  def muti(x:Int,y:Int):Int = {                   // 抽象类中已实现的方法
    x * y
  }
}
class AbstractTest1 extends AbstractTest{  //继承抽象类，必须实现抽象类中未实现的成员
  override val str: String = _
  override val integer: Int =_
  override var bool: Boolean = _
  override def add(a: Int, b: Int): Int = {        //实现父类未实现的方法
      3 + 5
  }
}
```

Scala 的特质相当于 Java 中的接口，不过特质中可以有实现的方法。

【例 2-14】 特质实战案例一

```
trait Person{
  def eat(str:String) = {
    println(str)
  }
}
trait Worker{
  def work{
    println("Working...")
  }
}
class Student extends Worker with Person
object Test{
def main(args: Array[String]) {
    val student = new Student
println(student.eat("吃饭了"))
println(student.work)
}
}
```

在 Scala 中实现特质被称为混入，混入的第一个特质用关键字 extends，混入更多的特质用 with 关键字。

【例 2-15】 特质实战案例二

```
trait Person {
  val name: String
  val age = 30
}
trait Worker {
  val age = 25
}
class Student extends Person with Worker {
  val name: String = "zhangsan"
  override val age = 15                    //override关键字重写age字段
}
object StudentTest {
  def main(args: Array[String]) {
    val s = new Student
    println("Name is :" + s.name + " ,age is : "+s.age)
  }
}
```

例 2-15 中特质 Person 和 Worker 中都有 age 字段，当 Student 混入这两个特质时，需要重写 age 字段，并且 override 关键字必须有，否则会报如下错误：

```
Error:(15, 7) class Student inherits conflicting members:
  value age in trait Person of type Int  and
  value age in trait Worker of type Int
(Note: this can be resolved by declaring an override in class Student.)
class Student extends Person with Worker {
     ^
```

2.1.5 Scala Option 类详解

Scala 的 Option［T］是容器对于给定的类型的零个或一个元件。Option［T］可以是一些［T］或 None 对象，它代表一个缺失值。例如，Scala 映射 get 方法产生，如果给定的键没有在映射定义的一些（值），如果对应于给定键的值已经找到，或 None。选项 Option 类型常用于 Scala 程序，可以比较这对 null 值 Java 可用这表明没有任何值。例如，java. util. HashMap 中的 get 方法将返回存储在 HashMap 的值，或 null，如果找到没有任何值。

【例 2-16】 Option 实战

```
object OptionTest {
  def main(args: Array[String]) {
    val capitals = Map("CHINA"-> "BEIJING", "FRANCE" -> "PARIS")
    println("capitals.get( \\"CHINA" ) : " + capitals.get( "CHINA" ))
    println("capitals.get( \"India\" ) : " + capitals.get( "India" ))
  }
}
//运行结果如下：
capitals.get( "CHINA" ) : Some(BEIJING)
capitals.get( "India" ) : None                    //当找不到时返回结果None
```

2.1.6 object 的提取器

object 中提供了 apply 方法。apply 的作用是在创建对象的时候不直接用 new，而是直接用对象加参数，这时调用了伴生对象的 apply 方法。

【例 2-17】 apply 方法实战

```
scala> val arr= Array(1,2,3)
 arr: Array[Int] = Array(1, 2, 3)
  scala> val arr= Array.apply(1,2,3)
  arr: Array[Int] = Array(1, 2, 3)
   //这两种方式的结果是一样的
```

注意：对于抽象类也是可以直接调用伴生对象的 apply 方法。它的具体工作原理是先调用其伴生对象的 apply 方法，在这个方法中调用的是抽象类的子类的伴生对象的 apply 方法。

【例 2-18】 提取器

```
object ExtractorTest {
  def main(args: Array[String]) {
    println ("Apply method : " + apply("Zara", "gmail.com"));
    println ("Unapply method : " + unapply("Zara@gmail.com"));
    println ("Unapply method : " + unapply("Zara Ali"));
  }
  // 注入
  def apply(user: String, domain: String) = {
    user +"@"+ domain
  }
  // 提取器
  def unapply(str: String): Option[(String, String)] = {
    val parts = str.split("@")
    if (parts.length == 2){
      Some(parts(0), parts(1))
    }else{
      None
    }
  }
}
```

提取器在 Scala 的模式匹配中有很多应用，将在第 4 章中结合模式匹配进行实战解析提取器。

2.1.7 Scala 的样例类实战详解

样例类很简单也很重要，在 Scala 中得到了大量的应用。尤其是与后面的模式匹配结合使用。

样例类定义：

case class Person//类似于抽象类，只是关键字为 case。

【例 2-19】 样例类和样例对象实战

```
abstract class Amount
//继承了普通类的两个样例类
case class Dollar(value: Double) extends Amount
case class Currency(value: Double, unit: String) extends Amount
//样例对象
case object Nothing extends Amount
```

注意：样例类不能继承样例类。

2.2 Scala 综合案例及 Spark 源码解析

【例 2-20】 在 Spark 的 Master 和 Worker 中的应用

```
// 样例类Heartbeat
case class Heartbeat(workerId: String, worker: RpcEndpointRef) extends DeployMessage

case Heartbeat(workerId, worker) => {
   // 这里用到了模式匹配
   idToWorker.get(workerId) match {
     case Some(workerInfo) =>
       workerInfo.lastHeartbeat = System.currentTimeMillis()
     case None =>
        ......
   }
}
```

【例 2-21】 Spark 源码

```
private[deploy] class Master(
    override val rpcEnv: RpcEnv,
    address: RpcAddress,
    webUiPort: Int,
    val securityMgr: SecurityManager,
    val conf: SparkConf)
  extends ThreadSafeRpcEndpoint with Logging with LeaderElectable
```

这段代码中用到了 Scala 的主构造器，RpcEnv 是抽象类，RpcAddress 是样例类，SecurityManager、SparkConf 是普通类，ThreadSafeRpcEndpoint、Logging、LeaderElectable 是三个特质。

【例 2-22】 Spark 中样例类经典实用案例

```
/**
 * Address for an RPC environment, with hostname and port.
 */
private[spark] case class RpcAddress(host: String, port: Int) {

  def hostPort: String = host + ":" + port

  /** Returns a string in the form of "spark://host:port". */
  def toSparkURL: String = "spark://" + hostPort
```

```
  override def toString: String = hostPort
}

private[spark] object RpcAddress {

  /** Return the [[RpcAddress]] represented by `uri`. */
  def fromURIString(uri: String): RpcAddress = {
    val uriObj = new java.net.URI(uri)
    RpcAddress(uriObj.getHost, uriObj.getPort)
  }

  /** Returns the [[RpcAddress]] encoded in the form of "spark://host:port" */
  def fromSparkURL(sparkUrl: String): RpcAddress = {
    val (host, port) = Utils.extractHostPortFromSparkUrl(sparkUrl)
    RpcAddress(host, port)
  }
}
```

通常定义的样例类是 case class Test 或者加上有参构造器，例 2-22 中定义的样例类 RpcAddress 有自己的属性和方法，这个实例可以极大提升对样例类的了解和使用。

2.3 小结

Scala 是面向对象和函数的编程语言，其在面向对象方面更彻底。在 Scala 中一切皆对象，“+”操作符也是对象的函数，所以可以通过对象 . + () 来调用。本章详细解析了 Scala 面向对象涉及的重要概念如 class、object、构造器、字段、方法、抽象类、接口等。最后通过 Spark 源码的解析加深对 Scala 面向对象的理解。

第3章 Scala函数式编程彻底精通及Spark源码阅读

Scala是一门函数式编程和面向对象完美结合的语言，要学好Scala就需要学习Scala函数式编程艺术。如果大家熟悉Java和JavaScript这两门编程语言，将会对理解Scala函数式编程有很大的帮助。本章会带领大家学习Scala函数式编程的主要思想以及函数式编程在Spark源码中的应用。

3.1 函数式编程概述

Scala作为一门函数式编程和面向对象完美结合的语言。使用Scala一般都会使用其函数式编程的风格，使代码更加简洁有弹性，同时Scala语言推崇变量都是不可变的、用val定义的变量类似于Java中的final关键字，因此要对Scala有深刻的认识和较好的使用，就要学习Scala函数式编程的思想。

从计算机硬件角度考虑，近年来计算机硬件行业发展迅速，其核心就是计算机的计算中心CPU的快速发展，使得计算机处理速度量化，计算机CPU扩大计算速度的架构也开始改变，通过多个Cores的出现并极大化地使用Cores的数量，来提高计算机处理能力。如何充分利用其多核处理，就需要在编写代码的过程中尽量使用可并行计算的代码，而函数式编程的并行操作性是其具有的一种特性，在Scala中函数式编程没有可变变量Var，那么就不会有内存共享的问题，也不会产生副作用函数。

(1) 函数式编程思维

函数式思维可以总结为以下几点。

① 函数表达式化。初学Scala或者有学习使用其他语言的基础，再来学习Scala语言就需要转换思维。可以试着理解一切都是表达式，将命令行改成表达式，去掉可变的变量（推崇使用Var）。

② 数据与行为分离。都知道面向对象是将数据与行为进行绑定的，而函数式中追求函

数和数据拥有同样的地位，没有谁地位高的说法，因此可以将函数作为值或者参数赋值给常量及变量，也可以当做返回值，在实际的编码过程中，不能让数据跑掉这一点在 Spark 程序编写中同样适用，即数据不变代码变，同样的思维，也要考虑行为消除副作用，数据不能改变，如下面函数：

$$f(x)=x+1$$

函数 $f(x)$ 中如果变量 x 改变，函数结果值改变。

③ 函数式编程的高阶逻辑。使用函数式之后，会逐渐摒弃循环、赋值等逻辑，考虑更加高阶的函数使用。通常情况下不知道函数式列表的遍历过程，这也是函数式适合并发的原因之一，比如要知道一个列表 List 中元素小于 10 的具体是哪些，只需用使用 List().filter(_ < 10)即可达到这一目的，不需用使用循环去遍历。为什么直接使用 filter 这些函数就可以避免使用循环的麻烦，是因为函数 filter 实现了这一高阶逻辑，使得可以不用拘泥于细节的实现，只要表达主观意愿，告诉 Computer 就行。

④ 函数式的组合子逻辑。函数式是自底向上的模式，正好和面向对象相反，首先需要定义基本操作，然后将这些操作汇总封装起来，层层嵌套封装达到实现复杂逻辑的目的。在 Scala 中同样提供了基本的逻辑运算子，如 filter、foreach、Map 等来实现高阶逻辑，这一点类似于数据库 Oracle、SQL server 中的 select * from table _ name where…，可以看出掌握组合子函数式逻辑思维对开发 Scala 程序非常重要。

⑤ 形而上的思维。遵循函数式编程的特点，Scala 同样具有数据不可变的思维：val x = 10，其实并非是将 10 赋值给变量 x，而是把 x 符号匹配到 10。

一切皆表达式：if x then a else b，是一个三元表达式。

函数是一等公民：函数可以作为参数，也可以作为结果返回，更可以由一个函数演化成另一个函数。

⑥ 形而下的思维。总的来说，是先有朴素的函数式语言，然后才有今天发现函数式编程的好处，启用了函数式语言的某些 feature，目的是为了把问题解构成更小的粒度，因此这些 feature 背后没什么逻辑。函数式思维形而下的特征如下。

a. 用递归替换循环。

b. 难以尾递归的时候考虑使用延续函数（continuation）。

c. 高阶函数、部分应用、Lambda 演算。

d. 用泛型、接口、可区别联合类型替换类继承。

e. 用二叉树替换普通链表后可以支持高并发计算。

(2) 函数式编程的好处

① 主要是函数的不变性带来的，没有可以改变的状态，函数的引用就是透明的和没有副作用的存在了，另外学过《操作系统》的人都知道任务处理程序执行流的最小单元就是线程，而函数式编程中不变性使得线程之间并不共享状态，就不会出现因资源争用产生死锁的情况，这对于并发编程来说极其有利，当然更加有利于 SMP（对称多处理器）架构下能够更好更多地使用 CPU 的 Cores 进行任务的并发处理，提高程序的执行效率，因而提高生产力。

② 函数不依赖外部的状态也不修改外部的状态，对于函数的调用，更加不会依赖于调用的时间和函数具体所在的位置，这使得代码更加容易进行推理，不容易出错，同时也利于调试程序、单元测试等。

(3) 函数式编程的特性

① 递归。从上面函数式编程的思维和好处一致阐述了函数式变量值不可变，也许在这里会产生疑问，因为变量不可变那么对函数的操作不就是变量值变化了吗？对于变量值不变，但对于值的操作并非修改原来的值，而是修改新产生的值，这使得函数式编程更多地倾向于使用递归。

如在下面代码中函数 func _ features _ 1 有两个参数 x，y，下面方法 new _ feature 不是改变类 func _ features _ 1 的值，而是返回一个新的实例。

同样由于变量的不可变特性，对于循环而言纯函数无法实现，因为在几种循环 for、while、doWhile 中都需要可变的状态才能跳出循环的条件，因此在函数式语言中更多地使用递归进行迭代。

下面以 $f(x)$ 计算 n 的阶乘：

$$f(x)=\prod_{k=1}^{n} k$$

而函数 $f(x)$ 计算 n 的阶乘的定义如下：

$$f(x)=n!=\begin{cases}1 & n=0\\(n-1)!\times n & n>0\end{cases}$$

对函数 $f(x)$ 使用 Java 实现，使用了递推的定义需要对每个中间结果 tmp 进行计算，对于中间结果 tmp 的保存用到了一个累加器，如例 3-1 所示。

【例 3-1】 使用 Java 实现计算 n 的阶乘示例

```
public static int func_factorial(int n ){
  int tmp = 1;
  for(int i = 1;i<= n;i++)
  {
    tmp = tmp *i;
  }
  return tmp;
}
```

使用 Scala 进行实现，递归地进行计算，Scala 如例 3-2 所示。

【例 3-2】 使用 Scala 实现递归计算 n 的阶乘示例

```
object diGui {

  def main(args:Array[String]) ={
    println(funFactorial(5))
  }
//n阶乘的具体实现
  def funFactorial(n:Int):Int ={
    if(n == 0)   return 1
    else n * funFactorial(n- 1)
  }
}
```

运行结果如图 3-1 所示。

可以从例 3-2 中看到 Scala 实现中，没有使用循环，没有使用可变的状态，并且函数体

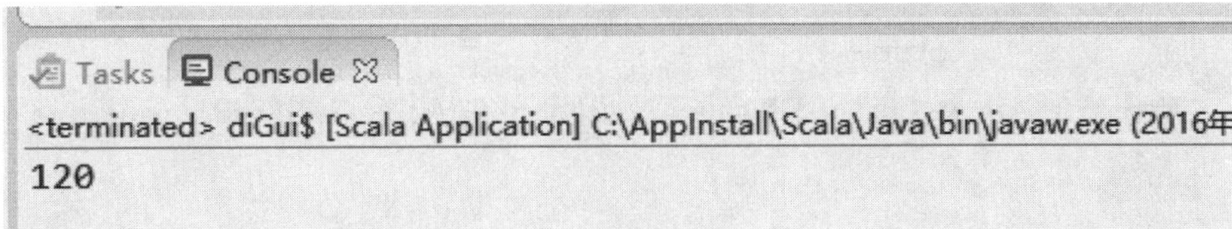

图 3-1　n的阶乘运行结果

代码更加的简洁有条理性，也不需要在累积器中保存每次运行产生的中间结果，使用函数参数来保存中间结果。

② 偏函数的使用。偏函数有很多定义，在维基百科中定义为：指定义域 X 中可能存在某些值在值域 Y 中没有对应的值。可以通俗理解为，如果有一个函数 $f(x)$ 具有多个参数，可以使其中某几个参数的值不变，也可以理解为只对函数定义域中一个子集进行定义的函数。

③ Curring（柯里化）。Curring 的概念是由俄罗斯人 Moses 提出，由美国物理逻辑学家 Haskell Curry 将其真正的丰富化，Curring 也是为了纪念 Haskell Curry 的贡献而命名。Curring 将带有 M 个元组参数的函数转换成 M 个一元函数链的方法。其实 Curring 给 Lambda 演算的一元函数添加一个或者多个变量的情况下即可使用 Curring。

Curring 的过程在此简单地使用实例的方式介绍一下，以方便理解。

假设有函数（其中 $x=3$，$y=3$，$z=5$，求值）：

$f(x,y,z)=x*y+z(x=3,y=3,z=5)$

第一步将 $x=3$ 这一条件应用到函数中，函数 $f(x,y,z)$ 变为新函数 $g(y,z)$：

$g(y,z)=3y+z(y=3,z=5)$

第二步将 $y=3$ 带入函数中，得到新函数 $h(z)$：

$h(z)=g(3,z)=z+9(z=5)$

第三步用 5 代替 z，得到结果：

$h(5)=g(3,5)=h(3,3,5)=9+5=14$

从上面三步过程可以看出，对于 n 元函数求值，有 n 元就会发生 n 次替换，且每次给函数参数赋值的过程均是按顺序进行的，函数式编程中对于 n 元函数的计算也是顺序进行的，这和数学中单纯地将 $x=3$，$y=3$，$z=5$ 直接带入函数 $f(x,y,z)$ 中不同，每一步需要一个参数，同时每一步产生一个新的函数，最后得到一个嵌套的函数链。使用 Curring 就可以很好地达到化简函数的作用，如例 3-3 所示。

【例 3-3】 函数 Curring

```
object myCurring {
  def main(args: Array[String]): Unit = {
    //定义函数Curring
    def funCurring(x:Int) = (y:Int) =>(z:Int) => x * y + z
    val step1 = funCurring(3)
    val step2 =step1(3)
    val step3 = step2(5)
    println(step3)
  }
}
```

程序运行结果如图 3-2 所示。

④ 高阶函数。高阶函数是可以把函数作为参数进行传递的函数，函数是一等公民，可以

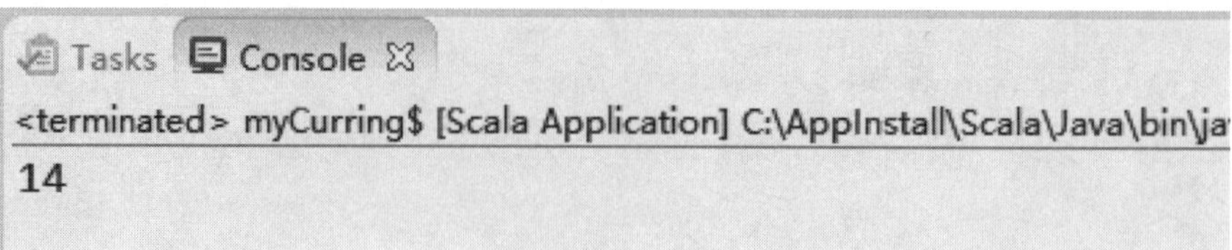

图 3-2　函数 Curring 运行结果

将函数作为参数，也可以将函数作为值进行赋值，也可以应用于泛型系统中，如例 3-4 所示。

【例 3-4】 高阶函数的应用

```
object Function_Hight {
  def func1(n: Int): Int = {
   val myFunc = (i: Int, x: Int) => i * x
   myFunc.apply(n, 2)
  }
 //将函数作为参数，参数类型Int，返回结果类型Int
def func2(x: Int => Int) = x
  //高阶函数
  def func3(f: (Int, Int) => Int) = f
  //匿名函数应用
  val func4 = (x: Int) => x + 1
//函数func5是一个匿名函数
//def func5(x:Int):Int = { if(x<1) x else x * func5(x-1)}
//函数字面量的多个占位符使用
val func6 = (_ : Int) + ( _ :Int)

  def func7(a:Int, b:Int, c:Int) = a+b+c

  val func8 = func7 _
  //偏应用函数
  val func9 = func7(1, _:Int, 3)
  //定义函数闭包
  def func10(x:Int) = (y:Int) => x+y
  //定义重复参数函数
  def func11  (args: Int*) = for (arg <- args) println(arg)

  def func12(x: Int): Int = {
    if (x == 0) {
      throw new Exception("The Exception")
    }
    else{
      func12(x-1)
    }
  }
  /**
   * 调用函数
   */
  def hello1(m: Int): Int = m
```

```
  def hello2(m: Int, n: Int): Int = m * n
  def main(args: Array[String]) {
    println(func1(78))
    println(func2(hello1)(2))
    println(func3(hello2)(2, 3))
    println(func4(3))
    println(func4(1))
    func6(7, 2)
    func8(1, 89, 3)
    func9.apply(8)
    func10(12)(2)
    //func11(Array(1, 2, 3, 4):Int*)
    func11(List(1, 2, 3): _*)
  }
}
```

程序运行结果如图 3-3 所示。

```
Tasks  Console
<terminated> Function_Hight$ [Scala Application] C:\AppInstall\Scala\Java\bin\javaw.exe (2016年
156
2
6
4
2
1
2
3
```

图 3-3　程序运行结果

3.2　函数定义

在 Scala 中函数的定义和方法的定义都是以关键字 def 开始的，后面是函数名，函数参数以及参数类型，返回值类型，和函数执行体构成。具体函数在使用的时候，如果在函数的函数体中只使用了一次函数的输入参数的值，此时可以将函数输入参数名称省略掉，用“ _ ”代替。函数定义的基本格式如下。

```
def 函数名（参数 A：A 类型，参数 B：B 类型，...）：返回值类型 = {
        函数执行体
}
```

有返回值的函数定义，在函数定义的过程中一般都会按照基本格式进行定义，包括了关键字 def、函数名、参数列表、返回值类型以及函数执行体，也可以归纳为始终带有返回值的函数定义如例 3-5 所示。

【例 3-5】 Scala 中常规定义函数示例

```
object FuncReturnType {

  def main(args:Array[String]):Unit ={
    println(add(2,6))
  }
  //定义函数add，始终待返回值
def add(x:Int,y:Int):Int={
    x+y
  }
```

运行结果如图 3-4 所示。

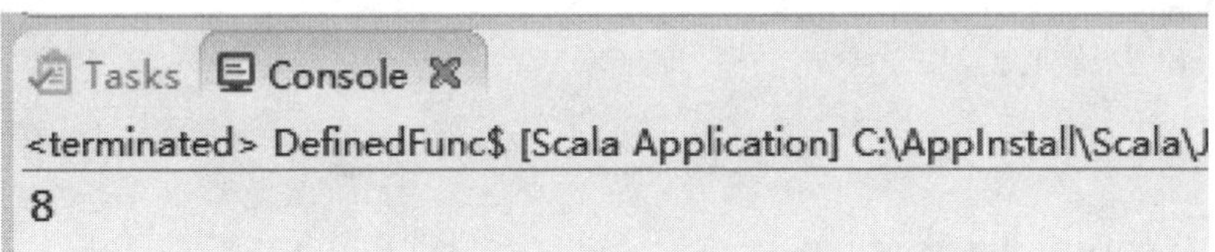

图 3-4　示例 3-5 运行结果

或者定义为返回类型为空的函数，如下所示。

```
def returnUnit():Unit={
    println("another way to return void")
  }
```

省略非 Unit 返回值的情形下，省略返回值，让程序根据代码块，根据代码块最后一个表达式的类型推导出返回值的类型，如下所示。

```
def add(x:Int)={ //something }
```

当编写的程序中函数的返回值为 Unit 的时候，可以省略函数返回值类型以及函数执行体前面的“＝”，函数不会产生其他副作用，如下所示。

```
def add(){
    // 返回值为Unit
    println("return void")
}
```

当定义的函数体中只有一条语句，可以省略花括号，此时函数包含了关键字 def、函数名、函数参数列表、“＝”以及“＝”后面的函数执行体，如下所示。

```
object MyFunc{
//定义函数max返回两整数中最大的数
def max(x: Int, y: Int) = if (x > y) x else y
def main(args:Array[String]):Unit ={
     println(max(123,232))
    println("Hello, world!")
  }
}
```

单例对象 MyFunc 运行结果如图 3-5 所示。

Scala 是一门非常灵活简洁的语言，介绍了函数定义的内容后，可以从 Spark 源代码的角度对 Scala 的函数定义有一个深刻的认识，在 Spark 最重要的数据集 RDD. scala 中有非常

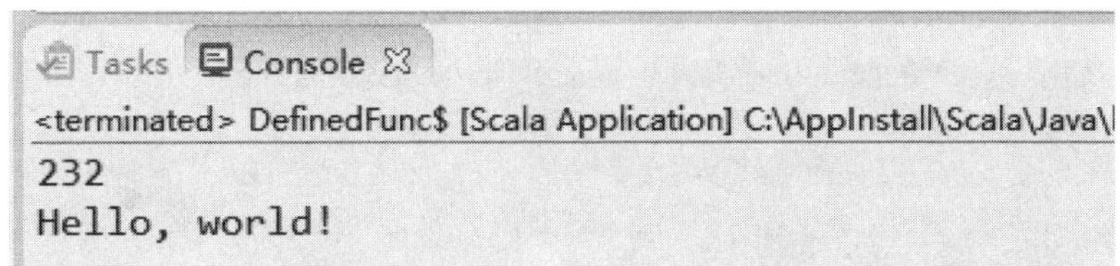

图 3-5　程序 MyFunc 运行结果

多的案例可以供学习者学习，如下所示。

```
private[spark] def computeOrReadCheckpoint(split: Partition, context: TaskContext): Iterator[T] =
  {
   if (isCheckpointedAndMaterialized) {
     firstParent[T].iterator(split, context)
   } else {
     compute(split, context)
   }
  }
```

在 RDD. scala 中定义了 computeOrReadCheckpoint 函数，函数返回值类型为 Iterator [T]，函数带有两个参数 split，context 分别为 Partition 和 TaskContext 类型，其中 split 是 RDD 中 Partition 的索引 index，而 context 是运行 Task 包含的信息等，该函数实现了判断如果是广播的数据，就读取正在广播的 Task 进行计算，如果不是则直接计算一个 RDD Partition，返回的是 Iterator [T]。

3.3　函数式对象

函数式对象是不具有改变任何状态的对象的类，在本节内容中主要介绍 Scala 中先决条件检查、自指向、辅助构造器、Scala 的标识符等内容。

(1) 先决条件

在面向对象编程中为了确保数据在整个程序生命周期内的有效性，一般会将数据封装到对象内，而对于函数式对象是不具有改变任何状态的类，例如不可变的对象就需要保证对象在创建时数据的有效性，此时会用到先决条件，使用先决条件的格式如下所示。

```
final def require(requirement: Boolean, message: =>Any): Unit
```

其中 requirement 是限制条件，布尔类型返回 false 或者 true，message 是不满足条件时的错误信息，可以是任意类型，require 返回值类型为 Unit。

先决条件是对传递给方法或者构造器的值进行的限制，也就是说调用者必须满足这一条件，才能正确地调用。下面以实例说明，创建一个 Person 伴生类，带有 4 个参数，主要用来实现对人 Person 工作属性的实现，即人靠什么进行工作，从人体器官的角度看，要用大脑思考，双手进行操作，双脚用来走路，来实现 Person 类，在伴生对象中进行调用的时候传入实参，人不能有两个大脑或者没有大脑，手和脚类似，因此要对实参的值使用先决条件进行限制。如例 3-6 所示。

【例 3-6】 先决条件示例

```
class Person(head:Int,hand:Int,foot:Int,name:String)extends Animals{
//先决条件，boolean类型
require(head > 0&& head <2)
//当传入的值不满足，程序运行异常，异常信息为"error input…"
 require(hand >=0&& hand <3, "error input of hand,please input hand retring ")
 require(foot >=0&& foot <3,foot)
//定义Working方法
 def Working(fangFa:String) = {
   val total = head + hand + foot
   println(name + "is a " + fangFa + " and working by\n\t " + head + " head to think and \n\t " + hand+" handsto cooperate something and\n\t " + foot + " foots to walk !")
   println(name + " are use total Organ:" + total  )
   total
 }
}
//在伴生对象中调用
object FunctionObject {
 def main(args: Array[String]): Unit = {
    val perple = new Person(1,2,2,"Tom")
    perple.Working("Programmer")
 }
}
```

程序运行结果如图 3-6 所示。

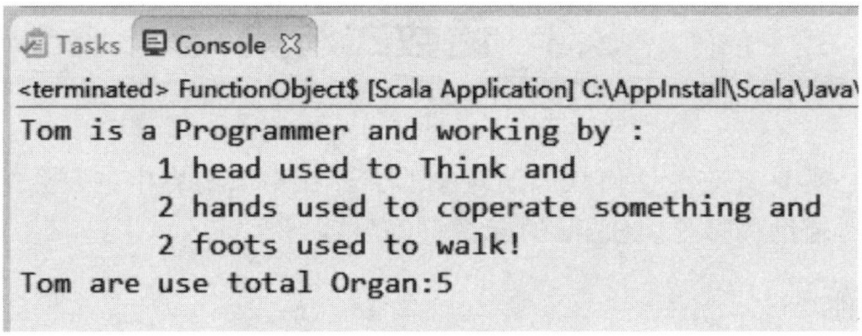

```
Tasks  Console
<terminated> FunctionObject$ [Scala Application] C:\AppInstall\Scala\Java\
Tom is a Programmer and working by :
        1 head used to Think and
        2 hands used to coperate something and
        2 foots used to walk!
Tom are use total Organ:5
```

图 3-6 先决条件程序运行结果

（2）自指向

This 关键字指向当前执行方法被调用的对象实例，或者如果使用构造器的话，就是正被构建的对象实例，如例 3-7 所示。

【例 3-7】 自指向示例

```
package guid
class SubConstructor(n:Int,m:Int){
 require(m!=0,m) //先决条件
 def this(a:Int) = this(a,10)
 def compare(a:Int,b:Int) = if(b ==0 )a%b else a
 val X = compare(n,m)

 def A = n*X
```

```
    def B = m*X
  //定义方法<
  def< (that:SubConstructor) = this.A * that.B < that.A * this.B

    def selfDiction(that:SubConstructor):SubConstructor = if (this.<(that)) that else this
  }
  //单例对象
  object SubConstructor{
    def main(args: Array[String]): Unit = {
      val subAdd = new SubConstructor(4,6)
      val add = subAdd.compare(7, 5)
      //println(subAdd)
      println(add)
   println(subAdd.X)
    }
  }
```

程序运行结果如图 3-7 所示。

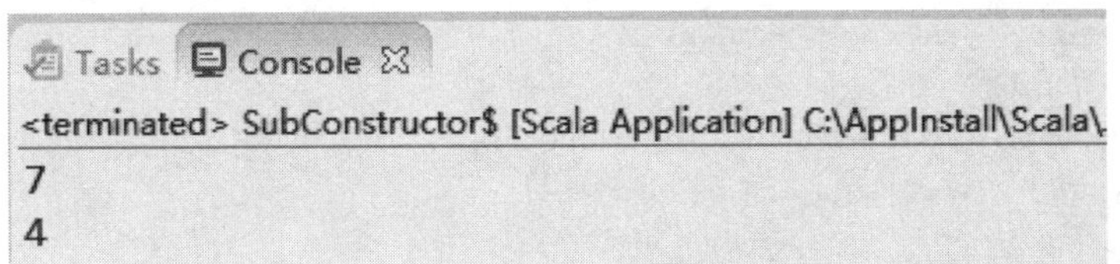

图 3-7 自指向示例运行结果

在方法“<”中，传入了一个 SubContractor 类型的 that 参数，其中 this. A 和 this. B 指向“<”方法被调用的对象，当然这里的 this 关键字也可以省略不写，而在方法 selfDiction 中不能省略 this 关键字，因为在类 SubContractor 中添加该方法返回较大的值。在方法 selfDiction 中第一个 this 可以直接省略，第二个 this 是执行方法调用的对象实例，不能省略。

(3) 辅助构造器

在定义类的时候“()”后面的内容会被定义为主构造器（Primary Constructor），那就是说主构造器是和类定义混合在一起的，在一个类中如果没有明确定义主构造器的话，则默认构造一个无参的主构造器，辅助构造器（Auxiliary Constructor）扩充了主构造器的内容，在 Java 中构造器的名称和类名保持一致，但 Scala 中是以 this 为名称的，并且其定义开始于 def this(…)。

在 Scala 中可以有若干个辅助构造器，当一个辅助构造器在执行的时候会首先调用相同类下的其他构造器，可以是主构造器也可以是已经创建的其他辅助构造器，而辅助构造器的调用起始于是主构造器的调用的结束，如例 3-8 所示。

【例 3-8】 辅助构造器示例

```
class constructorAnx {
  private var name =""
  private var age = 0
  //定义辅助构造器
  def this(name:String){
```

```
    //调用主构造器
    this()
    this.name = name
    println(name)
  }
  //定义另一个辅助构造器
  def this(name:String,age:Int){
    //调用前一个辅助构造器
    this(name)
    this.name = name
    this.age = age
    println(name + " : "+ age)
  }
}
//伴生对象
object constructorAnx{
  def main(args: Array[String]): Unit = {
    //主构造器
    val bigDate = new constructorAnx
    //辅助构造器
    val hadoop = new constructorAnx("Hadoop")
    val spark = new constructorAnx("Spark",4)
  }
}
```

例 3-8 运行结果如图 3-8 所示：

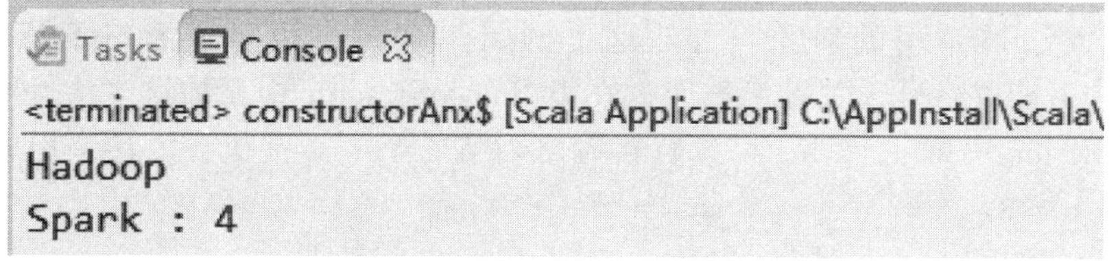

图 3-8　辅助构造器事例运行结果

（4）Scala 的标识符

Scala 的标识符包含了字母、数字、下划线，首字符为字母或者下划线，Scala 命名遵循驼峰式标识符的习惯，如 HashMap 和 SparkContext 等，需要注意的是下划线“_”在虽然标识符内合法，但是可能导致 Scala 标识符名称冲撞，因为“_”还有需要其他的用法。

采用驼峰式命名的方式命名字段、函数名、方法等以小写字母开始，如 reduceByKey、textFile 等，为了避免 Scala 编译器报错，标识符结尾的位置不使用下划线，如 val name。Scala 与 Java 的区别是常量名，在 Java 中，常量名习惯全部大写如 PI，Scala 总常量名通常只是首字母大写，这也说明了 Scala 惯例的常量也是驼峰式命名方式。

操作符标识符由若干个操作字符组成，通常是一些七位可打印的 ASCII 字符，这些字符不包括字母、括号、数字、下划线以及回退字符等，操作标识符也可以属于数学符号（Sm）或者其他符号的 Unicode 集，例如：+，-，->，!，<>等。Scala 内部的操作符标识符可以任意长，在 Scala 标识符“=>”不会被拆分，如 x=>y，而在 Java 中会被拆成四个词汇字符。

混合标识符由字母和数字组成，后面跟着下划线和一个操作符标识符，例如 hello _ -用于被定义的“-”操作符的方法名。字面量标识符由反引号包含的任意字符串组成，如：‘＜spark＞’，使用字面量标识符时可以将任意不是非法的字符加入反引号中作为标识符，当然也可以包括 Scala 的保留字。

3.4 本地函数

在 Java 以及 C＋＋等面向对象语言中，要让一个方法不被其他的对象访问，通常使用关键字 private 来限制其作用范围，Scala 中一般在某个方法中定义本地函数，其不被其他函数所访问，且只在包含它的代码块中可见。

通常用到的本地函数的场景不多，当需要辅助对象中的某个方法时，才会定义本地函数。使用本地函数的好处是给代码提供了较好的封装性，且减少可能的命名空间的污染，避免了因为函数命名的冲突，当然封装性类似于使用 private 关键字的效果，如例 3-9 所示。

【例 3-9】 本地函数示例

```
object FunctionOps {
  class AA{
    // ...
  }
  class BB{
    //定义函数A
    def A{
      def B{  //定义本地函数B
        // ...
      }
    }
    //该B函数不同于A的本地函数B
def B{
      println("本地函数！ ")
    }
  }
}
```

本地函数的另一个方面体现在函数内部可以嵌套函数，本地函数的作用域是本地函数可以访问包含该本地函数的参数。这里还是以扩展例 3-9 为基础进行讲解，也就是说在函数 A 的作用域内，其本地函数 B 也是可见的，当然本地函数 B 也可以访问函数的内容，包括参数，如下所示。

```
class BB{
 //定义函数A
 def A(x:Int,y:Int){
   //定义本地函数B
   def B{
     println(x*y)
     //创建BB的对象
     val b = new BB
```

```
        //调用函数A
        val a = b.A(3,6)
        println(a)
      }
    }
    //定义函数B
    def B{
      println("本地函数！")
    }
  };
```

可以看出定义本地函数和定义变量一样，都是函数可见性，本地函数在包含它的代码块中可见，也可以理解为在它定义的作用域中可见，体现了函数的嵌套和作用域原则。

3.5　头等函数

头等函数（Frist-class-value）是 Scala 函数式编程的一个重要特征，函数可以被当做值进行传递，可以被当做参数进行传递，也可以将其当做结果返回，此时又称为高阶函数。

在 3.6 节会讲到 Scala 函数字面量，这里对函数值进行简单的介绍，函数值都是某个扩展了 Scala 包的一些 FunctionN 特质之一的类的实例，例如 Function1 是具有一个参数的函数，FunctionN 是具有 N 个参数的函数，且每个 FunctionN 特质有一个 apply 方法用来调用函数。函数值是作为对象存在于运行期限内，他是一个对象，可以保存于变量中，通常也可以对其进行调用，而函数字面量在函数实例化的时候转换为函数值，函数字面量存在于源代码中。如下为函数字面量和函数值的简单例子。

```
//函数字面量
scala> (A:Long) => A +1
res15: Long => Long = <function1>
//将函数值保存到变量中
scala> val myScala = (A:Long) => A+1
myScala: Long => Long = <function1>
函数值的调用
scala> myScala(10)
res18: Long = 11
scala>
```

如果将作为对象的函数值保存于变量中，可以对其重新进行赋值，可以让函数字面量包含多条语句，使用函数字面量和花括号就可以组成代码块，当该代码块被调用时，代码块中的语句会逐条被执行，而函数的返回值就是最后一行表达式的值，如下所示。

```
scala> var myScala = (A:Int) => A+1
myScala: Int => Int = <function1>
//重新赋值函数值
scala> myScala = (A:Int) => A+99
myScala: Int => Int = <function1>
//调用
scala> myScala(1)
res19: Int = 100
```

```
//函数字面量包含代码块
scala> myScala = (A:Int) =>{
     | println("Hello Scala")
     | println("Hello Spark")
     | A-10
     | }
myScala: Int => Int = <function1>
//调用
scala> myScala(2)
Hello Scala
Hello Spark
res20: Int =-8
scala>
```

3.6 函数字面量和占位符

3.6.1 Scala 占位符

在函数字面量中占位符“ _ ”可以表示为一个或者多个参数，但是需要满足的是每个参数只能出现一次，并且如果有多个占位符时，需要和参数依次对应起来，即第一个参数对应第一个占位符，第二个参数对应第二个占位符，一一对应。如例 3-10 为使用占位符的例子。

【例 3-10】 使用占位符代替参数

```
scala> List(1,2,4,5).map((x:Int) => x+2)
res12: List[Int] = List(3, 4, 6, 7)

scala> List(1,2,4,5).map(x:Int => x+2)
<console>:1: error: identifier expected but integer literal found.
       List(1,2,4,5).map(x:Int => x+2)
                               ^
//省略（）会报错，因为编译器将Int =>x+2看出了一个整体，导致类型错误
scala> List(1,2,4,5).map(x => x+2)
res13: List[Int] = List(3, 4, 6, 7)
//使用占位符代替
scala> List(1,2,4,5).map(_+2)
res14: List[Int] = List(3, 4, 6, 7)
// 循环打印
scala> List(1,2,4,5).map(_+2).foreach(println)
3
4
6
7
scala>
```

3.6.2 函数字面量

在 Scala 中，函数字面量可以作为值进行传递，当然也可以定义或者调用，还可以把它

们写成匿名字面量的方式，称之为函数字面量，在函数编译函数字面量的同时将其实例化为函数值，类似于类和对象，函数值和字面量的区别在于函数字面量存在于源代码中，而函数值在函数运行期间有效。任何函数值都是某个扩展了 Scala 包的若干 FunctionN 特质之一的类的实例，如 Function0 是没有参数的函数，Function1 是有一个参数的函数等。每一个 FunctionN 特质有一个 apply 方法用来调用函数。

对函数字面量相加一个整数，例如：

```
scala> val functionValue = (x:Int) => x+10
functionValue: Int => Int = <function1>
//将匿名函数传递给functionValue
scala> val aa = functionValue(3)
aa: Int = 13
```

在函数签名中可以看到将匿名函数传递给变量 functionValue，将 Int 类型的 functionValue 转化为 Int 的函数，然后在=>右边具体封装了函数的执行代码块，这里由于函数只有一行表达式，因此省略了“{ }”，因为函数值可以作为对象使用，所以可以存入变量 functionValue 中，然后对其进行类似函数的调用。

Scala 还有一处非常强大的地方是类型推导，因为 scala 编译器可以推断出函数字面量的参数类型，因此在实际编写程序的时候为了减少代码冗余和简洁性，通常会省略类型以及参数的括号，如下所示。

```
scala> val array = Array(1,2,3,4,5,6)
array: Array[Int] = Array(1, 2, 3, 4, 5, 6)
//根据类型推导，省略了item类型，然后循环打印出>2的元素
scala> val out = array.filter(item => item > 2).foreach(println(_))
3
4
5
6
out: Unit = ()
```

为了让代码更加简洁，可以使用通配符“ _ ”当做单个参数的占位符，可以对上面的数组 array 进一步处理，更好地体现出字面量的短格式和代码的简洁，如下所示。

```
scala> val out = array.map(_+2).filter(_ > 2).foreach(println)
3
4
5
6
7
8
out: Unit = ()
scala>
```

3.6.3 部分应用函数

部分应用函数（partially applied function）也成为偏应用函数，只提供了或者指定了部分参数的函数就是部分应用函数，它是一种表达式，在 Scala 中当进行函数调用的时候不需要指定所有的参数，只需要提供部分参数即可，那么传入需要的参数，实际是在把函数应用到参数上。

如例 3-11 所示定义偏应用函数。

【例 3-11】 偏应用函数示例

```
//Scala通过两个参数类型推导出函数返回值类型String
scala> (_:Int) + (_:String)
res8: (Int, String) => String = <function2>
//定义函数functionOps
scala> def functionOps = (_:Int) + (_:String)
functionOps: (Int, String) => String
//完全函数调用
scala> functionOps(1,"Spark")
res9: String = 1Spark
//函数调用
scala> val partial = functionOps(1,"Spark")
partial: String = 1Spark
//部分应用函数
scala> val partial = functionOps(1,_:String)
partial: String => String = <function1
scala>
```

在定义了函数之后，要对函数进行调用，此时可以指定一个参数，可以指定多个参数，也可以一个参数都无需指定，此时可以使用占位符“ _ ”代替整个参数列表，以这种方式使用“ |_ ”就是在写部分应用函数的表现，如果将偏应用函数的调用保存为一个变量的话，此变量可以作为函数使用。如下所示。

```
scala> def HelloScala(x:Int,y:Int,z:Int) = x + y + z
HelloScala: (x: Int, y: Int, z: Int)Int

scala> HelloScala(_:Int,3,5)
res10: Int => Int = <function1>
//指定第一个参数
scala> HelloScala(4,5,_:Int)
res11: Int=> Int = <function1>
//一个参数都不指定
scala> HelloScala _
res12: (Int, Int, Int) => Int = <function3>
//将函数调用保存给变量hello
scala> val hello = HelloScala _
hello: (Int, Int, Int) => Int = <function3>
scala> hello(2,45,_:Int)
res13: Int => Int = <function1>
//对变量hello的调用
scala> hello(2,45,3)
res14: Int = 50
scala>
```

在 val hello＝HelloScala _ 中，变量 hello 指向一个函数值对象，Scala 编译器通过偏应用函数 HellloScala _ 自动生成一个类的实例就是该函数值的对象，而编译器产生的类有一个带有三个参数的 apply 方法，然后 Scala 编译器会将 hello（2,45,3）转化为为对函数值的 apply 方法的调用。那么通过提供部分参数来表达一个偏应用函数。

3.7 闭包和 Curring

(1) 闭包

当函数超出作用范围或者函数执行完成后，其内部的变量依旧可以被外界访问，就称该函数为闭包，也就是说依照函数字面量中运行期创建的函数值（对象）被称为闭包。一个函数中，如果没有自由变量的函数字面量就为封顶项，而如果有自由的函数字面量就为开放项，如（item _ A：Int)=>item _ A+10 就是一个封顶项，(item _ A：Int)=>item _ A+item _ B 是一个开放项。

下面看一个简单闭包的定义。

```
scala> var more = 1
more: Int = 1
//包含外部变量
scala> (x:Int) => x+more
res11: Int => Int = <function1>
//函数字面量
scala> (x:Int) => x+1
res12: Int => Int = <function1>
```

可以看出变量 more 不在函数的定义中，那么对于这类变量的引用称之为自由变量，其属于其他地方的变量，函数字面量本身没有指定，而对于参数 x 称之为绑定变量，因为他是函数传入的参数，如例 3-12 所示。

【例 3-12】 Scala 闭包解析

```
scala> def hiScala(humen:String) =(Lod:String) => println(Lod + " :" +humen)
hiScala: (humen: String)String => Unit

scala> val helloScala = hiScala("yesu")
helloScala: String => Unit = <function1>
scala> helloScala("Lod")
Lod :yesu
```

可以从示例 3-12 中看出给函数 hiScala 传入参数 yesu，并将其传递给 helloScala 变量，在 hiScala 执行完后，给变量 helloScala 传入参数，helloScala 转过来又访问传给 hiScala 的内容，正常情况下已经超出了函数作用范围，但是 Scala 中闭包却实现了这一过程，由此说明闭包实现了当函数超出作用范围后执行完成后，其内部的变量依旧可以被外界访问。

实质上其内部的原理是编译器会常见一个不可见的函数对象，把 hiScala 中的 humen 参数当做该对象的成员，此时调用 hiScala 是调用了对象内部的返回值，以此访问了该对象的内容。

(2) Curring

柯里化（Curring）是以逻辑学家 Curring 的名字来命名的，Curring 是把接受多个参数的函数转换成为接受一个单一参数（也可以为多参数）的函数，可以认为有多少个参数就会转换多少个函数。

首先，定义一个函数 fun，该函数带有两个参数 a，b，如下所示：

```
scala> def fun(a:Int,b:Long) = a+b
fun: (a: Int, b: Long)Long
```

下面定义一个函数 fun _ 1，同样带有两个参数 a，b，如下所示：

```
scala> def fun_1(a:Int) = (b:Long) => a-b
fun: (a: Int)Long => Long
```

那么调用函数 fun _ 1 计算两个数相减的结果，如下所示：

```
scala> fun_1(54)(34)
res10: Long = 20
```

例 3-13 就是一个柯里化之后的函数的调用过程。

【例 3-13】 Curring 函数示例

```
scala> def func_Curring(x:String)(y:String) {println(x +"+" + y)}
func_Curring: (x: String)(y: String)Unit

scala> func_Curring("Spark")("Hadop")
Spark+Hadop

scala> def functionCurring(a:Int)(b:Int)(c:Int) = println(a*b*c)
functionCurring: (a: Int)(b: Int)(c: Int)Unit

scala> def functionCurring(a:Int)(b:Int)(c:Int) { println(a*b*c)}
functionCurring: (a: Int)(b: Int)(c: Int)Unit
//返回值为Unit
scala> functionCurring(2)(3)(4)
24
```

Scala 的 Curring 主要表现在接受到的是一个参数列表，该函数的参数列表包含的不是多个参数，而是一个个的参数，当然这里的每个参数也可以有多个参数。例如 Curring 定义函数不是形如 def func(x:Int,y:Int,z:Int)＝x＋y＋z，而是 def func(x:Int)(y:Int,yy:Int)(z:Int)＝x＋y＋z，那么对函数的调用可写为 func(2)(3)(4)、func(2){3}{4}或者是 func{2}{3}{4}。

下面验证 Curring 的函数式编程风格，如例 3-14 所示。

【例 3-14】 Curring 参数列表验证

```
scala> def func(x:Int)(y:Int,yy:Int)(z:Int) = x+y+z
func: (x: Int)(y: Int, yy: Int)(z: Int)Int

scala> func _
res17: Int => ((Int, Int) => Int => Int) = <function1>
```

可以看到在定义的函数 func 中参数（y：Int，yy：Int）中定义了两个参数，然后调用了函数 func，这里实质上是创建了一个偏应用函数，那么函数中就表示有一个或者多个函数的参数没有被绑定，在函数的签名中，显示了函数转换的过程，最后转换的类型是一个具体的函数。也可以改变函数参数的类型，如例 3-15 所示。

【例 3-15】 Curring 示例

```
scala> def func(x:Int)(y:Int,yy:String)(z:Int) = x+y+yy+z
func: (x: Int)(y: Int, yy: String)(z: Int)String

scala> def func(x:Int)(y:Int,yy:String)(z:Int) = x+y+z+yy
func: (x: Int)(y: Int, yy: String)(z: Int)String
//不同类型的参数
scala> func(1)(2,"Spark")(4)
res0: String = 7Spark
```

(3) 偏函数解析

首先偏函数不同于偏应用函数，偏函数是一个数学概念，把只对函数定义域的一个子集进行定义的函数称为偏函数，从定义上不难看出偏函数定义域中可能存在某些值在值域中没有对应的值。偏函数定义包含两种方式，第一种方式是直接定义，即明确申明为 PartialFunction，第二种方式是通过模式匹配进行定义。

偏函数在 scala 中用 scala. PartialFunction[-T，+S]来表示，其中-T 为输入参数类型，在 Scala 的参数系统中它被称为逆变，+S 为输出类型同样的在 Scala 参数类型系统中被称为协变。偏函数将一个多参数的函数取该函数的部分参数进行定义得到一个新的函数。当不知道某些值的具体操作过程或者存在多种处理方式时，那么采用偏函数先对明确的部分进行定义，后续也可对函数定义域进行修改操作。

Scala 提供了定义偏函数（PartialFunction）的语法快捷，使用模式匹配的方式查看偏函数的使用，如例 3-16 所示。

【例 3-16】 偏函数实例

```
scala> val myPartialFunc:PartialFunction[Int ,String] ={
     | case x if x>0 => "Nice!"
     | case _ => "The other thing!"
     | }
myPartialFunc: PartialFunction[Int,String] = <function1>
//偏函数调用
scala> myPartialFunc(4)
res0: String = Nice!
//匹配到其他情况
scala> myPartialFunc(-34)
res2: String = The other thing!
scala>
```

使用 orElse 补充对其他域的定义：

```
scala> val myPartial:PartialFunction[Int ,String] ={
     | case x if x>0 => "Nice!"
     | }
myPartial: PartialFunction[Int,String] = <function1>

scala> partialFunc(2)
res5: String = Nice!
//使用orElse补充其他域的定义
scala> val partialFunc:(Int => String) = myPartial orElse{case _ =>"bad!"}
partialFunc: Int => String = <function1>
```

```
//调用partialFunc
scala> partialFunc(-2)
res6: String = bad!
```

偏函数出现在很多场景，实际在使用的时候偏函数只关心某一部分的值域，可能会有好几个函数，每个函数关心其定义的值域，然后将这些函数的操作组合起来形成较为复杂的函数，例如下面 people 只关心输出为“child”和“man”的域。

```
scala> val people:PartialFunction[Int,String] ={
     | case age if age <= 20 =>"Child "
     | case age if age <= 60 && age> 20 => "man"
     | }
people: PartialFunction[Int,String] = <function1>
scala> println(people.isDefinedAt(15))
true
//判断people中是否满足输入参数可以被匹配
scala> println(people.isDefinedAt(61))
false
//使用orElse匹配其他情况
scala> val oldman:PartailFunction[Int,String] = people orElse{
     | case age >60 =>"oldman"
     | }
oldman
scala>
```

3.8 高阶函数

高阶函数是将函数传递给函数的参数的函数，高阶函数是 Scala 语言最富魅力的特性之一。函数式参数带来的结果是只要符合函数签名就可以将函数传递给函数的参数，换句话说如果函数可以把参数设置为函数，那这个参数会接收一切符合函数签名的函数的实现。另外高阶函数还有一个特征，就是函数的返回值可能是函数。如例 3-17 所示。

【例 3-17】 Scala 中高阶函数。

```
scala>val hiScala = (name:String) => println(name)
hiScala:String => Unit = <function1>
//定义了变量hiScala返回值为Unit
scala> def helloScala(myFunction:(String) => Unit,context:String){myFunction(context)}
helloScala:(func:String => Unit,context:String)Unit
scala> helloScala(hiScala,"Spark is Wonderful!")
Spark is Wonderful!
scala>def helloSpark(name:String) = (name:String) => println(name)
helloSpark:(name:String)String => Unit
scala> val spark = helloSpark("Scala")
spark:String =>Unit =<funtion1>
//函数调用
```

```
scala> spark("good!")
good
```

示例 3-17 中看出首先定义了一个变量 hiScala，该函数签名中返回值为 Unit，function1 是具体的函数，是将匿名函数赋值给 hiScala。在定义的函数 helloScala 中，有两个函数的参数，第一个参数 myFunction 是一个函数，其内部接收一个 String 类型的传入值，返回的函数值为 Unit，第二个参数为 String 类型，函数 helloScala 的函数执行体为 Unit，省略了“=”，函数 helloScala 的返回值为 Unit。

Scala 中将函数作为参数传递给变量，这也是 Scala 函数式编程不同于 Java 的一个特点，而 Java 程序员需要付出更多才能达到相同的效果，一般会将动作放在一个实现某接口的类中，然后将该类的一个实例传递给另一个方法，而 Scala 就无需关心这些，只需要关心传入的参数本身，然后对参数进行处理，所以在某种程度上极大地降低了算法的复杂度且增加了代码的简洁性。

(1) 函数值

在 Scala 中，可以在函数内部创建函数，将函数赋值于引用，前面讲到可以将函数作为一等公民传递给函数的参数，而其内部实现了对函数值的处理和实现，将函数创建为特殊类的实例。从一个代码实例的角度进行探讨，先从提取范围内的值进行循环的代码开始，将其封装于方法中，如示例 3-18 所示。

【例 3-18】 函数值实战

```
object funcValue {
  def resultValue(num:Int,total:Int =>Int) :Int ={
    var hello = 0
    for(i <- 1 to num){
      hello += total(i)
    }
    hello
  }
  def total(x:Int) = x +2
  def main(args: Array[String]): Unit = {
    val result = resultValue(23,x => if(x>0) x+3 else 0)
    println(result)
    println(total(43))
  }
}
```

程序运行结果如图 3-9 所示。

```
Tasks  Console
<terminated> FuncValue$ [Scala Application] C:\AppInstall\Scala\Java\b
The result is :345
45
```

图 3-9　函数值程序运行结果

在示例 3-18 中单例对象 funcValue 中定义了方法 resultValue ()，并带两个参数，参数

num 表示 for 循环中值的作用范围，第二个参数 total 是一个函数，将该函数作为参数传递给 resultValue，函数 total 类型为函数，接收 Int 的值，返回 Int 型的值，方法 resultValue () 返回结果也是一个 Int 类型。

在方法 resultValue () 中，对 1 到 num 中的值进行循环，对每个元素调用给定的函数 total，该函数接收一个 Int 类型，即范围 num 中的一个元素，之后返回一个 Int 类型的值，为对该元素的计算结果。方法 resultValue 的调用者定义计算和操作，通过求和后，在 main 方法中进行打印。

对 0 到 23 之间的元素求和即可表示为：

println(resultValue(23,x=>x)

方法 resultValue 的两个参数中，参数 23 为循环执行的范围，第二个参数为匿名函数，(省略函数名称有具体实现的函数)，匿名函数 total 中只是实现了将其参数返回的作用，根据 Scala 的类型推断特征，方法 resultValue 中从参数列表中推断出 total (i) 中参数 i 的类型为 Int，但如果指定方法的返回类型为其他类型编译器就会报错。

分别对 for 循环内的奇数和偶数进行求和：

```
println(resultValue(23,x =>if(x%2 ==0) 1 else 0))
println(resultValue(23,x =>if(x%2 !=0) 1 else 0))
```

Scala 中函数值作为参数不限于个数，也就是说可以接受任意多个函数值作为参数传递，并且函数值不限于具体的位置。函数值的使用，可以更好地让代码不重复自己 (DRY 原则)，变得更加有弹性，将公用的代码封装到一个函数中，差异部分作为实参，该实参属于方法调用。

```
private[spark] def getRDDStorageInfo(filter: RDD[_] => Boolean): Array[RDDInfo] = {
  assertNotStopped()
  val rddInfos = persistentRdds.values.filter(filter).map(RDDInfo.fromRdd).toArray
  StorageUtils.updateRddInfo(rddInfos, getExecutorStorageStatus)
  rddInfos.filter(_.isCached)
}
```

这里可以参考下 Spark 源码中的应用程序上下文，SparkContext 中的方法 getRDDStorageInfo 中传递一个参数 filter，参数类型是一个 RDD，返回 Boolean 型的函数值，该方法在 Spark 包中有效，至于函数的具体实现在此不多赘述。

(2) 函数参数

Scala 函数参数默认是 Val 的，即在 Scala 函数列表中不能够显式地申明变量为 val，也不能够重新对其进行赋值操作 (非 Var)。由于 Scala 函数式编程的灵活性，函数可以作为头等公民将其赋值给变量，传递给函数参数，当然在定义函数的时候，函数参数的列表可以有若干个，且要申明函数参数的类型，负责编译器会报错，因为在一般情况下，Scala 编译器会推断出函数返回值的类型，但是无法推断函数的参数类型，因此需要在定义函数的时候对参数列表中参数申明参数类型。

对于重复参数，可以在参数的类型后面加上符号 “*”，这也就意味着可以给函数传入可变的函数参数列表，但类型必须相同，如例 3-19 所示。

【例 3-19】 重复参数示例

```
//重复参数，可传入可变的参数列表
scala> def functionOps(args:String*) = for(arg <- args) println(arg)
```

```
functionOps: (args: String*)Unit
//打印
scala> functionOps("Hadoop","Spark","Flink")
Hadoop
Spark
Flink
//非重复参数
scala> def paraFunction(args:String) = for(arg <- args) print(arg)
paraFunction: (args: String)Unit
//打印
scala> paraFunction("Spark")
Spark
scala>
```

可以看出，函数 functionOps 可以被多个 String 类型的函数参数调用，可以是零个，也可以是多个。而使用“ * ”申明重复的参数实质上是一个同类型的数组，即 String * 的参数 x 的类型相当于 Array[String]。

如果这里想对重复的函数参数传入整个参数数组，将每个数组中的元素当做一个参数传入，而不是将整个参数列表当做一个参数传入，需要使用“ _ * ”，将其添加在需要传入的参数名后面，告诉编译器把数组中的所有元素当做参数。下面在例 3-19 的基础上进行演示，定义一个参数 paramater，带有三个 String 类型的数组，将其传入给函数 functionOps，执行函数体，如下所示。

```
//定义重复参数函数functionOps
scala> def functionOps(args:String*) = for(arg <- args) println(arg)
functionOps: (args: String*)Unit
//定义任意类型的数组paramater，作为函数的参数
scala> val paramater = Array("Hadoop","Spark",23)
paramater: Array[Any] = Array(Hadoop, Spark, 23)
//将数组参数传入
scala> functionOps(paramater:_*)
<console>:30: error: type mismatch;
 found   : Array[Any]
 required: Array[_ <: String]
Note: Any >: String, but class Array is invariant in type T.
You may wish to investigate a wildcard type such as `_>: String`. (SLS 3.2.10)
          functionOps(paramater:_*)
                    ^
//如果传入的参数数组中类型不一致，编译器报错，提示数组类型的先决条件需要满足
//String类型协变
scala> val paramater = Array("Hadoop","Spark","Flink")
paramater: Array[String] = Array(Hadoop, Spark, Flink)
//将参数数组paramater中的每一个元素传入作为函数的参数
scala> functionOps(paramater:_*)
Hadoop
Spark
Flink
scala>
```

函数参数可以有多个，可以将函数作为参数进行传递，可以使用占位符代替参数，简化

代码，如例 3-20 所示。

【例 3-20】 Scala 函数参数示例

```
scala> def DT(myFunc:(String) =>Unit,data:String){myFunc(data)}
DT: (myFunc: String => Unit, data: String)Unit
//函数调用
scala> DT((name:String) =>println(name),"Spark")
Spark
//省略类型String
scala> DT(name => println(name),"Spark")
Spark

scala> DT(println(_),"Spark")
Spark

scala> DT(println,"Spark")
Spark
scala>
```

(3) 多参数函数值

函数值在 Scala 中是真正的对象，因为 Scala 内部实现了对其的处理过程，将函数值创建为一些特殊类的实例。函数值存在于函数的运行期内，这一般与函数字面量不同，通常在实际编码过程中定义和使用有多个参数的函数值。

下面的单例对象 MulParaFunctionOps 中定义了一个方法 mulParaFunctionOps，带有三个参数，作为函数值的 func，Int 类型的数组，以及 Int 类型的 item，在该方法内部将参数 item 的值赋给变量 tmp，对参数 Array [Int] 的数组内元素进行 foreach 循环遍历操作，具体操作为对数组每个元素调用方法作为函数值的参数 func，并将其赋给变量 tmp，而方法作为函数值的参数 func 实现了对其传入的两个参数求和，完成了对数组 Array [Int] 的每个元素的调用操作，后将变量 tmp 的值返回。如例 3-21 所示。

【例 3-21】 多参数函数

```
object MulParaFunctionOps {
//定义多参数的函数mulParaFunctionOps
  def mulParaFunctionOps(func:(Int,Int) => Int,array:Array[Int],item:Int):Int ={
//遍历数组，执行函数func
var tmp = item
     array.foreach { x => tmp = func(tmp,x) }
     tmp
  }
  //定义函数func
  def func(x:Int,y:Int) = x+y
  def main(args: Array[String]): Unit = {
    //定义变量arr
    val arr = Array(12,34,5,7,76)
    //求和
    val sum = mulParaFunctionOps((tmp,x) => tmp + x,arr,1)
    println(sum)
  }
}
```

程序运行结果如图 3-10 所示。

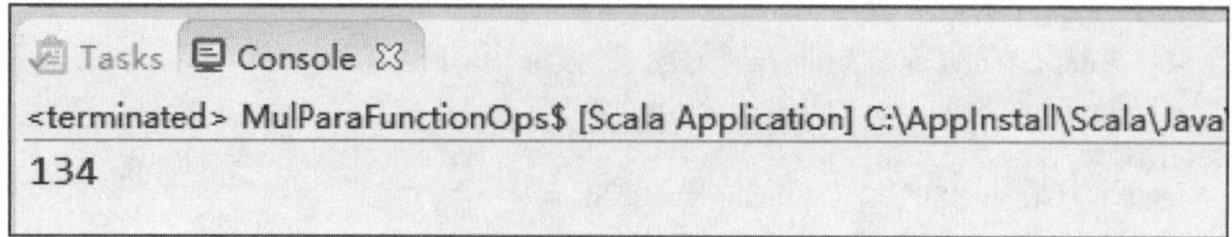

图 3-10 程序运行结果

在 main 方法中定义了变量 arr，它是一个 Int 类型的数组，然后实现对数组元素的求和，在方法 mulParaFunctionOps 中，第一个参数是执行元素求和的函数值，存在于运行期，第二个参数是整型数组，在方法调用的时候将定义的变量数组 arr 传入，第三个参数是初始值。

(4) 重用函数值

使用重用函数值主要用于代码的重用，在 Scala 中使用函数值可以创建很好的重用性代码用来消除代码冗余和代码重复，然而在定义了某个方法后，可能需要将该方法作为实参传入另一个方法中不利于代码的重用和维护，此时就需要使用重复函数值，创建函数值的引用，以此达到重用代码的作用。

下面创建一个伴生类 ReUseFuncValue，带有一个 val 的参数 bigData，定义方法 reUseFuncValue，然后在伴生对象 ReUseFuncValue 中实例化伴生类的对象并调用其方法。如例 3-22 所示。

【例 3-22】 重用函数值示例

```
//伴生类ReUseFuncValue
class ReUseFuncValue(val bigData:String =>String) {
 //定义方法reUseFuncValue
def reUseFuncValue(technology:String){
   println("It's a time of BigData!")
   bigData(technology)
  }
}
//伴生对象
object ReUseFuncValue{
def main(args: Array[String]): Unit = {
   //定义变量
val reuseFunctionValue = new ReUseFuncValue({technology =>
     println("new Tecknology of BigData is :" + technology)
     technology})
val reuseFunctionValue2 = new ReUseFuncValue({tecknology =>
     println("and :" + tecknology)
     tecknology})
   //调用
   reuseFunctionValue.reUseFuncValue("Spark")
   reuseFunctionValue.bigData("Hadoop")
   reuseFunctionValue2.reUseFuncValue("Flink")
  }
}
```

从例 3-22 的伴生对象 ReUseFuncValue 中可以看出在实例化伴生类 ReUseFuncValue 的对象时将函数值作为参数传入其中，对 ReUseFuncValue 的实例使用了相同的代码，可以看出 reuseFunctionValue 和 reuseFunctionValue2 代码相同，若对其进行修改，就都需要修改，不利于代码的重用。

程序运行结果如图 3-11 所示。

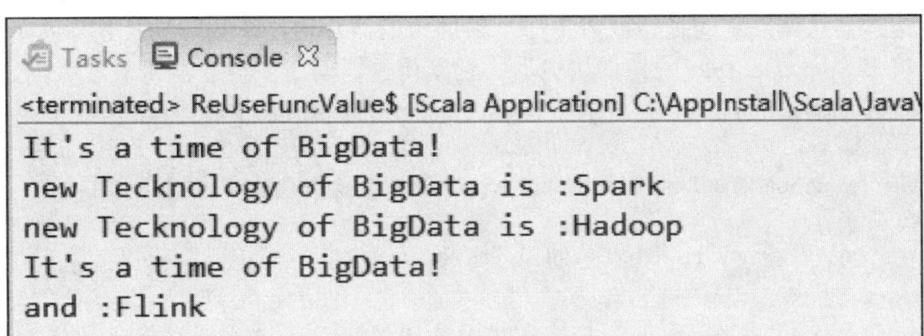

图 3-11　重用函数值程序运行结果

那么可以试着创建一次，以便代码的重用。方法是把函数值直接赋给 val 的变量，然后进行重用，如下所示。

```
val totalReuse = { technology:String =>
    println("new Tecknology of BigData is :" + technology)
    technology}
//代码重用
val reuseFunctionValue = new ReUseFuncValue(totalReuse)
val reuseFunctionValue2 = new ReUseFuncValue(totalReuse)
  //调用
  reuseFunctionValue.reUseFuncValue("Spark")
  //reuseFunctionValue.bigData("Hadoop")
  reuseFunctionValue2.reUseFuncValue("Flink")
```

运行结果如图 3-12 所示。

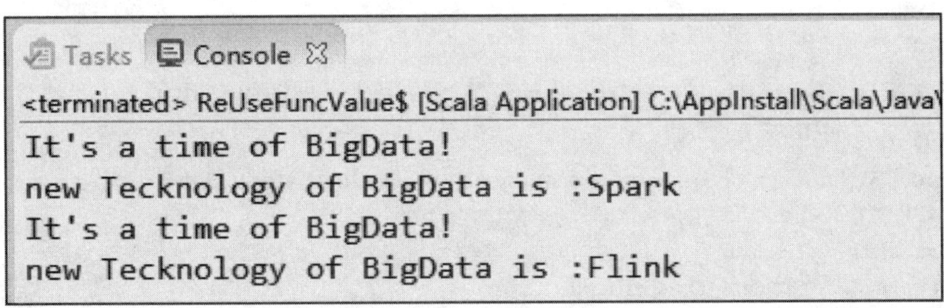

图 3-12　程序运行结果

3.9　从 Spark 源码角度解析 Scala 函数式编程

先决条件在 Spark 中也有广泛使用，使函数的参数在满足一定条件的情况下被执行或调用，如在 SparkContext 中 range 方法中所示。

```
assertNotStopped()
// when step is 0, range will run infinitely
require(step != 0, "step cannot be 0")
val numElements: BigInt = {
  val safeStart = BigInt(start)
  val safeEnd = BigInt(end)
```

辅助构造器从调用主构造器或者上一个辅助构造器开始，在 Spark 上下文 SparkContext 中构造了很多 SparkContext 的辅助构造器，辅助构造器调用主构造器如下所示。

```
@deprecated("Passing in preferred locations has no effect at all, see SPARK-8949", "1.5.0")
@DeveloperApi
def this(config: SparkConf, preferredNodeLocationData: Map[String, Set[SplitInfo]]) = {
this(config)
  logWarning("Passing in preferred locations has no effect at all, see SPARK-8949")
}
```

在 SparkContext 中的 sequenceFile 方法中使用了高阶函数，能够将函数作为函数进行传递的函数，该函数也使用了 Curring 的函数式编程风格，可以产生两个函数，第一个函数返回第二个函数的函数值，如下所示。

```
def sequenceFile[K, V]
      (path: String, minPartitions: Int = defaultMinPartitions)
      (implicit km: ClassTag[K], vm: ClassTag[V],
       kcf: () => WritableConverter[K], vcf: () => WritableConverter[V]): RDD[(K, V)] = {
    withScope {
     assertNotStopped()
     val kc = clean(kcf)()
     val vc = clean(vcf)()
     val format = classOf[SequenceFileInputFormat[Writable, Writable]]
     val writables = hadoopFile(path, format,
       kc.writableClass(km).asInstanceOf[Class[Writable]],
       vc.writableClass(vm).asInstanceOf[Class[Writable]], minPartitions)
     writables.map { case (k, v) => (kc.convert(k), vc.convert(v)) }
    }
   }
```

同样在 RDD. Scala 中的方法 groupBy 有类似的用法，同时又使用了 Scala 语言的泛型系统和隐式转换，如下所示。

```
def groupBy[K](f: T => K, p: Partitioner)(implicit kt: ClassTag[K], ord: Ordering[K] = null)
      : RDD[(K, Iterable[T])] = withScope {
    val cleanF = sc.clean(f)
    this.map(t => (cleanF(t), t)).groupByKey(p)
  }
```

一般情况下程序被解构成若干小函数，依次实现一种功能，使得编程更加灵活，但是也会造成命名空间容易被污染，所以使用本地函数其仅在包含它的代码块中可见，如在 RDD. Scala 中的 getNarrowAncestors 方法所示。

```
private[spark] def getNarrowAncestors: Seq[RDD[_]] = {
    val ancestors = new mutable.HashSet[RDD[_]]

    def visit(rdd: RDD[_]) {
```

```
    val narrowDependencies = rdd.dependencies.filter(_.isInstanceOf[NarrowDependency[_]])
   val narrowParents = narrowDependencies.map(_.rdd)
   val narrowParentsNotVisited = narrowParents.filterNot(ancestors.contains)
    narrowParentsNotVisited.foreach { parent =>
        ancestors.add(parent)
       visit(parent)
    }
  }
   visit(this)
   // In case there is a cycle, do not include the root itself
   ancestors.filterNot(_ == this).toSeq
  }
```

Spark 自从被推出，越来越受到业界的认可和推崇，其框架的简洁和强大的计算能力以及技术栈已经使得 Spark 成为大数据技术中最受欢迎的明星，从一定程度上来讲这得益于 Scala 语言的灵巧和简洁的特性，包括函数式编程等。因此掌握函数式编程不仅对于学习 Spark 来说是必要的，也对于学好 Scala 有非常重要的意义。

3.10 小结

本章主要对 Scala 函数式编程做了较为详细的介绍，包括函数的定义、匿名函数、函数作为一等公民可以当做参数进行传入，可以作为值赋给变量以及高阶函数等一些重要的特性来使用。从 Spark 源码解读的角度出发，在讲解 Scala 函数式编程的特性和使用的情况下，结合 Spark 源码在函数式的使用进行解析，方便广大读者对于学好 Scala 函数式编程有一个更加清晰和明确的认识。

第 4 章

Scala模式匹配、类型系统彻底精通与Spark源码阅读

模式匹配和类型系统在 Scala 占有举足轻重的地位。Scala 模式匹配类似于 Java 的 switch 语句，但比它更好，更不容易犯错。Scala 类型系统涉及泛型、协变和逆变等。本章将介绍 Scala 模式匹配、类型系统的具体内容。这里介绍的是常用的而且在 Spark 源码中经常看到的 Scala 模式匹配。比如，在 Spark 的事件驱动模型源码中，随处可见 Scala 模式匹配的踪影。

4.1 模式匹配语法

模式匹配包括替代的序列，每句开始使用关键字 case。每个备选中包括模式和一个或多个表达式，如果模式匹配将被计算。一个箭头符号“=>”分开的表达模式。

语法结构：

```
var match {
    Case ... =>执行语句
    Case ... =>执行语句
    Case _ =>执行语句         //下划线表示通配符
}
```

【例 4-1】 模式匹配语法实战

```
data match {
        case "Spark" => println("Wow!!!")        //匹配字符串类型时打印
case "Hadoop" => println("Ok")
        case _ => println("Something others")    //不匹配任何类型时输出Nothing
}
```

这是最简单的模式匹配，其中需要注意的是：在每一个匹配模型后面不再有分号，也没有 break（Scala 中没有 break 和 continue 语句），只要匹配到一个就不再往下运行。case_ => println("Nothing")下划线是通配符，匹配任何类型。

4.2 模式匹配实战

4.2.1 模式匹配基础实战

下面通过多个实战示例彻底地精通模式匹配。

【例 4-2】 守卫

```
def bigData(data:String){
    data match {
          case "Spark" => println("Wow!!!")
          case _ if data == "Flink" => println("Nice!!!")
          case _ => println("Something others")
    }
}
```

例 4-2 中，添加了 if 判断，当 data 为“Flink”时执行 println（" Nice!!!" ），这种语法成为守卫。

【例 4-3】 变量应用

```
def bigData(data:String){
    data match {
          case "Spark" => println("Wow!!!")
          case  data1 if data1 == "Flink" => println(data_)
          case _ => println("Something others")
    }
}
```

例 4-3 中，添加了 data1 变量，传入的值会被绑定到 data1 上，如果符合条件，就打印出来。

【例 4-4】 类型匹配

```

class Person
  case class Worker extends Person        //样例类前文讲过
  case class Student extends Person
  def matchTpye(p:Person){
     p match {
          case stu:Student => println("I am a student" + stu)
          case worker:Worker => println("I am a worker"+ worker)
          case _ => println("Nothing")
     }
  }
```

例 4-4 中，先定义一个父类 Person，两个子类 Worker 和 Student。当传入的类型匹配 Student 或者 Worker 时，执行其后面的语句。这个示例结合了样例类。这样的例子在 Spark 中有大量的应用。

4.2.2 数组、元祖实战

【例 4-5】 匹配数组

```
def data(array:Array[String]){
  array match {
   case Array("Scala") => println("Scala ...")
   case Array(spark,hadoop,storm) => println(spark + ":" +hadoop + " : "+ storm)
   case Array("Spark",_*) => println("Spark...")
   case _ => println("Nothing")
  }
}
```

例 4-5 匹配 Scala 数组，里面涉及了一个和多个参数的匹配还有可变参数的匹配。

【例 4-6】 匹配元组

```
def dataTuple(tuple:Tuple2[Any,Any]){
   tuple match {
          case  (0,_) => println("0 is matched")   //匹配第一个值为0的元组
     case  (y,0) => println(y + "0")             //匹配第二个值为0的元组
     case  (x,y) => println(x +" : "+ y)        //匹配有两个值的元组
     case  _  => println("Nothing")
   }
}
```

例 4-6 匹配 Scala 元组中的 Tuple。

4.2.3 Option 实战

【例 4-7】 Option 实战

```
object OptionMatch{
 def main(args: Array[String]) {
  val capitals = Map("FRANCE"-> "PARIS", "KOERA" -> "SEOUL")

  println("show(capitals.get(\"KOERA\")) : " +
   show(capitals.get( "KOERA")))
  println("show(capitals.get(\"INDIA\")) : " +
   show(capitals.get( "INDIA")) )
 }

 def show(x: Option[String]) = x match {
  case Some(s) => s                    //当匹配到Option的子类Some时返回s
  case None => "Nothing"               //当匹配Option的子类None时返回“Nothing”
 }
}
//运行结果如下：
show(capitals.get( "KOERA")) : SEOUL
show(capitals.get( "INDIA")) : Nothing
```

例 4-7 先定义一个 Map 映射，capitals. get（" KOERA" ）的返回值为 SEOUL，匹配到了 case Some（s），capitals. get（" INDIA" ）返回值为 None 类型，所以就匹配到 case None。

4.2.4 提取器

提取器是指定义了 unapply 方法的 object。在进行模式匹配的时候会调用该方法。unapply 方法接受一个数据类型，返回另一数据类型，表示可以把入参的数据解构为返回的数据。

【例 4-8】 提取器在模式匹配中的应用

```
object ExtractorTest2 {
  def main(args: Array[String]) {
    val x = ExtractorTest2(5)                    //调用apply方法，x值为10
    println(x)
    x match
    {
      case ExtractorTest2(num) => println(x+" is bigger two times than "+num)   //unapply is invoked
      case _ => println("i cannot calculate")
    }
  }
  def apply(x: Int) = x*2
  def unapply(z: Int): Option[Int] = if (z%2==0) Some(z/2) else None
}
//在进行模式匹配时， case ExtractorTest2(num) 会调用unapply方法。运行结果如下
10
10 is bigger two times than 5
```

4.2.5 Scala 异常处理与模式匹配

【例 4-9】 模式匹配与异常处理

```
import java.io.{FileReader, FileNotFoundException, IOException}

object ExceptionTest {
  def main(args: Array[String]) {
    try {
      val f = new FileReader("C://1.txt")
    } catch {
      case ex: FileNotFoundException =>{
        println("Missing file exception")
      }
      case ex: IOException => {
        println("IO Exception")
      }
     case _ => println("Unknown Exception")
    }finally {                                   //finally模块一定会被执行
      println("Exiting finally...")
    }
  }
}
//运行结果如下
Missing file exception
Exiting finally...
```

例 4-9 是捕获异常，当出现异常并被捕获时，通过模式匹配从最开始路由到匹配项，则返回或输出相应异常，然后停止模式匹配。

4.2.6 sealed 密封类

前面已经实战讲解了样例类的模式匹配，如果在声明超类（父类）时，加上 sealed 关键字，则超类就成为密封类。密封类的好处是在做模式匹配时，编译器会自动帮助检查所有匹配项是否已完全包含。要想添加一个匹配项，必须要写到密封类所在的里面。

【例 4-10】 密封类

```
Sealed class Person
  case class Worker extends Person            //样例类前文讲过
  case class Student extends Person
  def matchTpye(p:Person){
     p match {
          case stu:Student => println("I am a student" + stu)
          case worker:Worker => println("I am a worker"+ worker)
          case _ => println("Nothing")
     }
  }
```

把 Person 类声明为密封类，则其子类都必须声明在与密封类相同的文件里。编译器会得知这个密封类有哪些子类作为选项进行匹配。

4.3 类型系统

4.3.1 泛型

泛型就是定义以类型为参数的类或接口（Scala 中为特质）的功能，在 Scala 中类和特质都可以带类型参数，用方括号来定义类型参数。

【例 4-11】 泛型类

```
class Person[T](val content:T){
   def getContent(id:T) = id + " _ " + content
val p = new Person[String]("Spark")
p.getContent("Scala")              //返回res3: String = Scala _ Spark
```

在创建对象时已经指定了 T 为 String 类型，因此在调用 getContent 方法时必须传入字符串，否则：

```
<console>:10: error: type mismatch;
 found   : Int(100)
 required: String
             p.getContent(100)
                          ^
```

自定义的类型 T 在创建对象时可以随意指定，一旦指定，里面的方法就必须是指定的类型，当然还可指定 Int 类型。

【例 4-12】 泛型函数

```
def getElem[T](list:List[T]) = list(list.length-1)    //定义函数名为getElem，其参数为
                                                       List，返回最后一位元素。List中可以存放任意
类型，但是类型必须统一。
```

```
getElem(List("Spark","Storm","Scala"))          //返回Scala
getElem(List(1,2,3,4,5))                        //返回5
```

4.3.2 边界

泛型为 Scala 扩展了类及函数的复用性，但有时后也需要对泛型做限制。类型限制有类型界定，视图界定，上下文界定等。通过实战示例一一展示如下。

【例 4-13】 类型界定

```
class Test[T <: Comparable[T]](val first:T,val second:T){
      def bigger = if(first.compareTo(second) < 0) second else first
}
val t = new Test(“first”,”second”)
t.bigger                          //返回second
```

这个示例定义了 T 的上界 Comparable，也就是说 T 必须是 Comparable 的子类。很显然不能指定 T 为 Int 或 File 类型，因为在 Scala 中他们都不是 Comparable 的子类。有上界，那么同样也存在着下界，表示为“>:”，如下所示：

```
class Consumer[T](t: T){
      defuse[U >: T](u : U) = {println(u)}
}
```

【例 4-14】 视图界定

类型界定中控制了 T 的上下界，但是当传入 Int 类型时，编译报错，因为 Int 不是上界 Comparable 的子类，解决办法是“视图界定”。视图界定的标识符“<%”。

```
class Test[T <% Comparable[T]](val first:T,val second:T){
      def bigger = if(first.compareTo(second) < 0) second else first
}
val t = new Test(1,2)
T.bigger                          //返回值为2
```

这个示例在做大小比较时有些繁琐，总是希望能够两个值直接进行比较，那么 Scala 就利用 Ordered 特质解决了这个问题。

```
class Test[T <% Ordered[T]](val first:T,val second:T){
      def bigger = if(first < second) second else first
}
```

注意：视图界定 T<% Ordered［T］，要求必须存在一个 T 到 Ordered［T］隐式转换，隐式转换下章将通过实战介绍。

4.3.3 协变与逆变

"＋"表示协变，而"－"表示逆变。

C［＋T］：如果 A 是 B 的子类，那么 C［A］是 C［B］的子类。

C［－T］：如果 A 是 B 的子类，那么 C［B］是 C［A］的子类。

C［T］：无论 A 和 B 是什么关系，C［A］和 C［B］没有从属关系。

【例 4-15】 协变

```
trait Person[+T]{
  def eat
}
val children = new Person[String]{
    override def eat = {println("吃饭。。。")}
}
val father:Person[Any] = children
val action:Any = father.eat
```

children 可以被赋值给 father，说明 Person［Any］和 Person［String］是同类型的，而 Scala 中 Any 是所有类的基类。所以 father 必然可以调用到 children 的成员。协变是正向扩展的，返回对象必然被声明对象兼容，也就是说 father. eat 返回 String 类型，声明 Action 为 Any 类型，Any 必然兼容 String。

【例 4-16】 逆变

```
trait Person[-T]{
  def eat
}
val father = new Person[Any]{
    override def eat = {println("吃饭。。。")}
  }
val children:Person[String] = father
val action:Any = children.eat
```

详解：逆变是反向收缩的。实际处理类型比声明类型范围粗略，则返回对象必然会成为声明对象的兄弟对象，不能处理。可以看到，声明的 Person［String］可以指向 Person［Any］，但是因为 Any 的范围比 String 的范围粗略，声明对象的类型不能是返回值类型的兄弟对象，否则报错。例如：

val action：String＝children. eat 会报类型不匹配：type mismatch；found：Unit required：String。

4.4 Spark 源码阅读

【例 4-17】 Spark 源码之模式匹配展示

```
case DriverStateChanged(driverId, state, exception) => {
    state match {
     case DriverState.ERROR | DriverState.FINISHED | DriverState.KILLED | DriverState.FAILED
                =>      removeDriver(driverId, state, exception)
     case     _        =>
       throw new Exception(s"Received unexpected state update for driver $driverId: $state")
    }
}
```

这段是 Spark 的 Driver 状态发生改变后会执行的一段代码。DriverStateChanged 是样例类；state 通过模式匹配与 DriverState. ERROR | DriverState. FINISHED | DriverState. KILLED | DriverState. FAILED 进行匹配，如果上面这些都没有匹配上，则执行下面的“_”通配符，抛出异常。

class HashMap [A, B] private [collection] ...

这行代码是 Scala 的 HashMap 数据结构，在 Scala 的各种数据结构中有大量的泛型、视图等的应用。

【例 4-18】 Spark 之 Worker 类中样例类的经典实用案例

```
/**
* Worker向Master注册，收到返回的注册信息
*/
private def handleRegisterResponse(msg: RegisterWorkerResponse): Unit = synchronized {
   msg match {
  //注册成功时的处理方式，RegisteredWorker是样例类
  case RegisteredWorker(masterRef, masterWebUiUrl) =>
        logInfo("Successfully registered with master " + masterRef.address.toSparkURL)
        registered = true
        changeMaster(masterRef, masterWebUiUrl)
        forwordMessageScheduler.scheduleAtFixedRate(new Runnable {
          override def run(): Unit = Utils.tryLogNonFatalError {
            self.send(SendHeartbeat)
          }
        }, 0, HEARTBEAT_MILLIS, TimeUnit.MILLISECONDS)
        if (CLEANUP_ENABLED) {
          logInfo(
            s"Worker cleanup enabled; old application directories will be deleted in: $workDir")
          forwordMessageScheduler.scheduleAtFixedRate(new Runnable {
            override def run(): Unit = Utils.tryLogNonFatalError {
              self.send(WorkDirCleanup)
            }
          }, CLEANUP_INTERVAL_MILLIS, CLEANUP_INTERVAL_MILLIS, TimeUnit.MILL-
    ISECONDS)
        }
     //注册失败时的处理方式，RegisterWorkerFailed是样例类
      case RegisterWorkerFailed(message) =>
        if (!registered) {
          logError("Worker registration failed: " + message)
          System.exit(1)
        }
     //Master的状态为Standby时的处理方式，是样例类MasterInStandby
      case MasterInStandby =>
        // Ignore. Master not yet ready.
    }
  }
```

4.5 小结

本章主要介绍了 Scala 模式匹配的语法结构，结合案例代码实战了解模式匹配的基础、数组、元组等概念，以及介绍了模型系统的泛型、边界和协变与逆变的概念。最后通过结合 Spark 源码分析了 Scala 的模式匹配应用。

第5章 Scala隐式转换等彻底精通及Spark源码阅读

Scala 隐式转换具有很强大的功能。Scala 隐式转换让 Scala 编程语言更加富有表现力，不需要将一些显而易见的类型转换写进代码。Scala 隐式转换也帮助实现不需要修改代码就可以扩展新功能。Scala 隐式转换用 implicit 修饰符表示，可以作用在诸如类、参数和值等 Scala 关键字，使之分别成为隐式类、隐式参数和隐式值。

5.1 隐式转换

5.1.1 隐式转换的使用条件

隐式转换是函数参数或者对象的隐式转换，隐式值必须在伴生对象中声明。

隐式转换使用的条件。

（1）传入参数类型与预期类型不匹配时，会找是否有将该参数类型转为预期参数类型的隐式值；如果调用该函数的对象可以找到匹配的隐式值，同样会发生隐式转换；

（2）当对象访问不存在的成员时，会找该对象是否有转换成其他对象的隐式值。

使用隐式转换的限制条件。

（1）implicit 关键字只能用来修饰方法、变量（参数）和伴随对象；

（2）隐式转换的方法（变量和伴随对象）在当前范围内才有效。如果隐式转换不在当前范围内定义（比如定义在另一个类中或包含在某个对象中），那么必须通过 import 语句将其导入。

5.1.2 隐式转换实例

隐式转换是 Scala 的重难点之一，本节通过实例来详解 Scala 隐式转换。

【例 5-1】 隐式函数

```

object ImplicitTest {
```

```
    implicit def a2RichA(a:A) = new RichA(a)        //定义一个名称为a2RichA的隐式方法（函数）
    def main(args: Array[String]): Unit = {
      val a = new A
      a.rich
    }
}
class Implicit{}
class A{}
dass RichA(a:A){
  def rich{
    println("Hello Scala Implicit!!!")
  }
}
// 运行结果如下
Hello Scala Implicit!!!
```

这个实例中定义了一个隐式函数 a2RichA。在实例化 val a＝new A 对象 A 之后，a. rich 调用 rich 方法，但是对象 A 并没有 rich 方法，编译器会查找当前范围内是否有可转换的方法，如果没有则编译失败。a2RichA 这个函数的名称可以随便命名，通常应该具有可读性。

【例 5-2】 Int 类型隐式转换成 String 类型

```
object Int2StringTest {
  implicit def int2String(i: Int) = i.toString()
  def main(args: Array[String]): Unit = {
    println(3.length)
  }
}
//运行结果如下
1
```

例 5-2 中 length 调用了 String 类型的 length 方法，Int 类型本身没有 length 方法，但是在可用范围内定义了可以把 Int 转换为 String 的隐式函数 int2String，因此函数编译通过并且运行出正确结果。在此实例中隐式函数的定义必须定义在使用之前，否则编译报错。

【例 5-3】 导入隐式函数

```
object FuncTest {
    implicit def int2String(i:Int) = i.toString()
}
object Int2StringTest {
    import book.FuncTest._                    //将隐式函数导入可用作用域
  def main(args: Array[String]): Unit = {
    println(3.length)
  }
}
//运行结果如下
1
```

例 5-3 中，import book. FuncTest. _ 会将 FuncTest 对象内部的成员导入到相应的作用域内，否则无法调用隐式函数。FuncTest 对象必须定义在调用类之前或者定义到不同的文件中。

5.2 隐式类

所谓隐式类：就是对类增加 implicit 限定的类，其作用主要是对类的加强。

【例 5-4】 隐式类实战

```
object PracticeTest {
  def main(args: Array[String]) {
  }
  implicit class Calc(x:Int){                    //定义隐式类
    def add(a:Int) = {
      a + x
    }
  }
  println("1.add(2):"+1.add(2))                  //隐式类应用
}
```

Calc 前面的 implicit，通过这个隐式类，就可以让 Int 型数据具有 add 方法。整数类型本身没有 add 方法，编译器在遇到 1. add（2）时不会立马报错，而检查当前作用域有没有用 implicit 修饰的，同时可以将 Int 作为参数的构造器，并且具有方法 add 的类。

5.3 隐式参数详解

【例 5-5】 隐式参数实战

```
object ArgImplicitTest {
  def main(args: Array[String]): Unit = {
    hiScala
  }
  implicit val name:String = "张三"
  //implicit val sex:String = "nv"

  def hiScala(implicit str:String){
    println("str is : "+ str)
  }
}
```

上面定义带有隐式参数 str 的函数 hiScala，并且定义了隐式值 name。函数的隐式值会在本实例域内找一个对应的字符串类型的隐式值。若没有则编译报错 could not find implicit value for parameter str：String；若有多个则编译报 ambiguous implicit values，就是有歧义，找到了多个而不知道用哪个。

5.4 隐式值

【例 5-6】 隐式值

隐式值就是定义在 object 对象中的隐式变量。

```
object VarImplicit{
  implicit val name :String = "lishi"
  def hiScala(implicit str:String){
    println("str is : "+ str)
  }

  def main(args: Array[String]) {
    hiScala
  }
}
//运行结果如下
str is : lishi
```

例 5-6 中定义了隐式变量 name，方法 hiScala 有隐式的参数，若在调用 hiScala 时没有显示传入参数，则会调用对应类型的隐式值。

注意：相同类型的隐式值在相同的作用范围内只能有一个。

5.5 Spark 源码阅读解析

【例 5-7】 Spark 源码：隐式参数应用

```
def accumulator[T](initialValue: T)(implicit param: AccumulatorParam[T]): Accumulator[T] =
  {
    val acc = new Accumulator(initialValue, param)
    cleaner.foreach(_.registerAccumulatorForCleanup(acc))
    acc
  }
```

Spark 中这段代码应用了 Scala 的隐式参数 implicit param：AccumulatorParam［T］，在函数调用时可以传递对应类型的参数，也可以不传。

【例 5-8】 Spark 源码：隐式函数应用

```
//Int类型隐式转换为IntWritable
implicit def intToIntWritable(i: Int): IntWritable = new IntWritable(i)
//Long类型隐式转换为LongWritable
  implicit def longToLongWritable(l: Long): LongWritable = new LongWritable(l)
//Float类型隐式转换为FloatWritable
  implicit def floatToFloatWritable(f: Float): FloatWritable = new FloatWritable(f)
//Double类型隐式转换为DoubleWritable
  implicit def doubleToDoubleWritable(d: Double): DoubleWritable = new DoubleWritable(d)
```

5.6 小结

本章主要展示了在什么条件下使用隐式转换以及隐式转换的一些实战案例，以及其他三个重要概念如隐式类、隐式参数和隐式值。在 Spark 中隐式转换使用广泛，本章最后列举了两个关于隐式参数和隐式函数的 Spark 源码案例。

第6章 并发编程及Spark源码阅读

许多开发者在创建和维护多线程应用程序时经历过各种各样的问题，他们希望能在一个更高层次的抽象上进行工作，以避免直接和线程与锁打交道。Scala 在 2.9.x 之后放弃了自己的 actor，采用 Akka 进行并发编程。因此本节主要讲 Akka 案例实战。

6.1 并发编程彻底详解

actor 是一个封装了状态和行为的对象，每个 actor 都通过 message 交流，从自己的 mailbox 中读取别的 actor 发送的消息。

actor 是一个容器，拥有 state，behavior，mailbox，children，supervisor strategy 这些内容。Akka 保证所有 actor 都只运行在自己的轻量级线程，并一次性处理一个消息，这样程序员就不用处理同步、竞态。

State：actor 拥有的一组变量，即 actor 的 state。state 是可恢复的。

Behavior：actor 消息行为模型。

Mailbox：所有收到的消息，会进入 actor 的 mailbox 队列，默认是 FIFO。

Children：每一个 actor，都是潜在的监控者。actor 会自动监控执行子任务的 actor. children 被放在 context 中，通过 context. actorof（…）或者 context. stop（child）操作 children. 这些操作都是异步的，所以相应非常快。

Supervisor Strategy：actor 一旦创建，监控策略是不可修改的。Akka 自动帮我们处理错误故障。

6.1.1 actor 工作模型

actor 就是对象发送与接收消息的过程。P_1 给 P_2 发送消息，收到消息后处理并返回，P_1 接收 P_2 返回的消息。图 6-1 中虚线表示 P_1 可能只发送消息或者说 P_2 只接收并处理消息，不返回消息。

消息可以是任何类型的对象，但必须是不可变的。目前 Scala 还无法强制不可变性，因此这一点必须作为约定。String，Int，Boolean 这些原始类型总是不可变的。除了它们以外，

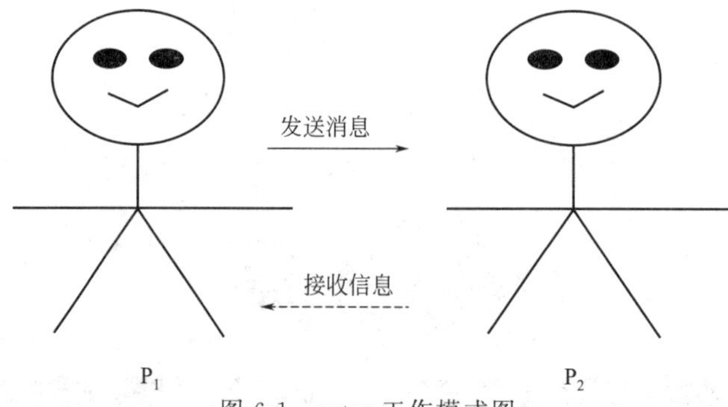

图 6-1　actor 工作模式图

推荐的做法是使用 Scala case class，它们是不可变的（如果不专门暴露数据的话），并与接收方的模式匹配配合得非常好。以下是一个例子。

```
//定义 case class
case class Register(user: User)
//创建一个 case class 消息
val message = Register(user)
```

其他的适合做消息的类型包括 scala. Tuple2，scala. List，scala. Map 它们都是不可变的，可以很好地进行模式匹配。

6.1.2　发送消息

向 actor 发送消息是使用下列方法之一。

①“!”，意思是“fire-and-forget”，例如异步发送一个消息并立即返回。也称为“tell”。

②“?”，异步发送一条消息并返回一个 Future，代表一个可能的回应。也称为“ask”。

每一个消息发送者分别保证自己的消息的次序。

Tell：Fire-forget，这是发送消息的推荐方式。不会阻塞地等待消息。它拥有最好的并发性和可扩展性。

actor!" hello"，如果是在一个 Actor 中调用，那么发送方的 actor 引用会被隐式地作为消息的 sender：ActorRef 成员一起发送。目的 actor 可以使用它来向原 actor 发送回应，使用 sender! replyMsg。

如果不是从 Actor 实例发送的，sender 成员缺省为 deadLetters actor 引用。

ask：Send-And-Receive-Future，ask 模式既包含 actor 也包含 future，因此它是作为一种使用模式，而不是 ActorRef 的方法。

```
import akka.pattern.{ ask, pipe }

case class Result(x: Int, s: String, d: Double)
case object Request

implicit val timeout = Timeout(5 seconds) // 下面的 `?` 会用到

val f: Future[Result] =
  for {
    x <- ask(actorA, Request).mapTo[Int] // 直接调用
    S <- actorB ask Request mapTo manifest[String] // 隐式转换调用
    D <- actorC ? Request mapTo manifest[Double] // 通过符号名调用
```

```
    } yield Result(x, s, d)

f pipeTo actorD //... 或...
pipe(f) to actorD
```

上面的例子展示了将 ask 与 future 上的 pipeTo 模式一起使用，因为这是一种非常常用的组合。请注意上面所有的调用都是完全非阻塞和异步的：ask 产生 Future，三个 Future 通过 for 语法组合成一个新的 Future，然后用 pipeTo 在 future 上安装一个 onComplete 处理器来完成将收集到的 Result 发送到其他 actor 的动作。

使用 ask 将会像 tell 一样发送消息给接收方，接收方必须通过 sender! reply 发送回应来为返回的 Future 填充数据。ask 操作包括创建一个内部 actor 来处理回应，必须为这个内部 actor 指定一个超时期限，过了超时期限内部 actor 将被销毁以防止内存泄露。

如果要以异常来填充 future，需要发送一个 Failure 消息给发送方。这个操作不会在 actor 处理消息发生异常时自动完成。

```
try {
  val result = operation()
  sender ! result
} catch {
  case e: Exception⇒
    sender ! akka.actor.Status.Failure(e)
    throw e
}
```

如果一个 actor 没有完成 future，它会在超时时限到来时过期，以 AskTimeoutException 来结束。超时的时限是按下面的顺序和位置来获取的。

显式指定超时：

```
import akka.util.duration._
import akka.pattern.ask
val future = myActor.ask("hello")(5 seconds)
```

提供类型为 akka. util. Timeout 的隐式参数，例如

```
import akka.util.duration._
import akka.util.Timeout
import akka.pattern.ask
implicit val timeout = Timeout(5 seconds)
val future = myActor ? "hello"
```

Future 的 onComplete，onResult，或 onTimeout 方法可以用来注册一个回调，以便在 Future 完成时得到通知，从而提供一种避免阻塞的方法。

注意：在使用 future 回调如 onComplete，onSuccess，and onFailure 时，在 actor 内部要小心避免捕捉该 actor 的引用。例如：不要在回调中调用该 actor 的方法或访问其可变状态。这会破坏 actor 的封装，会引用同步 bug 和 race condition，因为回调会与此 actor 一同被并发调度。不幸的是目前还没有一种编译时的方法能够探测到这种非法访问。

转发消息。你可以将消息从一个 actor 转发给另一个。虽然经过了一个“中转”，但最初的发送者地址/引用将保持不变。当实现功能类似路由器、负载均衡器、备份等的 actor 时会很有用。

6.1.3 回复消息

如果需要一个用来发送回应消息的目标，可以使用 sender，它是一个 Actor 引用。可以用 sender！replyMsg 向这个引用发送回应消息。也可以将这个 Actor 引用保存起来将来再作回应。如果没有 sender（不是从 actor 发送的消息或者没有 future 上下文），那么 sender 缺省为“死信”actor 的引用。

```
case request =>
 val result = process(request)
 sender ! result// 缺省为死信actor
```

6.1.4 actor 创建

要定义自己的 actor 类，需要继承 Actor 并实现 receive 方法。receive 方法需要定义一系列 case 语句（类型为 PartialFunction［Any，Unit］）来描述 Actor 能够处理哪些消息（使用标准的 Scala 模式匹配），以及实现对消息如何进行处理的代码。

```
class MyActor extends  Actor{
       override def receive: Receive = {
           case "test" ⇒log.info("received test")
           case _       ⇒log.info("received unknown message")
       }
}
```

请注意 Akka Actor receive 消息循环是“穷尽的（exhaustive）”，这与 Erlang 和 Scala 的 actor 行为不同。这意味着需要提供一个对它所能够接受的所有消息的模式匹配规则，如果希望处理未知的消息，就需要像上例一样提供一个缺省的 case 分支。

actor 在创建后将自动异步地启动。当创建 actor 时它会自动调用 actor 的 preStart 回调方法。这是一个非常好地用来添加 actor 初始化代码的位置。

```
override def preStart() = {
... //初始化代码
}
```

创建 actor 有两种方式，如下。

(1) 用缺省构造方法创建 actor（actor 是从系统创建的）

例如：

```
val system = ActorSystem("HelloWorld")
system .actorOf()
```

对 actorOf 的调用返回一个实例。这是一个 Actor 实例的句柄（handle），可以用它来与实际的 Actor 进行交互。The ActorRef 是不可变量，与它所代表的 Actor 之间是一对一的关系。The ActorRef 还是可序列化的（serializable），并且携带网络信息。这意味着可以将它序列化以后，通过网络进行传送，在远程主机上它仍然代表原结点上的同一个 Actor。

(2) 从其他的 actor 使用 actor 上下文（context）来创建

例如：

```
context.actorOf(Props[MyActor], name = "myactor")
```

name 参数是可选的，但建议为 actor 起一个合适的名字，因为它将在日志信息中被用于标识各个 actor。名字不可以为空或以“$”开头。如果给定的名字已经被赋给了同一个父 actor 的其他子 actor，将会抛出 InvalidActorNameException。

6.1.5 用上下文 context 创建 actor

【例 6-1】 创建 actor 之 context

```
object Greeter {
//定义两个样例类
  case object Greet
  case object Done
}
import akka.actor.Actor
class Greeter extends Actor {
 override  def receive = {                                //override也可以不写，但是通常我们都会写
    case Greeter.Greet =>
      println("Hello World!")
     //给发送者返回消息
      sender() ! Greeter.Done
  }
}
class HelloWorld extends Actor{
  override def preStart(): Unit = {
    // 用context对象创建greeter
    val greeter = context.actorOf(Props[Greeter], "greeter")  //Props是一个用来在创建actor时指定选项的配置类。
    //给greeter 发送消息
    greeter ! Greeter.Greet
  }
  def receive = {
    // 接收到返回的消息后停止此actor
    case Greeter.Done => context.stop(self);println("消息处理完毕。。。")
  }
}
object MainTest {
  def main(args: Array[String]): Unit = {
    akka.Main.main(Array(classOf[HelloWorld].getName))
  }
}
//运行结果如下
HelloWorld!
消息处理完毕。。。
```

在例 6-1 中，akka. Main. main（Array（classOf［HelloWorld］. getName））执行后会调用 HelloWorld 的 preStart 方法，在此方法中用 context 对象创建名为 greeter 的 Actor，

接着给 greeter 发送 Greeter. Greet 消息，Greeter 对象收到消息执行匹配的语句 println（" Hello World!" ），并给发送者（HelloWorld）返回消息，HelloWorld 类接收到返回的消息后执行 case Greeter. Done=>context. stop（self）；println（" 消息处理完毕…" ）语句来停止此消息发送与接收的模型。

6.1.6 用 ActorSystem 创建 actor

【例 6-2】 actor 消息发送与接收实战

先定义 MyActor 类继承 Actor 特质，用 ActorSystem 来创建可执行的 actor，示例代码如下。

```
class MyActor extends  Actor{
  override def receive: Receive = {
    case "test" => println("received test")
    case _      => println("received unknown message")
  }
}

object MyActorTest {
  def main(args: Array[String]) {
    val system = ActorSystem("MyActor")
    val myActor = system.actorOf(Props[MyActor],"MyActor")
    myActor ! "test"
  }
}
```

实战详解：

```
val system = ActorSystem("MyActor")
这一步是创建ActorSystem，初始化actor系统，参数随便写。

val myActor = system.actorOf(Props[MyActor],"MyActor")
创建myActor 实例，用于发送消息。Props是一个configuration class。

myActor ! "test"
感叹号是发送消息的方法，后面接消息体。
//运行结果如下
received test
```

MyActor 具体 actor 类，继承 actor 并实现 receive，receive 方法中是一系列的 case 语句，这是 Scala 的模式匹配。

6.1.7 用匿名类创建 actor

在从某个 actor 中派生出新的 actor 来完成特定的子任务时，可能使用匿名类来包含将要执行的代码会更方便。

【例 6-3】 匿名类 actor

```
def receive = {
  case m: DoIt ⇒
    context.actorOf(Props(new Actor {              //创建匿名类actor
      override def receive = {                     //重写receive方法
```

```
    case DoIt(msg) ⇒
      val replyMsg = doSomeDangerousWork(msg)
      sender ! replyMsg
      context.stop(self)
    }
    def doSomeDangerousWork(msg: ImmutableMessage): String = { "done" }
  })) forward m
}
```

用匿名类创建 actor 时，需要避免捕捉外层 actor 的引用，例如：不要在匿名的 actor 类中调用外层 actor 的方法。这样就破坏 actor 的封装，可能会引入同步 bug 和资源竞争，因为其他的 actor 可能会与外层 actor 同时进行调度。这种错误，在编译器不会被发现。

6.1.8 actor 生命周期

actor 基本的生命周期非常的直观。

① 和其他普通类一样，需要一个构造函数；

② preStart 函数在其之后被调用。在这里，可以初始化一些资源，然后在 postStop 中清除；

③ servicing 或 receive 方法里面的消息处理占用了绝大多数的时间。

【例 6-4】 actor 生命周期

```
import akka.actor.{ActorLogging, Actor}
import akka.actor.{ActorSystem, Props}
import akka.event.LoggingReceive
//定义actor
class BasicLifecycleLoggingActor extends Actor with ActorLogging {
  log.info("Inside BasicLifecycleLoggingActor Constructor")
  log.info(context.self.toString())
//复写preStart方法
  override def preStart() = {
    log.info("BasicLifecycleLoggingActor preStart method is invoked...")
  }
  def receive = LoggingReceive {
    case "hello" => log.info("hello")
  }
//复写postStop方法
  override def postStop() = {
    log.info("BasicLifecycleLoggingActor postStop method is invoked...")
  }
}
object LifecycleApp extends App {
  def main(args: Array[String]) {
    val actorSystem = ActorSystem("LifecycleActorSystem")
    val lifecycleActor = actorSystem.actorOf(Props[BasicLifecycleLoggingActor], "lifecycleActor")
    lifecycleActor ! "hello"    //发送消息
    Thread.sleep(2000)
    actorSystem.shutdown()
  }
}
//运行结果
```

```
Inside BasicLifecycleLoggingActor Constructor
Actor[akka://LifecycleActorSystem/user/lifecycleActor#1496463646]
BasicLifecycleLoggingActor preStart method is invoked...
hello
BasicLifecycleLoggingActor postStop method is invoked...
```

从运行结果可以看出，在 actor 被创建，接收消息，然后被销毁的过程如下所述。

① BasicLifecycleLoggingActor 的构造方法；

② preStart 方法被调用；

③ receive 方法被调用；

④ postStop 方法被调用。

Actor 从被创建到被销毁一共经过了四步。

6.1.9 终止 actor

从程序中看出，ActorSystem 关闭的时候，postStop 方法被调用。其实还有其他几种情况可以调用 postStop。

【例 6-5】 ActorSystem. stop（）终止 actor

```
object LifecycleApp {
  def main(args: Array[String]) {
    val actorSystem = ActorSystem("LifecycleActorSystem")
    val lifecycleActor = actorSystem.actorOf(Props[BasicLifecycleLoggingActor], "lifecycleActor")
    lifecycleActor ! "hello"
    Thread.sleep(2000)
    actorSystem.stop(lifecycleActor)
  }
}
```

程序在等待 2 秒后执行 actorSystem. stop（lifecycleActor）终止 actor。

【例 6-6】 context. stop（self）

```
  def receive = LoggingReceive {
    case "hello" => log.info("hello")
    case "stop" => context.stop(self)          //当接收到stop消息后调用context.stop(self)来终止actor
  }

  object LifecycleApp {
  def main(args: Array[String]) {
    val actorSystem = ActorSystem("LifecycleActorSystem")
    val lifecycleActor = actorSystem.actorOf(Props[BasicLifecycleLoggingActor], "lifecycleActor")
    lifecycleActor ! "hello"
    lifecycleActor ! "stop"                    //发送 stop消息
  }
}
```

【例 6-7】 POISONPILL 终止 actor

在前面的例子里面，通过在 LifecycleApp 发送带有 stop 的消息到 Actor，Actor 收到这个消息，并通过 context. stop 来终止自己。其实可以通过发送 PoisonPill 消息来达到同样的功能。PoisonPill 和前面的那个 stop 消息类似，它也会被扔进到一个普通邮箱里面排队，只有当轮到它的时候才会进行处理。

```
object LifecycleApp {
  def main(args: Array[String]) {
    val actorSystem = ActorSystem("LifecycleActorSystem")
    val lifecycleActor = actorSystem.actorOf(Props[BasicLifecycleLoggingActor], "lifecycleActor")
    lifecycleActor ! "hello"
    lifecycleActor ! PoisonPill
  }
}
```

【例 6-8】 Kill 终止 actor

```
object LifecycleApp {
  def main(args: Array[String]) {
    val actorSystem = ActorSystem("LifecycleActorSystem")
    val lifecycleActor = actorSystem.actorOf(Props[BasicLifecycleLoggingActor], "lifecycleActor")
    lifecycleActor ! "hello"
    lifecycleActor ! Kill
  }
}
```

actorSystem. shutdown（）是直接终止的系统，而通过发送 stop、POISONPILL、Kill 这三种发送消息和 actorSystem. stop（lifecycleActor）方式终止 actor 后，程序并没有完结。

当终止了 actor 后，如果继续给这个 actor 发送消息，那么就进入了 Terminated 状态。在进入 Terminated 状态后，从日志中可以看到好几个 deadletters，所有发送到已终止的 Actor 的消息都会被转发给一个叫做 DeadLetterActor 的内部 Actor。

DeadLetter Actor 会处理自己邮箱里的消息，并把每条消息都封装成一个 DeadLetter，然后再发布到 EventStream 里面去。

另一个叫做 DeadLetterListener 的 Actor 会去消费所有这些 DeadLetter 消息并把它们作为一条日志消息发布出去。

【例 6-9】 处理 DeadLetter 消息

```
import akka.actor.{DeadLetter, ActorLogging, Actor}
import akka.event.LoggingReceive
import akka.actor.ActorSystem
import akka.actor.Props

class BasicLifecycleLoggingActor extends Actor with ActorLogging {
  log.info("Inside BasicLifecycleLoggingActor Constructor")
  log.info(context.self.toString())
  override def preStart() = {
    log.info("BasicLifecycleLoggingActor preStart method is invoked...")
}
def receive = LoggingReceive {
    case "hello" => log.info("hello")
    case "stop" => context.stop(self)
}
  override def postStop() = {
```

```
    log.info("BasicLifecycleLoggingActor postStop method is invoked... ")
  }
}
object LifecycleApp {
  def main(args: Array[String]) {
    val actorSystem = ActorSystem("LifecycleActorSystem")
    val lifecycleActor = actorSystem.actorOf(Props[BasicLifecycleLoggingActor], "lifecycleActor")
    //创建消息处理的actor
    val deadLetterListener = actorSystem.actorOf(Props[MyCustomDeadLetterListener])
    actorSystem.eventStream.subscribe(deadLetterListener, classOf[DeadLetter])

    lifecycleActor ! "hello"
    lifecycleActor ! "stop"
    lifecycleActor ! "hello"
  }
}
class MyCustomDeadLetterListener extends Actor {
  def receive = {
//处理终止actor后的消息
    case deadLetter: DeadLetter => println(s"FROM CUSTOM LISTENER $deadLetter")
  }
}
//运行结果
Inside BasicLifecycleLoggingActor Constructor
Actor[akka://LifecycleActorSystem/user/lifecycleActor#634495506]
BasicLifecycleLoggingActor preStart method is invoked...
hello
FROM CUSTOM LISTENER DeadLetter(hello,Actor[akka://LifecycleActorSystem/deadLetters],Actor[akka://LifecycleActorSystem/user/lifecycleActor#634495506])
BasicLifecycleLoggingActor postStop method is invoked...
Message [java.lang.String] from Actor[akka://LifecycleActorSystem/deadLetters] to Actor[akka://LifecycleActorSystem/user/lifecycleActor#634495506] was not delivered. [1] dead letters encountered. This logging can be turned off or adjusted with configuration settings 'akka.log -dead-letters' and 'akka.log-dead-letters-during-shutdown'.
```

从打印结果中可以看出，在终止了 actor 后，如果继续给被终止的 actor 发送消息，那么就会封装成一个 DeadLetter，然后再发布到 EventStream 里面去。

6.1.10 actor 实战

【例 6-10】 actor 与样例类实战

```
object MyActor1 {
  case class Greeting(from: String)          //定义带参数的样例类
  case object Goodbye
  def main(args: Array[String]) {
    val system = ActorSystem("test")         //使用系统创建actor方式创建actor
    val myActor1 = system.actorOf(Props[MyActor1],"MyActor1")
    myActor1 ! Greeting("Hello")             //给Greeting发送消息
```

```
    }
  }
  class MyActor1 extends Actor with ActorLogging {
    import MyActor1._
    override def receive = {
      case Greeting(greeter) => println(s"I was greeted by $greeter.")
      case Goodbye => println("Someone said goodbye to me.")
      case _ => println("Something else ...")
    }
  }
  //运行结果如下
  I was greeted by Hello.
```

这个实例主要是结合了样例类来发送消息。在 spark 中结合样例类发送消息到处可见。

样例类：

```
case class Greeting(from: String)
```

样例对象：

```
case object Goodbye
```

以样例类的构造器作为消息体发送出去：

```
myActor1 ! Greeting("Hello")
```

匹配消息体为 Greeting（" "）的消息。后面打印的字符串中 $ greeter. 是变量，这是因为在开头添加了 s。这里同时还使用了模式匹配的变量，把 Greeting 构造器的参数绑定到了 greeter 上，从而后面可以直接使用。

```
case Greeting(greeter) => println(s"I was greeted by $greeter.")
```

【例 6-11】 消息发送与接收实战

```
import akka.actor._
//定义四个样例对象，用于消息发送与接收
case object SendMessage
case object ReceMessage
case object StartMessage
case object StopMessage
//定义消息发送类Send
class Send(rece: ActorRef) extends Actor {
  var count = 0
  def incrementAndPrint { count += 1; println("Send") }
  //重写receive方法
  override def receive = {
    //根据以上面定义的样例对象作为消息体进行模式匹配
    case StartMessage =>
      incrementAndPrint
      rece ! SendMessage
    case ReceMessage =>
      if (count > 9) {
        sender ! StopMessage
        println("Send stopped")
        context.stop(self)                    //终止Actor
```

```
      } else {
        incrementAndPrint
        sender ! SendMessage                    //给sender发送消息
      }
    }
  }
//定义接收消息类
class Rece extends Actor {
  override def receive = {
    case SendMessage =>
      println("  Rece")
        sender ! ReceMessage
    case StopMessage =>
      println("Rece stopped")
      context.stop(self)
      context.system.shutdown()
  }
}
//Scala给我们提供了App类，直接继承就可以运行程序
 object SendReceTest extendsApp{
  val system=ActorSystem("SendReceSystem")
  val rece=system.actorOf(Props[Rece],name="Rece")
  val send=system.actorOf(Props(newSend(rece)),name="Send")
  send!StartMessage
}
//运行结果如下
Send
  Rece
Send
  Rece
Send
......
Send stopped
Rece stopped

```

例 6-11 定义了两个 actor：Send 和 Rece。

Send 接收 StartMessage 和 ReceMessage。StartMessage 是一个启动消息，由 main 对象发送，ReceMessage 来自 Receactor，如果次数还未达到，它继续发送 SendMessage。

Rece 接收 StopMessage 和 SendMessage。如果接收到 SendMessage，它就发送一个 ReceMessage，如果是 StopMessage，停止 ActorSystem。

6.2 小结

本章主要讲解了 Akka 的 actor 的生命周期和两种创建方式，终止 actor 的各种方式。Akka 的通讯机制在 Spark 中被封装成了 RPC 通讯，且应用很广泛。因此彻底精通 Akka 对学习 Spark 的 RPC 通讯将有很大帮助。

第7章 源码编译

Spark 源码编译就是把用 Java 或 Scala 等高级语言编写的 Spark 程序源代码变成计算机可以识别的二进制代码，并以安装包的形式提供给应用程序开发者。

虽然可以从 Spark 官网下载编译后的安装包，但还是需要自己编译源码。首先 Spark 能同 Hadoop 进行交互，而发行 Hadoop 的厂商比较多且有各自的版本，Spark 官方提供的安装包不一定和开发者所使用的 Hadoop 版本相同，如果不相同在程序运行时有可能出现错误；其次官方提供的安装包有时不能包含开发者所需的依赖包。Spark 的源码编译有三种方式，如下所示。

① Sbt（Simple Build Tool）方式。这种方式可以分别在 Windows 系统和 Linux 系统上进行编译。

② Maven 方式。这种编译方式是根据源码中 pom. xml 文件来编译的，可以分别在 Windows 系统和 Linux 系统上进行编译。

③ make-distribution. sh 脚本方式。这种方式是使用源码目录中的脚本文件 make-distribution. sh，这种方式实际上也是采用 Maven 方式编译。这种方法只能在 Linux 系统上编译。

本章介绍在 Windows10 和 Ubuntu 两种操作系统中编译 Spark 1. 6. 0 源码。7. 1 节介绍在 64 位 Windows 10 系统下分别用 Sbt、Maven 方式编译源码 Spark 1. 6. 0 源码，7. 2 节介绍在 64 位 Ubuntu kylin-15. 10-desktop 系统下分别用 Sbt、Maven、打包方式编译 Spark 1. 6. 0 源码。

7.1 Windows 下源码编译

本节介绍在 Windows 10 操作系统中编译 Spark 1. 6. 0 源码的方法和过程，以及在编译的过程中要注意的几个问题，基本步骤是先到 Spark 的官网下载 Spark 1. 6. 0 的源码，然后再用 Sbt 或 Maven 进行编译。

7.1.1 下载 Spark 源码

编译 Spark 1. 6. 0 源码前先要下载源码到本机。既可以从官网下载源码，也可以从

Github 上下载。本小节分别介绍在 Windows 10 系统中从官网和 Github 上下载 Spark 1.6.0 源码。

从 Spark 官网下载 Spark1.6.0 源码，详细步骤如下所示。

① 在网页浏览器中打开 Spark 官网，网址是 http://spark.apache.org；

② 鼠标单击网页右侧的 Download Spark 按钮，单击后进入源码下载页面；

③ 在下载页面上选择 1.6.0（Jan 04 2016）版本和 Source Code [can build several Hadoop versions]，然后选择 Direct Download 选项，如图 7-1 所示；

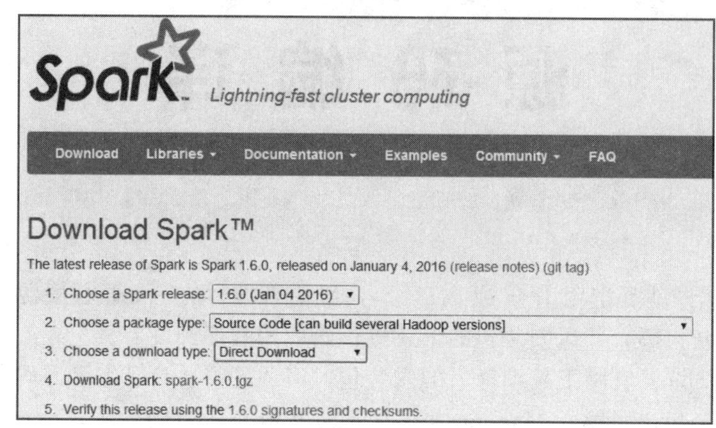

图 7-1　Spark 1.6.0 源码下载选项

④ 单击 Download Spark：spark-1.6.0.tgz 后，出现新建下载任务对话框，选择下载保存的路径，单击下载按钮即可下载源码到本机上。

从 Github 上下载源码的步骤如下所述。

① 在网页浏览器中打开 Github 的官网，网址是：https://github.com；

② 在搜索框中输入 Spark 关键字进行搜索；

③ 在搜索结果中找到 apache/spark；

④ 单击 apache/spark 即可下载最新版本 Spark 1.6.0 源码。

7.1.2 Sbt 方式

在 64 位 Windows 10 系统下编译 Spark 1.6.0 源码需要安装的软件有：Java、Scala、Git、Sbt。Git 用于在用 Sbt 或者 Maven 编译源码的过程中从 Github 下载依赖的软件包。本节采用 Java1.8 版本和 Scala 2.10.4 版本编译源码。下面介绍 Java 1.8、Scala 2.10.4、Git、Sbt 的安装和配置，以及用 Sbt 编译 Spark 1.6.0 源码的方法。

注意编译的过程中需要保持网络连接。

（1）安装和配置 Java 1.8

本节介绍在 Windows 10 系统中安装和配置 Java 1.8 的方法，其过程如下所述。

① 用网页浏览器打开 Oracle 的官网，网址是：http：//www.oracle.com/index.html。

② 在页面的选项卡中选择 Downloads 选项，然后选择 Java SE 选项，点击左侧的链接 JavaDOWNLOAD 即可进入下载页面，如图 7-2 所示。

③ 进入下载页面后，选择 Accept License Agreement 单选框，再选择 Java 的版本，作者选择的是 Java SE Development Kit 8u73，再选择计算机的版本，作者选择的是 Windows

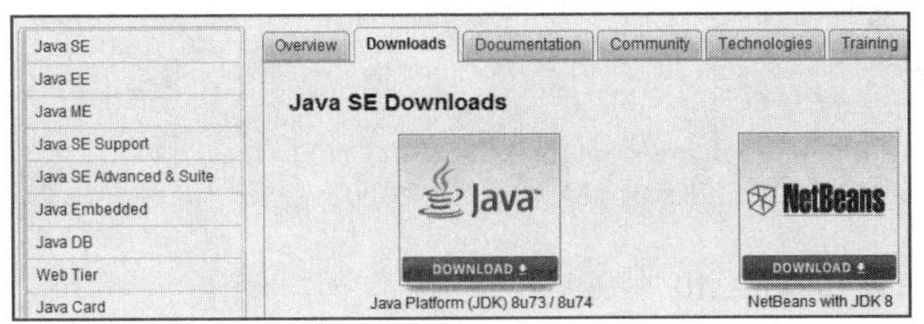

图 7-2　下载 Java

x64，单击即可下载，如图 7-3 所示。

Java SE Development Kit 8u73

You must accept the Oracle Binary Code License Agreement for Java SE to download this software.
Thank you for accepting the Oracle Binary Code License Agreement for Java SE; you may now download this software.

Product / File Description	File Size	Download
Linux ARM 32 Hard Float ABI	77.73 MB	jdk-8u73-linux-arm32-vfp-hflt.tar.gz
Linux ARM 64 Hard Float ABI	74.68 MB	jdk-8u73-linux-arm64-vfp-hflt.tar.gz
Linux x86	154.75 MB	jdk-8u73-linux-i586.rpm
Linux x86	174.91 MB	jdk-8u73-linux-i586.tar.gz
Linux x64	152.73 MB	jdk-8u73-linux-x64.rpm
Linux x64	172.91 MB	jdk-8u73-linux-x64.tar.gz
Mac OS X x64	227.25 MB	jdk-8u73-macosx-x64.dmg
Solaris SPARC 64-bit (SVR4 package)	139.7 MB	jdk-8u73-solaris-sparcv9.tar.Z
Solaris SPARC 64-bit	99.08 MB	jdk-8u73-solaris-sparcv9.tar.gz
Solaris x64 (SVR4 package)	140.36 MB	jdk-8u73-solaris-x64.tar.Z
Solaris x64	96.78 MB	jdk-8u73-solaris-x64.tar.gz
Windows x86	181.5 MB	jdk-8u73-windows-i586.exe
Windows x64	186.84 MB	jdk-8u73-windows-x64.exe

图 7-3　选择 Java 版本

④ 安装 Java。鼠标双击下载的 Java 安装包开始安装，并选择安装的路径，作者安装的路径是：C：\ UserProgramFile \ java \ java8。

⑤ 配置环境变量。在计算机桌面上鼠标右键单击“我的计算机”图标，选择“属性”，选择“高级系统设置”，选择“环境变量（N）…”，在系统变量选项卡中单击“新建”，在“变量名（N）”文本框中输入 JAVA _ HOME，在“变量值（V）：”文本框中输入 C：\ UserProgramFile \ java \ java8，如图 7-4 所示。

图 7-4　设置 JAVA _ HOME 系统变量

⑥ 把 JAVA _ HOME 配置到系统的 Path 中。选中系统变量中 Path 行，鼠标单击“编辑（I）…”，在“变量值（V）：”文本框中输入%JAVA _ HOME% \ bin，如图 7-5 所示。

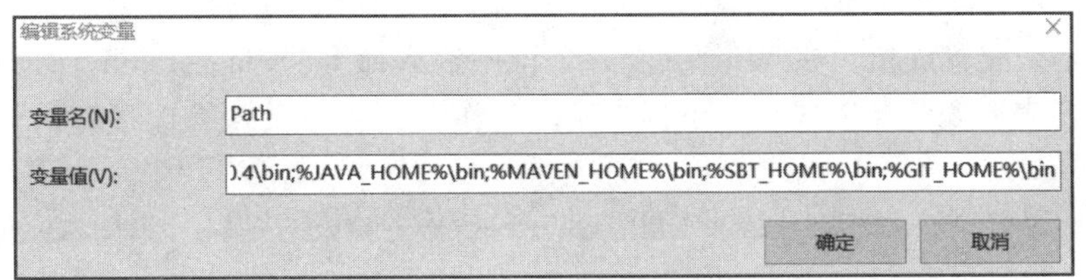

图 7-5　JAVA _ HOME 系统变量配置到 Path 路径中

⑦ 检查是否配置成功。在 Windows 命令行中输入命令 java-version，会打印出安装的

Java 的版本，表明安装配置成功，下列中所示的作者安装的版本是 1.8.0。

```
>java -version
java version "1.8.0_73"
Java(TM) SE Runtime Environment (build 1.8.0_73-b02)
Java HotSpot(TM) 64-Bit Server VM (build 25.73-b02, mixed mode)
```

(2) 安装和配置 Scala 2.10.4

本小节介绍在 Windows 10 系统中安装和配置 Scala 2.10.4 的方法，其过程如下所示。

① 在网页浏览器中打开 Scala 的官网，网址是：http://www.scala-lang.org。

② 鼠标单击左侧的 DOWNLOAD 按钮进入下载页面。

③ 在下载页面上选择 All downloads 按钮，进入 Scala 的所有版本页面，选择 Scala 的版本，作者选择的是 Scala 2.10.4。

④ 安装 Scala 2.10.4。鼠标双击下载的文件 scala-2.10.4.msi 即可安装，作者安装的位置是：C:\UserProgramFile\scala\scala2.10.4。

⑤ 配置环境变量。在计算机桌面上鼠标右键单击“我的计算机”图标，选择“属性”，选择“高级系统设置”，选择“环境变量（N）...”，在系统变量选项卡中单击“新建”，在“变量名（N）”文本框中输入 SCALA_HOME，在“变量值（V）:”文本框中输入 C:\UserProgramFile\scala\scala 2.10.4，如图 7-6 所示。

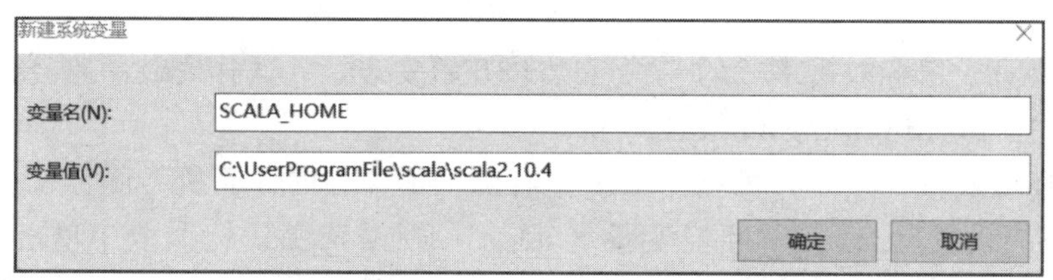

图 7-6 设置 SCALA_HOME 系统变量

⑥ 把 SCALA_HOME 配置到系统的 Path 中。选中系统变量中 Path 行，鼠标单击“编辑（I）...”，在“变量值（V）:”文本框中输入%SCALA_HOME%\bin，如图 7-7 所示。

图 7-7 SCALA_HOME 系统变量配置到 Path 路径中

⑦ 检查是否配置成功。在 Windows 命令行中输入命令 scala-version，会打印出安装的 Scala 的版本，表明安装配置成功，下列中所示的作者安装的版本是 2.10.4。

```
>scaa-version
Scala code runner version 2.10.4--Copyright 2002-2013, LAMP/EPFL
```

(3) Git 的安装和配置

需要先下载 Git，然后再进行安装和配置环境变量。Git 的安装和配置详细步骤如下所示。

① 在网页浏览器中打开 Git 官网，官网网址是 http://git-scm.com。

② 下载 Git。单击网页左侧的 Download 下载选项，选择与本机对应的版本，作者选择 64-bit Git for Windows Setup，选择下载到本机的保存路径，单击确定后开始下载。

③ 安装。鼠标双击下载的安装包后开始安装，作者安装在 C:\UserProgramFile\sbt，安装过程中都选择默认选项。

④ 配置系统变量。在计算机桌面上鼠标右键单击“我的计算机”图标，选择“属性”，选择“高级系统设置”，选择“环境变量（N）...”，在系统变量选项卡中单击“新建”，在“变量名（N）”文本框中输入 GIT_HOME，在“变量值（V）:”文本框中输入 C:\UserProgramFile\git\Git，如图 7-8 所示。

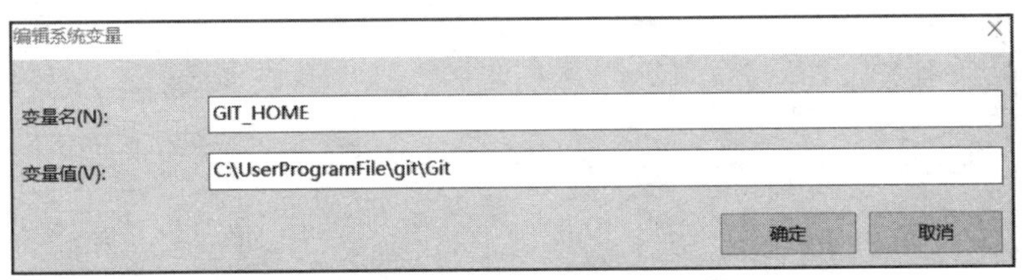

图 7-8 设置 GIT_HOME 系统变量

⑤ 把 GIT_HOME 配置到系统的 Path 中。选中系统变量中 Path 行，鼠标单击“编辑（I）…”，在“变量值（V）:”文本框中输入%GIT_HOME%\bin，如图 7-9 所示。

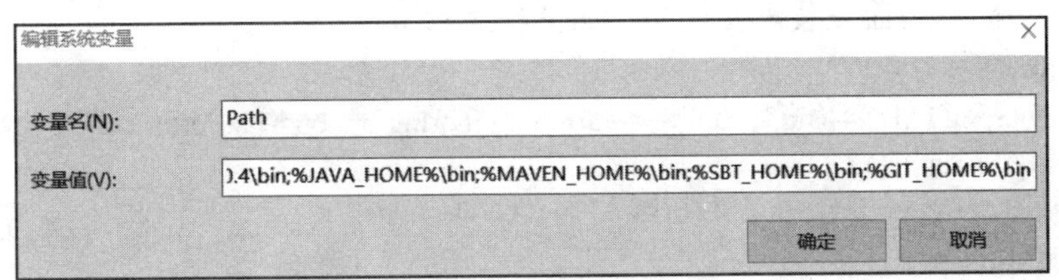

图 7-9 GIT_HOME 系统变量配置到 Path 路径中

⑥ 检查是否配置成功。在 Windows 命令行中输入命令 git--version，会打印出安装的 Git 的版本，表明安装配置成功，下列中所示的作者安装的版本是 2.7.0。

```
1. >git–version
2. git version 2.7.0.windows.1
```

(4) 安装并用 Sbt 编译 Spark 1.6.0 源码

安装并用 Sbt 编译 Spark 1.6.0 源码的步骤如下所示。

① 用网页浏览器打开 Sbt 官网 http://www.scala-sbt.org。

② 下载。鼠标单击网页上的 DOWNLOAD 按钮，选择与本机对应的版本，作者下载的是 Allplatforms 选项下的 SBT-0.13.9.ZIP。

③ 安装。下载之后鼠标双击进行解压，解压之后鼠标双击进行安装，作者安装在 C:\UserProgramFile\sbt。

④ 设置系统变量。在计算机桌面上鼠标右键单击“我的计算机”图标，选择“属性”，选择“高级系统设置”，选择“环境变量（N）...”，在用户变量选项卡中单击“新建”，在“变量名（N）”文本框中输入 SBT_HOME，在“变量值（V）:”文本框中输入 C:\UserProgramFile \ sbt \ sbt-0. 13. 9 \ sbt，如图 7-10 所示。

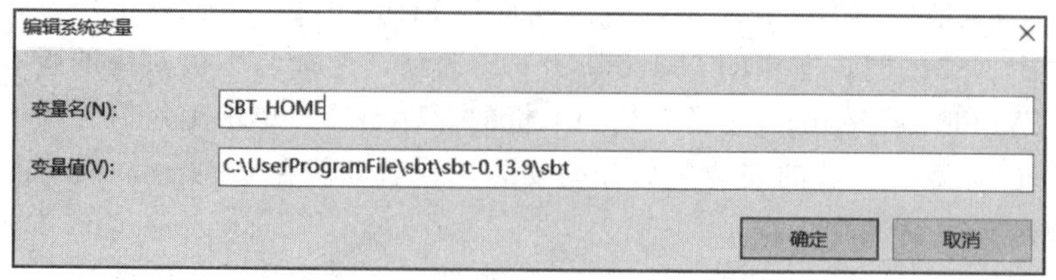

图 7-10　配置 SBT _ HOME 系统变量

⑤ 把系统变量值配置到 Path 路径中。选中系统变量中 Path 行，鼠标单击“编辑(I) …”，在“变量值（V）:”文本框中输入%SBT _ HOME% \ bin，如图 7-11 所示。

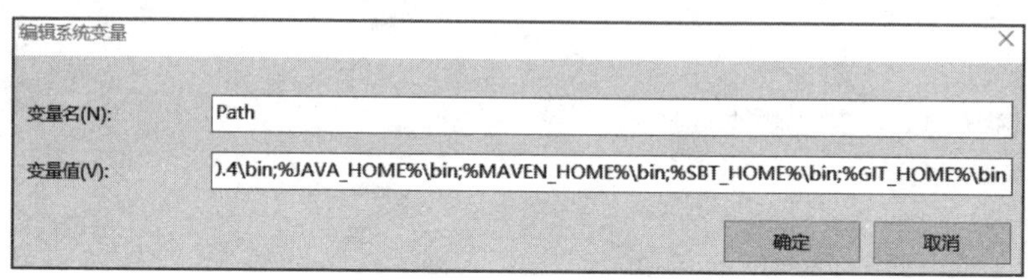

图 7-11　配置 SBT _ HOME 系统变量到 Path

⑥ 验证是否安装配置成功。打开 Windows 10 系统的命令行，输入命令 sbtsbt-version 用于查看安装的 Sbt 版本，验证是否安装配置成功，从下面可以看出安装的是 0. 13. 9 版本，和下载的版本相同，表面安装配置成功。如下程序所示：

```
>sbtsbt-version
Java HotSpot(TM) 64-Bit Server VM warning: ignoring option MaxPermSize=256m;
support was removed in 8.0
[info] Set current project to dell (in build file:/C:/Users/dell/)
[info] 0.13.9
```

⑦ 把下载的 Spark 1. 6. 0 源码解压，作者解压到 C：\ UserProgramFile \ sbt-spark 文件夹下。

⑧ 打开源码文件夹，找到文件夹中的 pom. mxl 文件，并用记事本打开。

⑨ 修改 pom. xml 文件中的 Java 版本和 Hadoop 版本。把＜java. version＞1. 7＜/java. version＞改为＜java. version＞1. 8＜/java. version＞，把＜hadoop. version＞2. 4. 0＜/hadoop. version＞改为＜hadoop. version＞2. 6. 0＜/hadoop. version＞。

⑩ 打开 Windows 10 系统的命令行，输入下面的命令进入到 Spark 1. 6. 0 源码目录。

```
>cd C:\UserProgramFile\sbt-spark\spark-1.6.0\spark-1.6.0
```

⑪ 设置编译时的变量用以制订最大内存缓存，在命令行输入下面的命令

```
>set SBT_OPTS=-Xmx2g -XX:MaxPermSize=512M -XX:ReservedCodeCacheSize=512m
```

⑫ 在命令行输入下面的命令

```
>sbt-Pyarn  assembly
```

按＜Enter＞键开始编译，其中-Pyarn 选项表示带 yarn 编译，编译时 Sbt 会自动的下载安装它所需要的程序包。

⑬ 当屏幕上出现的［success］的时候表示编译成功，本次编译用时 4762 秒，如图 7-12 所示。

```
[warn] Merging 'plugin.properties' with strategy 'first'
[warn] Merging 'plugin.xml' with strategy 'first'
[warn] Merging 'reference.conf' with strategy 'concat'
[warn] Merging 'rootdoc.txt' with strategy 'first'
[warn] Strategy 'concat' was applied to a file
[warn] Strategy 'discard' was applied to 1757 files
[warn] Strategy 'first' was applied to 820 files
[info] SHA-1: 5741d40f18aad12444634047142864196623ddaf
[info] Packaging C:\UserProgramFile\sbt-spark\spark-1.6.0\spark-1.6.0\
[info] Done packaging.
[success] Total time: 4762 s, completed 2016-2-26 0:42:25
```

图 7-12　用 Sbt 编译 Spark 1.6.0 成功

编译完成后的的文件是在 assembly \ target \ scala-2.10 文件夹中的 spark-assembly-1.6.0-hadoop2.6.0.jar，这就是编译 Spark1.6.0 源码得到的 Jar 包。

7.1.3　Maven 方式

用 Maven 方式编译 Spark 需要安装 Java、Scala、Maven，本节是用 Java 1.8 版本、Scala 2.10.4 版本。先下载并安装 Maven，然后用 Maven 编译 Spark 1.6.0 源码。

注意编译的过程中需要连接网络。

下载并用 Maven 编译 Spark 1.6.0

下载并用 Maven 编译 Spark1.6.0 过程如下所示。

① 在网页浏览器中打开 Maven 官网 https：//maven.apache.org。

② 鼠标单击网页左侧的 Download 按钮之后，再单击 Binary tar.gz archive 选项的 apache-maven-3.3.9-bin.tar.gz，选择下载位置后单击确定开始下载。

③ 鼠标双击下载后的软件包进行解压 apache-maven-3.3.9-bin.tar.gz，作者解压到 C：\ UserProgramFile \ maven \ 文件夹，解压后的文件是 apache-maven-3.3.9-bin。

④ 配置系统变量。在计算机桌面上鼠标右键单击“我的计算机”图标，选择“属性”，选择“高级系统设置”，选择“环境变量（N）...”，在用户变量选项卡中单击“新建”，在“变量名（N）”文本框中输入 MAVEN _ HOME，在“变量值（V）:”文本框中输入 C：\ UserProgramFile \ maven \ apache-maven-3.3.9-bin \ apache-maven-3.3.9，如图 7-13 所示。

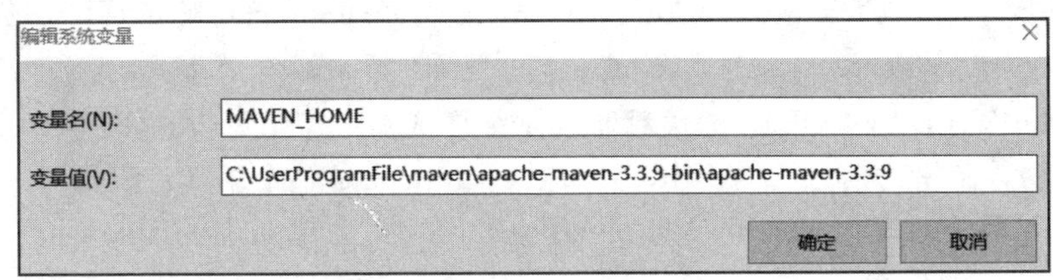

图 7-13　配置 Maven 系统变量

⑤ 把系统变量值配置到 Path 路径中。选中系统变量中 Path 行，鼠标单击“编辑（I）…”，在“变量值（V）:”文本框中输入%MAVEN _ HOME% \ bin，如图 7-14 所示。

⑥ 进入到 Windws 10 命令行，输入命令 mvn-v，检验是否配置成功。

图 7-14　MAVEN _ HOME 系统变量配置到 Path 路径中

```
>mvn–V
Apache Maven 3.3.9 (bb52d8502b132ec0a5a3f4c09453c07478323dc5; 2015-17-
11T00:41:47+08:00)
Maven home: C:\UserProgramFile\maven\apache-maven-3.3.9-bin\apache-maven-
3.3.9\bin\..
Java version: 1.8.0_71, vendor: Oracle Corporation
Java home: C:\UserProgramFile\java\java8\jre
Default locale: zh_CN, platform encoding: GBK
OS name: "windows 10", version: "10.0", arch: "amd64", family: "dos"
```

⑦ 把下载的 Spark 1.6.0 源码解压，作者解压到 C:\UserProgramFile\maven-spark 文件夹中，进入 Spark 1.6.0 源码文件夹，在 pom. xml 文件中修改 Java 版本和 Hadoop 版本，把<java. version> 1.7 </java. version> 改为 <java. version> 1.8 </java. version>，把<hadoop. version>2.4.0</hadoop. version>改为<hadoop. version>2.6.0</hadoop. version>。

⑧ 从 Windows 10 命令行进入到 Spark 1.6.0 源码目录，命令是：

```
>cd C:\UserProgramFile\maven-spark\spark-1.6.0\spark-1.6.0
```

⑨ 设置 Maven 使用的内存参数，命令是：

```
>set MAVEN_OPTS=-Xmx2g -XX:MaxPermSize=512M -XX:ReservedCodeCacheSize=512m
```

如果没有设置，可能会出现下面的错误。

```
[INFO] Compiling 203 Scala sources and 9 Java sources to
/Users/me/Development/spark/core/target/scala-
2.10/classes... [ERROR] PermGen space -> [Help 1]
[INFO] Compiling 203 Scala sources and 9 Java sources to
/Users/me/Development/spark/core/target/scala-
2.10/classes... [ERROR] Java heap space -> [Help 1]
```

⑩ 源码编译，输入命令 mvn-DskipTests clean package 进行源码编译。例如：

```
>mvn-DskipTests clean package
```

⑪ 当出现 BUILD SUCCESS 的信息时表示编译成功，图 7-15 表示用 Maven 编译 Spark 1.6.0 源码成功，图中的信息 Total time：59：13 min 表示整个编译过程用时 59 分 13 秒。

编译后的结果文件是 assembly \ target \ scala-2.10 文件夹中的 spark-assembly-1.6.0-hadoop2.6.0.jar，这个包含了编译 Spark1.6.0 得到的 Jar 包。

7.1.4　需要注意的几个问题

在 Windows 10 系统中编译 Spark 1.6.0 源码的过程中，下面的几个问题需要注意。

(1) 指定 Hadoop 的版本

由于不同版本的 HDFS（Hadoop Distribution File System）不兼容，如果要读写 HDFS

```
[INFO] Spark Project External Kafka ...................... SUCCESS [ 45.643 s]
[INFO] Spark Project Examples ............................ SUCCESS [02:45 min]
[INFO] Spark Project External Kafka Assembly ............. SUCCESS [  8.707 s]
[INFO] ------------------------------------------------------------------------
[INFO] BUILD SUCCESS
[INFO] ------------------------------------------------------------------------
[INFO] Total time: 59:13 min
[INFO] Finished at: 2016-02-14T22:54:43+08:00
[INFO] Final Memory: 412M/2039M
[INFO] ------------------------------------------------------------------------
```

图 7-15　用 Maven 编译 Spark 1.6.0 源码成功

数据，就需要针对指定 HDFS 版本编译 Spark 1.6.0 源码。通过设置 hadoop.version 的属性来指定 HDFS 的版本。如果没有设置，将会使用默认的 Hadoop2.2.0 版本。表 7-1 是 Hadoop 版本与配置参数表。

表 7-1　Hadoop 版本与配置参数表

Hadoop 版本	配置参数
1.x to 2.1.x	hadoop-1
2.2.x	hadoop-2.2
2.3.x	hadoop-2.3
2.4.x	hadoop-2.4
2.6.x and later2.x	hadoop-2.6

对于 Apache Hadoop versions 1.x、Cloudera CDH "mr1" 的发行版本其他不带 YARN 的版本，用以下方法编译。

```
# Apache Hadoop 1.2.1
>mvn-Dhadoop.version=1.2.1 -Phadoop-1 -DskipTests clean package
# Cloudera CDH 4.2.0 with MapReduce v1
```

如果 hadoop.version 版本不同，可以配置 yarn 和 yarn.version 选项。Spark 仅支持 YARN2.2.0 及以上版本。例如：

```
# Apache Hadoop 2.2.X
>mvn-Pyarn -Phadoop-2.2 -DskipTests clean package
# Apache Hadoop 2.3.X
>mvn-Pyarn -Phadoop-2.3 -Dhadoop.version=2.3.0 -DskipTests clean package
# Apache Hadoop 2.4.X or 2.5.X
>mvn-Pyarn -Phadoop-2.4 -Dhadoop.version=VERSION -DskipTests clean package
#Versions of Hadoop after 2.5.X may or may not work with the -Phadoop-2.4 profile (they
were released after this
version of Spark).
# Different versions of HDFS and YARN.
>mvn-Pyarn -Phadoop-2.3 -Dhadoop.version=2.3.0 -Dyarn.version=2.2.0 -DskipTests
clean package
```

(2) 带 Hive 和支持 JDBC 的 Spark 源码编译

官网提供的编译包并不支持 Hive，为了把 Hive 融合到 Spark SQL、JDBC 服务和 CLI，需要添加-Phive 和 Phive-thriftserver 选项。Spark 默认绑定到 Hive 0.13.1 版本，例如下面

的命令。

```
1. # Apache Hadoop 2.4.X with Hive 13 support
2. >mvn-Pyarn -Phadoop-2.4 -Dhadoop.version=2.4.0 -Phive -Phive-thriftserver -DskipTests
   clean package
```

(3) 用 Scala 2.11 版本编译

官网提供的 Spark1.6.0 的编译包的 Scala 语言的版本是 2.10.4，可以用-Dscala-2.11 选项用 Scala2.11 版本编译打包，例如：

```
1. ./dev/change-scala-version.sh 2.11
2. >mvn-Pyarn -Phadoop-2.4 -Dscala-2.11 -DskipTests clean package
```

注意 Spark 1.6.0 目前暂不支持 Scala2.11 版的 JDBC。

(4) 编译单独的子模块

如果只使用 Spark1.6.0 的子模块的话，可以用 mvn-pl 选项编译单独的子模块。例如下面的方法编译 Spark Streaming 子模块。

```
>mvn-pl :spark-streaming_2.10  clean  install
```

这里的 spark-streaming _ 2.10 是 streaming/pom.xml 文件中的 artifactId。

(5) 在 Windows 系统中编译 SparkR

在 Windows 系统中编译 SparkR，需要下面几个步骤。

① 安装 R 和 Rtools，要求 R 是 3.1 或以上版本，并配置 R 和 Rtools 到系统的环境变量中。

② 安装 JDK7 并把 JAVA _ HOME 配置到系统环境变量中。

③ 下载并安装 Maven 并把 MAVEN _ HOME 配置到系统环境变量中。

④ 设置 MAVEN _ OPTS 选项。

⑤ 打开 Windows10 的命令行，并输入下面的命令开始编译。

```
>run mvn–DskipTests –Psparkr package
```

7.2 Ubuntu 下源码编译

为了方便在 Linux 系统下编译和部署 Spark，Spark 1.6.0 源码目录自带有 Sbt 和 Maven 安装脚本，这两个脚本是在源码 Build 目录下的 Sbt 和 Mvn，这些脚本在源码编译过程中会自动下载和安装编译所需要的软件包，如 Maven，Scala，Zinc，因此不需安装 Sbt 和 Maven。本章用的 Linux 系统是 Ubuntu kylin-15.10-desktop。

本小节介绍在 Ubuntu 系统中分别用 Sbt 方式、Maven 方式、make-distribution.sh 脚本方式编译 Spark 1.6.0 源码，需要安装 Java 和 Scala，本节使用的 Java 1.8 版本和 Scala 2.10.4 版本。

先要下载 Spark 1.6.0 源码到本机中，然后介绍在 Ubuntu kylin-15.10-desktop 系统中用 Sbt、Maven、make-distribution.sh 脚本方式的编译源码的方法。

7.2.1 下载 Spark 源码

本节介绍在 Ubuntu 系统中下载 Spark 1.6.0 源码的方法，在其他类 Linux 的系统中也是一样的过程。作者的 Ubuntu 版本是 64 位的 Ubuntu kylin-15.10-desktop，用其下载源码的步骤如下所示。

① 在网页浏览器中打开 Spark 的官网，网址是 http：//spark.apache.org；

② 鼠标单击网页右侧的 Download Spark 按钮进入下载页面；

③ 在下载页面上选择 1.6.0（Jan 04 2016）版本和 Source Code [can build several Hadoop versions]，然后选择 Direct Download 选项，如图 7-16 所示；

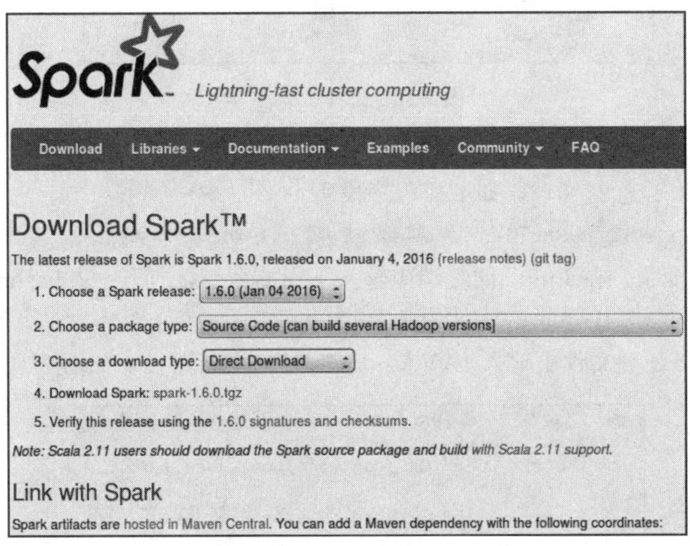

图 7-16 Spark 1.6.0 源码下载选项

④ 选择 spark-1.6.0.tgz 后，出现保存文件对话框，单击保存按钮后开始下载；

⑤ 把下载后的 spark-1.6.0.tgz 文件解压，解压之后即可用于编译了。

(1) 安装和配置 Java 1.8

本小节介绍在 Ubuntu kylin-15.10-desktop 系统中安装和配置 Java 1.8 的方法，其过程如下所示。

① 用网页浏览器打开 Oracle 的官网，网址是：http：//www.oracle.com/index.html。

② 在页面的选项卡中选择 Downloads 选项，然后选择 Java SE 选项，点击左侧的 Java DOWNLOAD 即可进入下载页面，如图 7-17 所示。

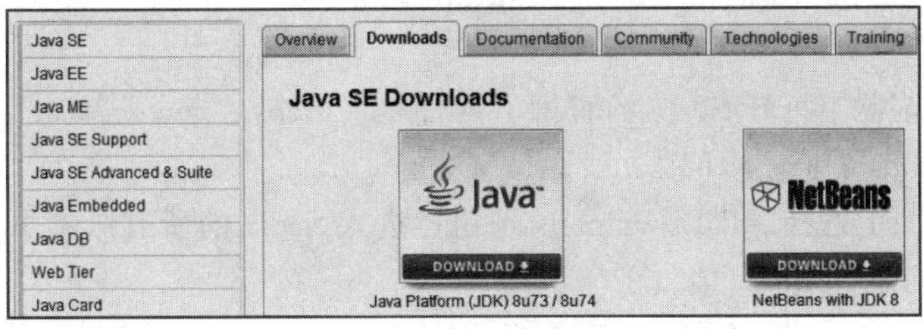

图 7-17 下载 Java

③ 进入下载页面后，选择 Accept License Agreement 单选框，再选择 Java 的版本，作者选择的是 Java SE Development Kit 8u73，再选择计算机的版本，作者选择的是 Linux x64，单击即可下载，如图 7-18 所示。

Java SE Development Kit 8u73

You must accept the Oracle Binary Code License Agreement for Java SE to download this software.
Thank you for accepting the Oracle Binary Code License Agreement for Java SE; you may now download this software.

Product / File Description	File Size	Download
Linux ARM 32 Hard Float ABI	77.73 MB	jdk-8u73-linux-arm32-vfp-hflt.tar.gz
Linux ARM 64 Hard Float ABI	74.68 MB	jdk-8u73-linux-arm64-vfp-hflt.tar.gz
Linux x86	154.75 MB	jdk-8u73-linux-i586.rpm
Linux x86	174.91 MB	jdk-8u73-linux-i586.tar.gz
Linux x64	152.73 MB	jdk-8u73-linux-x64.rpm
Linux x64	172.91 MB	jdk-8u73-linux-x64.tar.gz
Mac OS X x64	227.25 MB	jdk-8u73-macosx-x64.dmg
Solaris SPARC 64-bit (SVR4 package)	139.7 MB	jdk-8u73-solaris-sparcv9.tar.Z
Solaris SPARC 64-bit	99.08 MB	jdk-8u73-solaris-sparcv9.tar.gz
Solaris x64 (SVR4 package)	140.36 MB	jdk-8u73-solaris-x64.tar.Z
Solaris x64	96.78 MB	jdk-8u73-solaris-x64.tar.gz
Windows x86	181.5 MB	jdk-8u73-windows-i586.exe
Windows x64	186.84 MB	jdk-8u73-windows-x64.exe

图 7-18 选择 Java 版本

④ 安装 Java。解压下载的 Java 安装，并选择安装的路径，作者安装的路径是：/usr/software/java/jdk. 1. 8. 3 _ 073，如下面程序段所示。

```
>tar  zxvf  Downloads/jdk-8u73-linux-x-4.tar.gz  -C  /usr/software / java
```

⑤ 配置环境变量。在 Linux 的终端中输入下面的命令，打开～/. bashrc 配置文件。

```
>vim  ~/.bashrc
```

在～/. bashrc 配置文件中输入 JAVA _ HOME 的安装路径，如图 7-19 所示。

```
export JAVA_HOME=/usr/software/java/jdk1.8.0_73
export JRE_HOME=${JAVA_HOME}/jre
export SCALA_HOME=/usr/software/scala/scala-2.10.4

export CLASS_PATH=.:${JAVA_HOME}/lib:${JRE_HOME}/lib

export PATH=${SCALA_HOME}/bin:${JAVA_HOME}/bin:$PATH
```

图 7-19 设置 JAVA _ HOME 环境变量

⑥ 检查是否配置成功。在终端命令行中输入命令 java-version，会打印出安装的 Java 的版本，表明安装配置成功，如下所示作者安装的版本是 1. 8. 0。

```
>java-version
java version "1.8.0_73"
Java(TM) SE Runtime Environment (build 1.8.0_73-b02)
Java HotSpot(TM) 64-Bit Server VM (build 25.73-b02, mixed mode)
```

（2）安装和配置 Scala 2. 10. 4

本小节介绍在 Ubuntu kylin-15. 10-desktop 系统中安装和配置 Scala 2. 10. 4 的方法，其过程如下所示。

① 在网页浏览器中打开 Scala 的官网，网址是：http：//www. scala-lang. org。鼠标单击左侧的 Download 按钮进入下载页面。

② 在下载页面上选择 All Downloads 按钮，进入 Scala 的所有版本页面，选择 Scala 的版本，作者选择的是 Scala 2. 10. 4。

③ 安装 Scala2. 10. 4。解压下载的文件 scala-2. 10. 4. tgz 即可安装，作者安装的位置是：C：\ UserProgramFile \ scala \ scala2. 10. 4。如下面程序段所示。

```
>tar  zxvf Downloads/scala-2.10.4.tgz  -C  /usr/software /scala
```

④ 配置环境变量。在 Linux 的终端中输入下面的命令，打开～/. bashrc 配置文件。

＞vim ～/. bashrc

在～/. bashrc 配置文件中输入 SCALA _ HOME 的安装路径，如图 7-20 所示。

```
export JAVA_HOME=/usr/software/java/jdk1.8.0_73
export JRE_HOME=${JAVA_HOME}/jre
export SCALA_HOME=/usr/software/scala/scala-2.10.4

export CLASS_PATH=.:${JAVA_HOME}/lib:${JRE_HOME}/lib

export PATH=${SCALA_HOME}/bin:${JAVA_HOME}/bin:$PATH
```

图 7-20　设置 SCALA _ HOME 环境变量

⑤ 检查是否配置成功。在终端命令行中输入命令 scala-version，会打印出安装的 Scala 的版本，表明安装配置成功，如下所示作者安装的版本是 2. 10. 4。

```
>scala-version
Scala code runner version 2.10.4-- Copyright 2002-2013, LAMP/EPFL
```

7.2.2　Sbt 方式

由于 Spark 1. 6. 0 源码目录 bulid 中提供了 sbt 编译脚本，运行该脚本会自动下载 Sbt 软件，所以不需要手动显示安装 Sbt。

注意编译的过程中需要连接网络。

在 64 位 Ubuntu kylin-15. 10-desktop 系统下用 Sbt 编译 Spark 1. 6. 0 源码的步骤如下所示。

① 新建 spark-sbt 目录，用于存放编译后的结果文件。

```
>mkdir-p  /usr/software /spark-sbt
```

② 把下载的 Spark 1. 6. 0 的源码解压到目录/usr/software/spark-sbt 中。

```
>tar  zxvf  Downloads/spark-1.6.0.tgz  -C  /usr/software /spark-sbt
```

③ 进入到解压后的 spark-1. 6. 0 目录。

```
>cd /usr/software /spark-sbt/ spark-1.6.0
```

④ 用 vim 打开 pom. xml 文件。

```
>vim  pom.xml
```

修改 pom. xml 文件中的 Java 和 Hadoop 版本，把<java. version>1. 7</java. version>改为<java. version>1. 8</java. version>，把<hadoop. version>2. 4. 0</hadoop. version>改为<hadoop. version>2. 6. 0</hadoop. version>，保存并退出。

⑤ 输入编译命令开始编译。

```
>./build/sbt -Pyarn -Phadoop-2.6.0 -Dhadoop.version=2.6.0 assembly
```

⑥ 部分过程如图 7-21 所示。

⑦ 当出现［success］的信息时表示编译成功，如图 7-22 所示，表示整个编译用 987 秒。

编译完成后的文件是 ./assembly/target/scala-2. 10 下 spark-assembly-1. 6. 0-hadoop2. 6. 0. jar，这个包包含了 Spark 编译得到的 Jar 包。

```
[info] Set current project to spark-parent (in build file:/usr/software/spark-sbt-v2/spark-1.6.0/)
[info] Updating {file:/usr/software/spark-sbt-v2/spark-1.6.0/}test-tags...
[info] Resolving org.fusesource.jansi#jansi;1.4 ...
[info] Done updating.
[info] Updating {file:/usr/software/spark-sbt-v2/spark-1.6.0/}streaming-flume-sink...
[info] Updating {file:/usr/software/spark-sbt-v2/spark-1.6.0/}unsafe...
[info] Updating {file:/usr/software/spark-sbt-v2/spark-1.6.0/}launcher...
[info] Resolving org.fusesource.jansi#jansi;1.4 ...
[info] Done updating.
[info] Resolving org.spark-project.spark#unused;1.0.0 ...
[info] Updating {file:/usr/software/spark-sbt-v2/spark-1.6.0/}network-common...
[info] Resolving org.apache.hadoop#hadoop-client;2.6.0 ...
```

图 7-21 Sbt 部分编译过程

```
[warn] Merging 'rootdoc.txt' with strategy 'first'
[warn] Strategy 'concat' was applied to a file
[warn] Strategy 'discard' was applied to 1759 files
[warn] Strategy 'filterDistinctLines' was applied to 10 files
[warn] Strategy 'first' was applied to 1085 files
[info] SHA-1: 618363b9c8fc4c63a23188defeca37036af86fc8
[info] Packaging /usr/software/spark-sbt-v2/spark-1.6.0/examples/target/scala-2.10/spark-examples-1.6.0-hadoop2.6.0.jar
[info] Done packaging.
[info] Done packaging.
[success] Total time: 987 s, completed Mar 5, 2016 9:41:55 AM
```

图 7-22 Sbt 编译源码成功

7.2.3 Maven 方式

为了方便编译和部署，在 Spark 源码的 build/目录下提供了脚本 mvn 用于编译源码，可以自动下载 Maven 和编译过程中的依赖包到 build 目录下，因此不需要安装 Maven。

注意编译的过程中需要保持网络连接。

在 64 位 Ubuntu kylin-15.10-desktop 系统下用 Maven 编译 Spark 1.6.0 源码的步骤如下所述。

① 新建 spark-maven 目录，用于存放编译后的结果文件。

```
>mkdir-p /usr/software /spark-maven
```

② 把下载的 Spark 1.6.0.tgz 解压到目录/usr/software/spark-maven 中。

```
>tar  zxvf Downloads/spark-1.6.0.tgz -C /usr/software /spark-maven
```

③ 进入到解压后的 spark-1.6.0 目录。

```
>cd /usr/software /spark- maven / spark-1.6.0
```

④ 用 vim 打开 pom.xml 文件。

```
>vim pom.xml
```

修改 pom.xml 文件中的 Java 和 Hadoop 版本，把<java.version>1.7</java.version>改为<java.version>1.8</java.version>，把<hadoop.version>2.4.0</hadoop.version>改为<hadoop.version>2.6.0</hadoop.version>，修改后保存并退出。

⑤ 设置 MAVEN_OPTS 选项，避免在编译过程中出现内存不足的错误。

```
export MAVEN_OPTS="-Xmx2g -XX:MaxPermSize=512M -XX:ReservedCodeCacheSize=512m"
```

如果没有设置则可能会出现下面的错误。

```
1. [INFO] Compiling 203 Scala sources and 9 Java sources to
   /Users/me/Development/spark/core/target/scala-
2. 2.10/classes... [ERROR] PermGen space-> [Help 1]
3. [INFO] Compiling 203 Scala sources and 9 Java sources to
   /Users/me/Development/spark/core/target/scala-
4. 2.10/classes... [ERROR] Java heap space-> [Help 1]
```

⑥ 输入编译命令并按＜Enter＞键开始编译。

```
>./bulib/mvn-Pyarn -Phadoop-2.6 -Dhadoop.version=2.6.0 -DskipTests clean package
```

这里的选项-Pyarn 代表支持 yarn，-Phadoop 指定 hadoop 版本。

⑦ 图 7-23 是部分 Maven 编译过程。

```
[INFO] --- maven-shade-plugin:2.4.1:shade (default) @ spark-streaming-kafka_2.10 ---
[INFO] Excluding org.scala-lang:scala-library:jar:2.10.5 from the shaded jar.
[INFO] Excluding org.apache.avro:avro-mapred:jar:hadoop2:1.7.7 from the shaded jar.
[INFO] Excluding org.apache.avro:avro-ipc:jar:1.7.7 from the shaded jar.
[INFO] Excluding org.apache.avro:avro:jar:1.7.7 from the shaded jar.
[INFO] Excluding org.apache.avro:avro-ipc:jar:tests:1.7.7 from the shaded jar.
[INFO] Excluding org.codehaus.jackson:jackson-core-asl:jar:1.9.13 from the shaded jar.
[INFO] Excluding org.codehaus.jackson:jackson-mapper-asl:jar:1.9.13 from the shaded jar.
[INFO] Excluding org.apache.hadoop:hadoop-client:jar:2.6.0 from the shaded jar.
[INFO] Excluding org.apache.hadoop:hadoop-common:jar:2.6.0 from the shaded jar.
[INFO] Excluding commons-cli:commons-cli:jar:1.2 from the shaded jar.
[INFO] Excluding xmlenc:xmlenc:jar:0.52 from the shaded jar.
[INFO] Excluding commons-httpclient:commons-httpclient:jar:3.1 from the shaded jar.
[INFO] Excluding commons-collections:commons-collections:jar:3.2.2 from the shaded jar.
[INFO] Excluding commons-configuration:commons-configuration:jar:1.6 from the shaded jar.
[INFO] Excluding commons-digester:commons-digester:jar:1.8 from the shaded jar.
[INFO] Excluding commons-beanutils:commons-beanutils:jar:1.7.0 from the shaded jar.
```

图 7-23　Maven 编译部分过程

⑧ 当出现［INFO］BUILD SUCCESS 的信息提示时，表示编译成功，如图 7-24 所示，从图中可以看出整个编译过程用了 10 分 30 秒。

```
[INFO] Spark Project External Twitter .................... SUCCESS [  5.172 s]
[INFO] Spark Project External Flume Sink ................. SUCCESS [  3.923 s]
[INFO] Spark Project External Flume ...................... SUCCESS [  6.316 s]
[INFO] Spark Project External Flume Assembly ............. SUCCESS [  2.456 s]
[INFO] Spark Project External MQTT ....................... SUCCESS [ 12.464 s]
[INFO] Spark Project External MQTT Assembly .............. SUCCESS [  4.860 s]
[INFO] Spark Project External ZeroMQ ..................... SUCCESS [  5.111 s]
[INFO] Spark Project External Kafka ...................... SUCCESS [  7.689 s]
[INFO] Spark Project Examples ............................ SUCCESS [01:18 min]
[INFO] Spark Project External Kafka Assembly ............. SUCCESS [  5.475 s]
[INFO] ------------------------------------------------------------------------
[INFO] BUILD SUCCESS
[INFO] ------------------------------------------------------------------------
[INFO] Total time: 10:30 min
[INFO] Finished at: 2016-03-05T13:27:42+08:00
[INFO] Final Memory: 89M/1333M
[INFO] ------------------------------------------------------------------------
```

图 7-24　Maven 编译 Spark 1.6.0 成功

Maven 编译完成后，会在/assembly/target/scala-2.10 目录下生成 spark-assembly-1.6.0-hadoop2.6.0.jar 包，它包含了 Spark 1.6.0 编译得到的 Jar 包以及编译过程中的依赖包。

注意：如果在加密的文件系统上编译 Spark 源码（例如 home 目录是加密的），那么编译可能失败，出现文件名太长的错误。这时一种变通的方法，在 pom.xml 文件中给 scala-maven-plugin 添加下面的配置参数：＜arg＞-Xmax-classfile-name＜/arg＞＜arg＞128＜/arg＞，如下面程序段所示。

```
<arg>-unchecked</arg>
<arg>-deprecation</arg>
<arg>-feature</arg>
<arg>-Xmax-classfile-name</arg>
<arg>128</arg>
</args>
<jvmArgs>
```

并且在文件 project/SparkBuild.scala 中给 val 变量 sharedSettings 添加下面的内容：scalacOptions in Compile＋＋＝Seq（"-Xmax-classfile-name","128" ）。

如下面程序段所示。

```
retrieveManaged := true,
retrievePattern := "[type]s/[artifact](-[revision])(-[classifier]).[ext]",
publishMavenStyle := true,
scalacOptions in Compile ++= Seq("-Xmax-classfile-name", "128"),
resolvers += Resolver.mavenLocal,
otherResolvers <<= SbtPomKeys.mvnLocalRepository(dotM2 =>
Seq(Resolver.file("dotM2", dotM2))),
```

7.2.4 make-distribution. sh 脚本方式

为了构建类似于官网上发布版本行，可以用根目录下的 make-distribution. sh 脚本文件进行编译 Spark 1. 6. 0 源码。用 make-distribution. sh 编译 Spark 1. 6. 0 源码的步骤如下所述。

① 新建 spark-distribution 目录，用于存放编译后的结果文件。

```
>mkdir-p  /usr/software /spark- distribution
```

② 把下载的 Spark 1. 6. 0. tgz 解压到目录/usr/software/spark-distribution 中。

```
>tar  zxvf  Downloads/spark-1.6.0.tgz  -C  /usr/software /spark- distribution
```

③ 修改 Java 和 Hadoop 的版本，进入到解压后的 spark-1. 6. 0 目录。

```
>cd   /usr/software /spark- distribution / spark-1.6.0
```

用 vim 打开 pom. xml 文件。

```
>vim  pom.xml
```

修改 Java 和 Hadoop 版本，把<java. version>1. 7</java. version>改为<java. version>1. 8</java. version>，把<hadoop. version>2. 4. 0</hadoop. version>改为<hadoop. version>2. 6. 0</hadoop. version>，保存并退出。

④ 执行 make-distribution. sh 脚本，开始编译。

```
>make-distribution.sh
```

make-distribution. sh 可以带以下参数：

--tgz：在根目录下生成 spark-1. 6. 0 -bin. tar. gz，不加参数时不生成 tgz 文件，只生成/dist 目录。

--hadoop VERSION：打包时所用的 Hadoop 版本号，不加参数时为 1. 0. 4。

-Pyarn：是否支持 Hadoop YARN，不加参数时表示不支持。

-Phive：是否支持 Hive，不加参数时表示不支持。

--with-tachyon：是否支持分布式内存文件系统 Tachyon，不加参数时表示不支持。

⑤ 编译的部分过程如图 7-25 所示。

```
[INFO] --- maven-shade-plugin:2.4.1:shade (default) @ spark-core_2.10 ---
[INFO] Excluding org.apache.avro:avro-mapred:jar:hadoop2:1.7.7 from the shaded jar.
[INFO] Excluding org.apache.avro:avro-ipc:jar:1.7.7 from the shaded jar.
[INFO] Excluding org.apache.avro:avro:jar:1.7.7 from the shaded jar.
[INFO] Excluding org.apache.avro:avro-ipc:jar:tests:1.7.7 from the shaded jar.
[INFO] Excluding org.codehaus.jackson:jackson-core-asl:jar:1.9.13 from the shaded jar.
[INFO] Excluding org.codehaus.jackson:jackson-mapper-asl:jar:1.9.13 from the shaded jar.
[INFO] Excluding com.twitter:chill_2.10:jar:0.5.0 from the shaded jar.
[INFO] Excluding com.esotericsoftware.kryo:kryo:jar:2.21 from the shaded jar.
[INFO] Excluding com.esotericsoftware.reflectasm:reflectasm:jar:shaded:1.07 from the shaded jar.
[INFO] Excluding com.esotericsoftware.minlog:minlog:jar:1.2 from the shaded jar.
[INFO] Excluding com.twitter:chill-java:jar:0.5.0 from the shaded jar.
[INFO] Excluding org.apache.xbean:xbean-asm5-shaded:jar:4.4 from the shaded jar.
[INFO] Excluding org.apache.hadoop:hadoop-client:jar:2.6.0 from the shaded jar.
[INFO] Excluding org.apache.hadoop:hadoop-common:jar:2.6.0 from the shaded jar.
```

图 7-25　make-distribution. sh 部分编译过程

⑥ 当出现［Info］DUILD SUCCESS时表编译成功，如图 7-26 所示，整个编译过程用时 22 分 51 秒。

```
[INFO] Spark Project Assembly ............................ SUCCESS [02:16 min]
[INFO] Spark Project External Twitter .................... SUCCESS [ 11.558 s]
[INFO] Spark Project External Flume Sink ................. SUCCESS [  8.305 s]
[INFO] Spark Project External Flume ...................... SUCCESS [ 12.666 s]
[INFO] Spark Project External Flume Assembly ............. SUCCESS [  4.559 s]
[INFO] Spark Project External MQTT ....................... SUCCESS [ 34.457 s]
[INFO] Spark Project External MQTT Assembly .............. SUCCESS [ 11.719 s]
[INFO] Spark Project External ZeroMQ ..................... SUCCESS [ 16.889 s]
[INFO] Spark Project External Kafka ...................... SUCCESS [ 18.371 s]
[INFO] Spark Project Examples ............................ SUCCESS [02:44 min]
[INFO] Spark Project External Kafka Assembly ............. SUCCESS [ 11.728 s]
[INFO] ------------------------------------------------------------------------
[INFO] BUILD SUCCESS
[INFO] ------------------------------------------------------------------------
[INFO] Total time: 22:51 min
[INFO] Finished at: 2016-03-04T02:36:24+08:00
[INFO] Final Memory: 89M/1288M
[INFO] ------------------------------------------------------------------------
```

图 7-26　make-distribution. sh 编译成功

编译完成后，在 assembly/target/scala-2. 10 目录下生成 spark-assembly-1. 6. 0-hadoop2. 6. 0. jar 文件，在 examples/target/scala-2. 10 目录下生成 spark-examples-1. 6. 0-hadoop2. 6. 0. jar 文件，在 dist/lib 目录下生成 spark-assembly-1. 6. 0-hadoop2. 6. 0. jar，spark-examples-1. 6. 0-hadoop2. 6. 0. jar 文件。

7.2.5　需要注意的几个问题

在 Ubuntu kylin-15. 10-desktop 系统中编译 Spark 1. 6. 0 源码的过程中，需要注意下面的几个问题。

(1) Hadoop 的版本

由于不同版本的 HDFS 之间不兼容，如果要从 HDFS 上读写数据，就需要针对指定的 HDFS 版本进行编译 Spark 1. 6. 0 源码，这时可以通过设置 hadoop. version 的属性来指定 HDFS 的版本，如果没有设置，将会使用默认的 Hadoop2. 2. 0 版本。表 7-2 是 Hadoop 版本与配置参数表。

表 7-2　Hadoop 版本与配置参数表

Hadoop 版本	配置参数
1. x to 2. 1. x	hadoop-1
2. 2. x	hadoop-2. 2
2. 3. x	hadoop-2. 3
2. 4. x	hadoop-2. 4
2. 6. x and later 2. x	hadoop-2. 6

对于 Apache Hadoop versions 1. x、Cloudera CDH mr1 的发行版本以及其他不带 Yarn 的版本，可以用以下方法编译。

```
1. # Apache Hadoop 1.2.1
2. >mvn-Dhadoop.version=1.2.1 -Phadoop-1 -DskipTests clean package
3. # Cloudera CDH 4.2.0 with MapReduce v1
4. >mvn-Dhadoop.version=2.0.0-mr7-cdh4.2.0 -Phadoop-1 -DskipTests clean package
```

如果 Yarn 与 hadoop. version 版本不同，可以配置 yarn 和 yarn. version 选项。Spark 仅

支持 Yarn2.2.0 及以上版本。例如：

```
# Apache Hadoop 2.2.X
>mvn-Pyarn -Phadoop-2.2 -DskipTests clean package
# Apache Hadoop 2.3.X
>mvn-Pyarn -Phadoop-2.3 -Dhadoop.version=2.3.0 -DskipTests clean package
# Apache Hadoop 2.4.X or 2.5.X
>mvn-Pyarn -Phadoop-2.4 -Dhadoop.version=VERSION -DskipTests clean package
#Versions of Hadoop after 2.5.X may or may not work with the-Phadoop-2.4 profile (they were released after this
version of Spark).
# Different versions of HDFS and YARN.
>mvn-Pyarn -Phadoop-2.3 -Dhadoop.version=2.3.0 -Dyarn.version=2.2.0 -DskipTests clean package
```

(2) 支持 Hive 和 JDBC 的 Spark 源码编译

由于官网提供的编译包不支持 Hive，为了把 Hive 融合到 Spark SQL、JDBC 和 CLI，需要添加-Phive 和 Phive-thriftserver 选项。Spark 默认绑定到 Hive 0.13.1 版本，例如下面的命令：

```
# Apache Hadoop 2.4.X with Hive 13 support
>mvn-Pyarn -Phadoop-2.4 -Dhadoop.version=2.4.0 -Phive -Phive-thriftserver -DskipTests clean package
```

(3) 用 Scala 2.11 版本编译 Spark 1.6.0 源码

官网提供的 Spark1.6.0 的编译包的 Scala 语言是 2.10.4 版本，也可以用其他的版本来编译。方法是用-Dscala-2.11 选项来指定 Scala 2.11 版本编译打包，例如：

```
./dev/change-scala-version.sh 2.11
>mvn-Pyarn -Phadoop-2.4 -Dscala-2.11 -DskipTests clean package
```

Spark 1.6.0 目前暂不支持 Scala2.11 版的 JDBC。

(4) 编译单独的子模块

如果想只使用 Spark1.6.0 的子模块，可以用 mvn -pl 选项来指定要单独编译的子模块，例如下面的方法编译 Spark Streaming 子模块。

```
>mvn-pl :spark-streaming_2.10  clean  install
```

这里的 spark-streaming_2.10 是 streaming/pom.xml 文件中的 artifactId。

7.3 小结

编译 Spark 1.6.0 的方式有 Sbt 方式、Maven 方式、打包方式，既可以在 Windows 系统中编译源码，也可以在 Ubuntu 系统中编译源码。本章分别介绍了在 Windows10 系统中用 Sbt 方式、Maven 方式编译 Spark 1.6.0 的源码的步骤和需要注意的几个问题，也介绍了在 64 位 Ubuntukylin-15.10-desktop 系统中用 Sbt 方式、Maven 方式、打包方式编译 Spark 1.6.0 的源码的步骤和需要注意的问题。

第8章 Hadoop分布式集群环境搭建

为了更好地学习 Spark，需要学习者搭建 Hadoop 集群环境和 Spark 集群环境，因为 Spark 要用到 HDFS 分布式文件系统，熟练掌握 Hadoop 集群环境的搭建是学习的基础，可以通过集群环境的工作掌握 Hadoop 框架和 Spark 架构以及作业的执行细节。

本章较为细致地演示 Hadoop 单机模式搭建、Hadoop 伪分布式模式集群搭建以及 Wordcount 程序的运行、Hadoop 完全分布式环境的搭建以及 Wordcount 程序的运行。主要从零基础的前提下展开 Hadoop 集群搭建步骤，旨在让初学者更好地掌握 Hadoop 集群的搭建。

8.1 搭建 Hadoop 单机环境

8.1.1 安装软件下载

此处是在 ThinkPad T450 个人笔记本上进行 Hadoop 集群的安装演示，为了能够更好地学习 Hadoop，并基于 Hadoop 做大数据开发，建议个人硬件最低配置为内存至少 8G，CPU 处理器 Intel i3 以上处理器，硬盘容量至少 50G（不包含系统占容）。需要 Vmware 虚拟机，虚拟机可以使用 VMware-workstation，下载地址为 https://my.vmware.com/en/web/vmware/downloads，然后下载 UbuntuKylinubuntukylin-14.04.2-desktop-amd64.iso 镜像文件，下载地址为 http://www.ubuntu.com/download/alternative-downloads，如图 8-1 所示。

部署 Hadoop 集群，首先下载 Hadoop 安装包，Hadoop 安装版本 hadoop-2.6.0.tar，下载地址为 http://apache.fayea.com/hadoop/common/hadoop-2.6.0/，下载安装好正式开始 Spark 环境的搭建工作。下载 Hadoop 如图 8-2 所示。

8.1.2 Ubuntu 系统的安装

首先安装 Vmware 虚拟机，点击下载的 Vmware 安装文件 VMware-workstation -full-11.1.2，这里是安装到 D：\ Spark \ install \ Vmware 文件夹下，然后下一步，继续安装，然后输入产品序列号，安装完成后安装 Ubuntu Kylin 操作系统。

接下来在安装好的 Vmware 上安装 Ubuntukylin-14.04.2 系统。为了方便更多的初学者

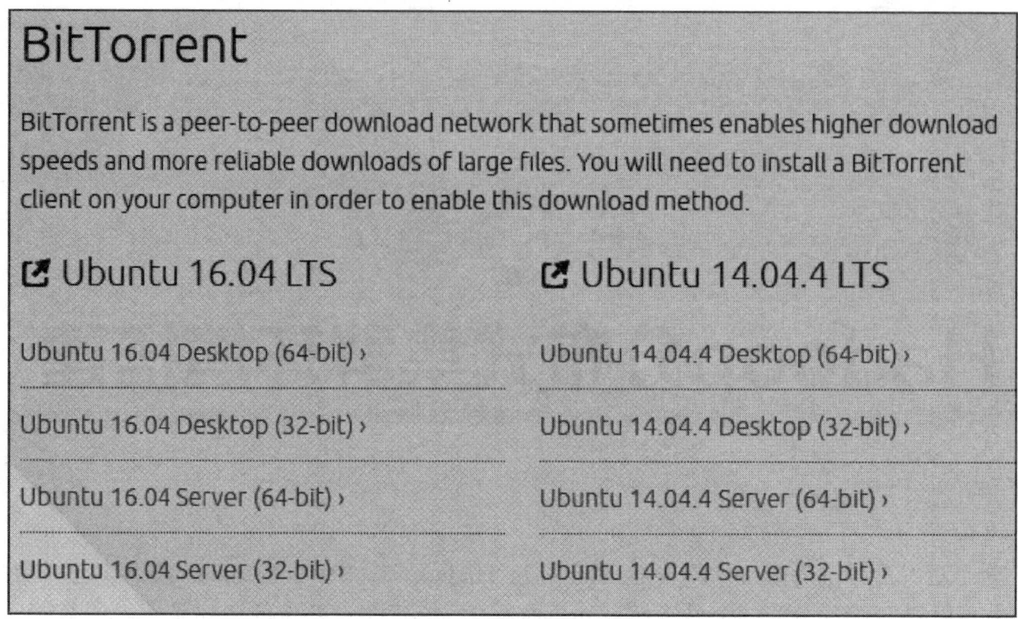

图 8-1　Ubuntu Kylin 下载

apache.fayea.com/hadoop/common/hadoop-2.6.0/

应用　建议网站　从 IE 中导入　百度云 网盘-Rocky_A　Study Area　w3school 在线教程

Index of /hadoop/common/hadoop-2.6.0

Name	Last modified	Size	Description
Parent Directory		-	
hadoop-2.6.0-src.tar.gz	2014-12-01 07:52	17M	
hadoop-2.6.0-src.tar.gz.mds	2014-12-01 07:52	1.1K	
hadoop-2.6.0.tar.gz	2014-12-01 07:52	186M	
hadoop-2.6.0.tar.gz.mds	2014-12-01 07:52	958	

图 8-2　Hadoop 安装包下载

进行环境搭建，选择 ubuntuKylin 系统进行安装演示，当然也可以选择其他的 Linux 操作系统 centos、redhat 均可。此时打开安装好的 VMware 虚拟机后，点击“新建虚拟机”，选择自定义安装的模式，下一步后选择 Linux，Ubuntu（64）位系统，如图 8-3 所示。

然后点击“下一步”，输入虚拟机名称，这里输入 Master，选择安装 Ubuntu Kylin 操作系统所在的位置，如图 8-4 所示。

继续“下一步”后，选择 CPU 的核心数，此时的 Core 数量决定了 Spark 集群能够使用多少个核心，决定了作业运行的效率，然后选择内存大小，这里选择 1500MB，如图 8-5 所示。

“下一步”后选择网络类型，这里选择“NAT”模式，可以为客户操作系统提供使用主机 IP 地址访问主机拨号连接或者外部以太网网络连接的权限。一直默认下一步到需要指定磁盘容量，这里选择 20GB，当然如果条件允许可以选择更大的空间，确定后编辑虚拟机 Master。

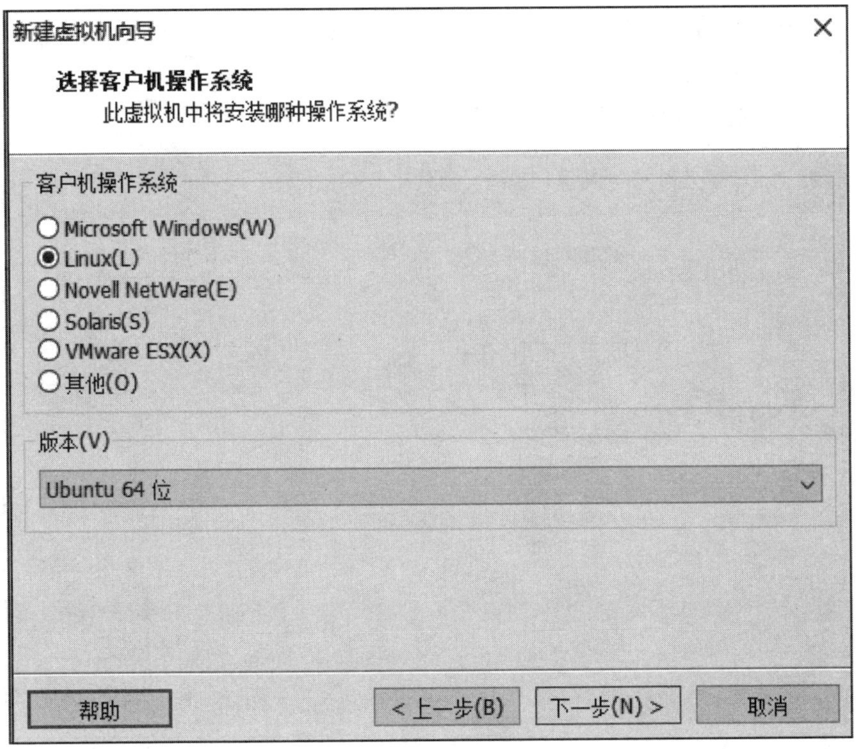

图 8-3　选择 Linux 系统

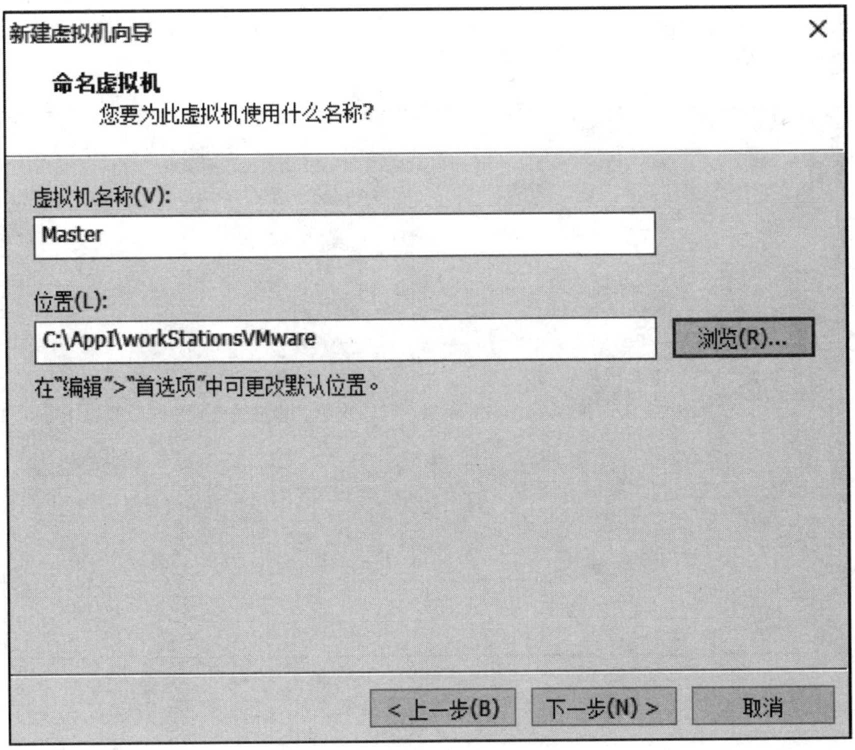

图 8-4　输入虚拟机名称

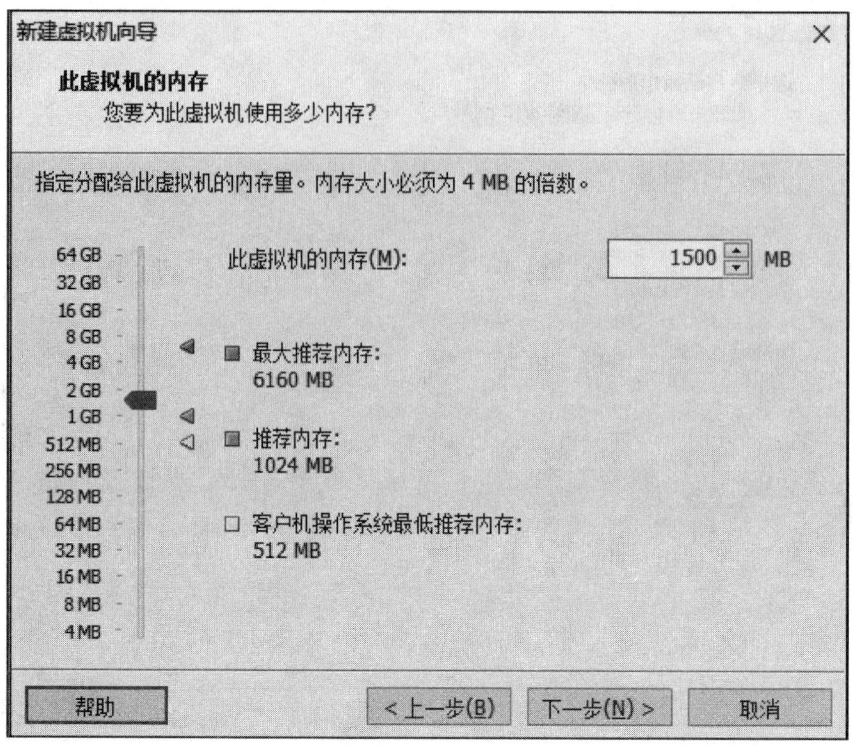

图 8-5 选择可用的内存大小

选择虚拟机的 Master 硬件配置和 iso 镜像的加载编辑虚拟机设置虚拟机内存，找到 Ubuntu 镜像文件 ubuntukylin-14.04.2-desktop-amd64 所在的位置，完成虚拟机的编辑。然后点击“确定”，如图 8-6 所示。

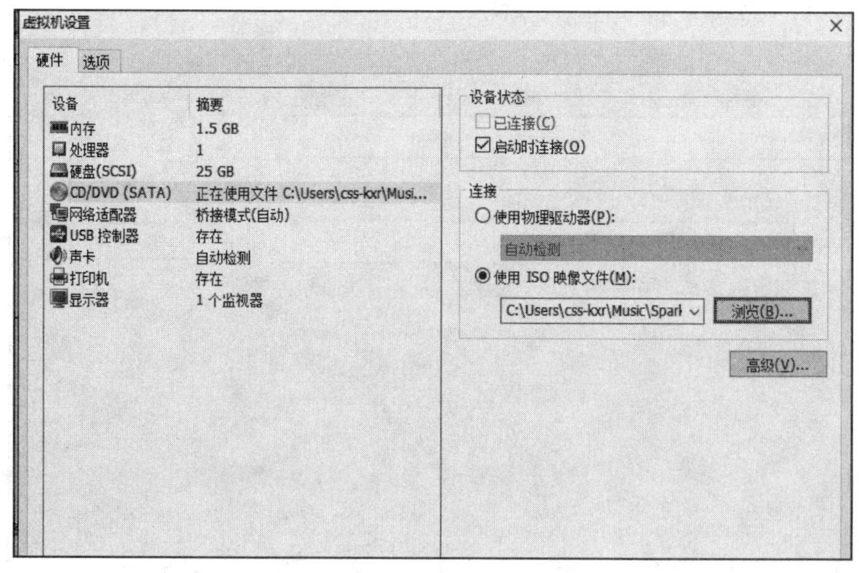

图 8-6 选择 ISO 镜像文件

完成以上步骤后，点击“开启虚拟机”正式开始 Ubuntu 的安装，如图 8-7 所示。

点击“开启虚拟机”正式开始 Ubuntu 的安装，等待虚拟机加载镜像文件后，开始进行安装，如图 8-8 所示，点击“install Ubuntu Kylin”。

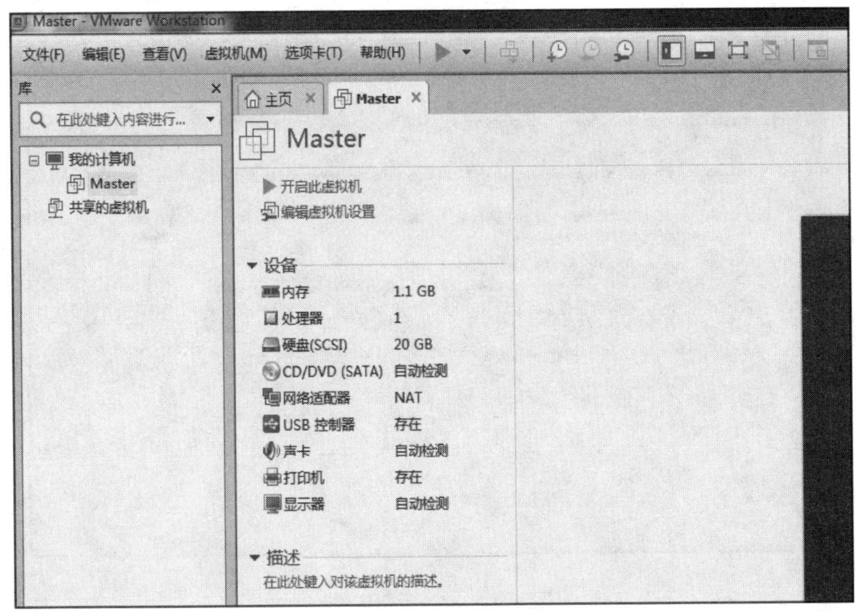

图 8-7　开启虚拟机

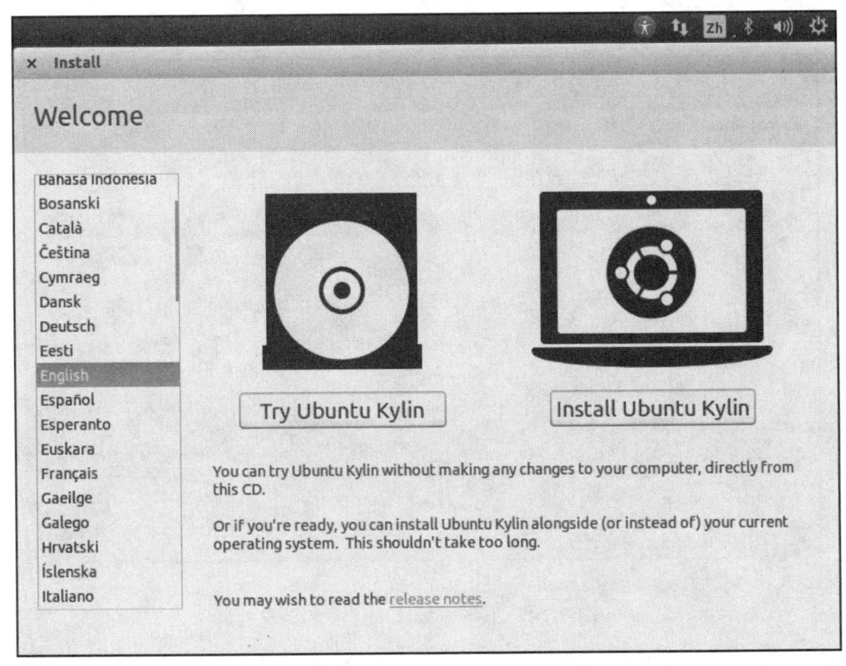

图 8-8　在虚拟机开始安装 Ubuntu Kylin

选择了安装的语言版本后，开始正式安装，然后就需要选择安装的类型，有清除磁盘、安全加密安装、LVM 安装和自定义安装的方式，在此选择自定义分区的安装方式，如图 8-9 所示。

选择自定义安装后就需要对安装的磁盘空间进行分区，点击“New Partition Table”按钮后，继续 continue 将一整个磁盘创建为一个空分区，此时就会创建 freespace 为 16106M 大小新空间，如图 8-10 所示。

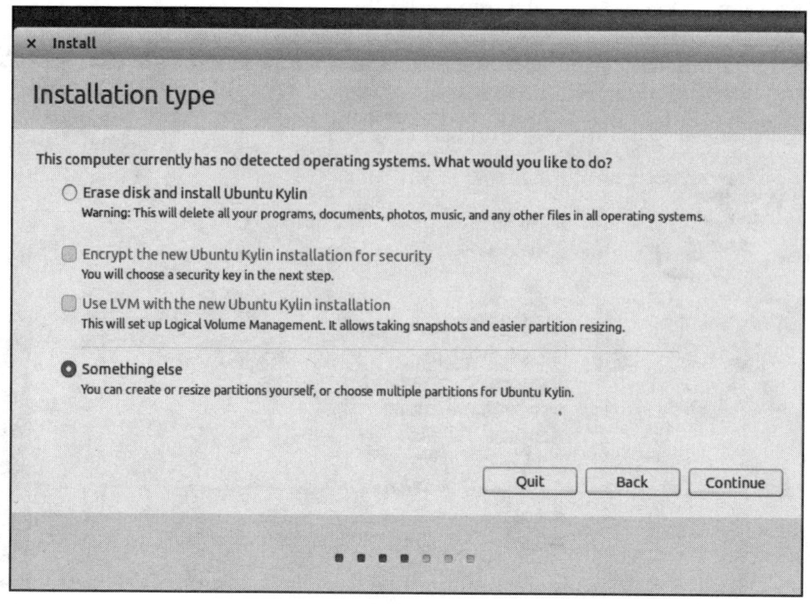

图 8-9　自定义安装类型

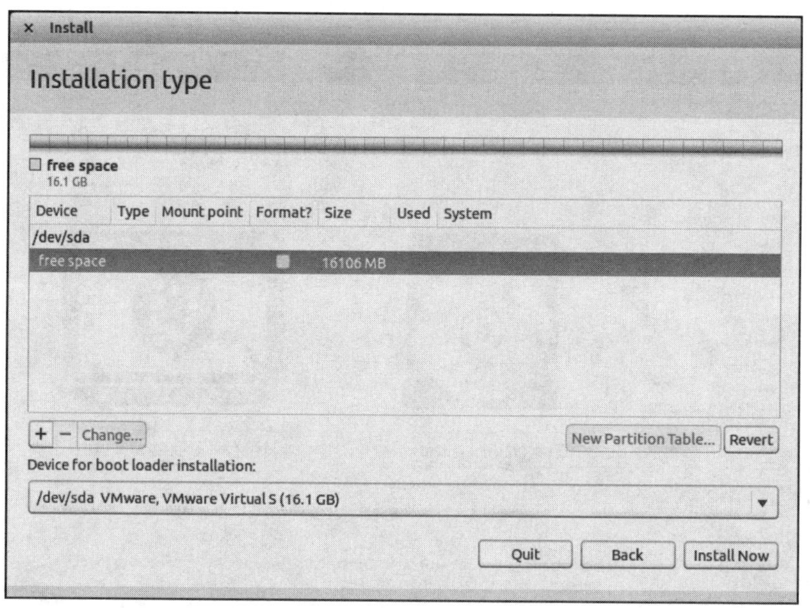

图 8-10　准备开始 Ubuntu 的分区

开始进行分区，在 Ubuntu 系统中为了不必要的分区麻烦，先对主分区进行创建，然后再创建逻辑分区，即 boot 分区作为主分区，然后创建逻辑分区“/”和 home，将 swap 分区放于最后，具体的分区如下所示。

（1）创建 boot 分区

将 boot 进行单独分区，那么为 boot 分配 102MB 的磁盘空间，实际分配的空间为 100MB，选择“Primary”，使 boot 成为主分区，如图 8-11 所示。

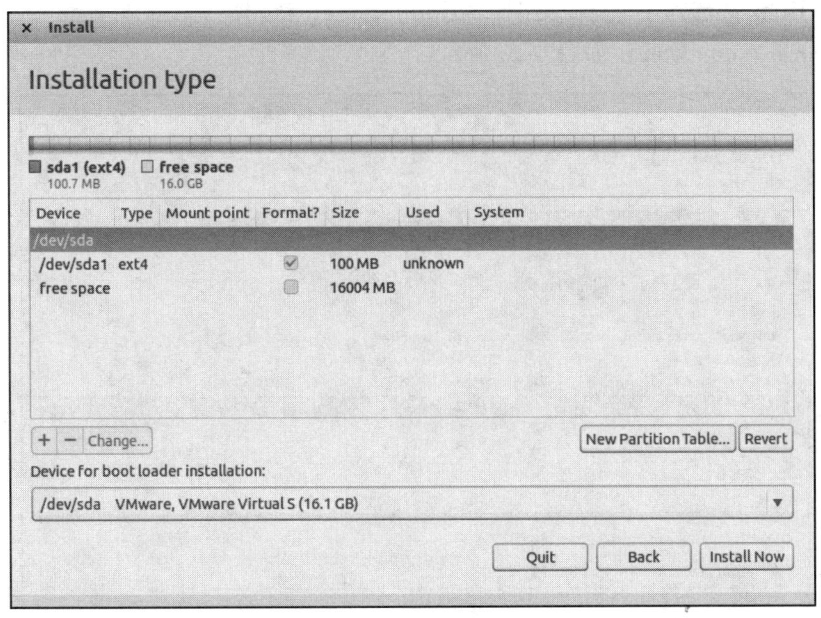

图 8-11 创建 boot 分区

(2) 创建“/”分区

创建好 boot 主分区之后，开始创建逻辑分区“/”，该分区是大多数使用中的存放文件的磁盘目录树根目录，可以分配 8MB 的空间大小，“Type for the new partition”选择 Logical 类型，如图 8-12 所示。

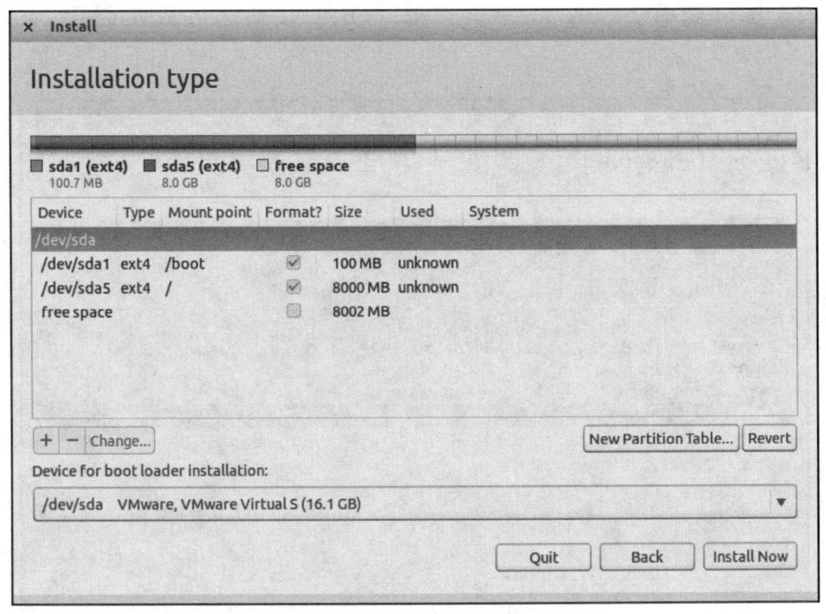

图 8-12 创建“/”分区

(3) 创建 home 分区

创建好了根分区，接下来创建逻辑分区 home，分配 6000MB 的空间大小，“Type for

the New Partition”选择 Logical 类型，“Mountpoint”选择“/home”，点击“OK”按钮后，就创建好了 home 分区，如图 8-13 所示。

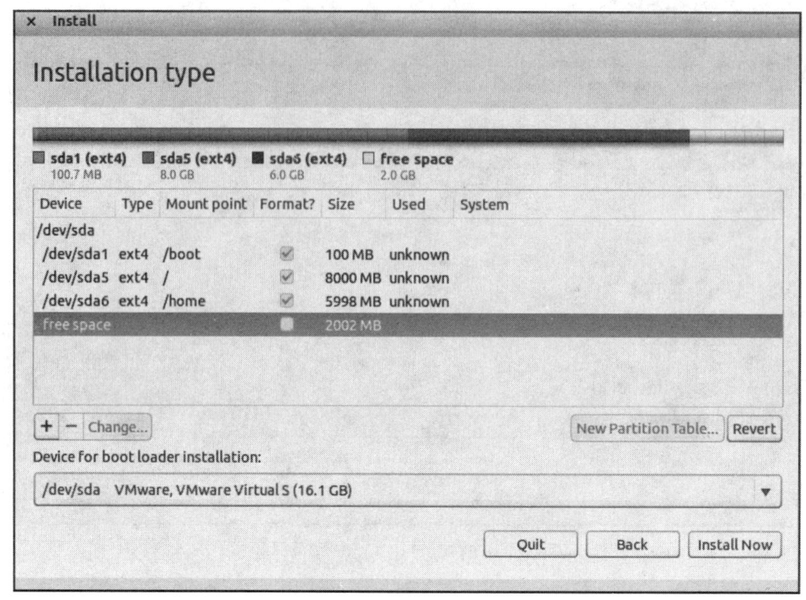

图 8-13 创建 home 分区

(4) 创建 swap 内存置换空间

创建内存置换空间 swap，不需要有挂载点，“Type for the new partition”选择 Logical 类型“useas”，所以选择“swap area”，点击“OK”按钮之后成功创建 swap 空间，如图 8-14所示。

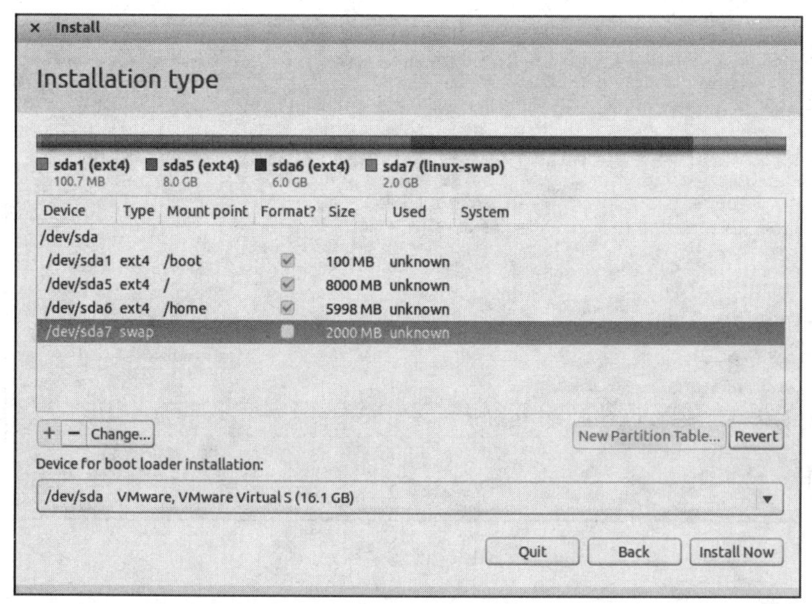

图 8-14 创建 swap 空间

创建好 Ubuntu 系统分区之后，点击“installNow”按钮开始安装系统，选择位置为

Shanghai，然后选择系统键盘为“EnglishUS”，点击“continue”后需要设置用户名、系统名称以及用户登录密码信息，如图 8-15 所示。

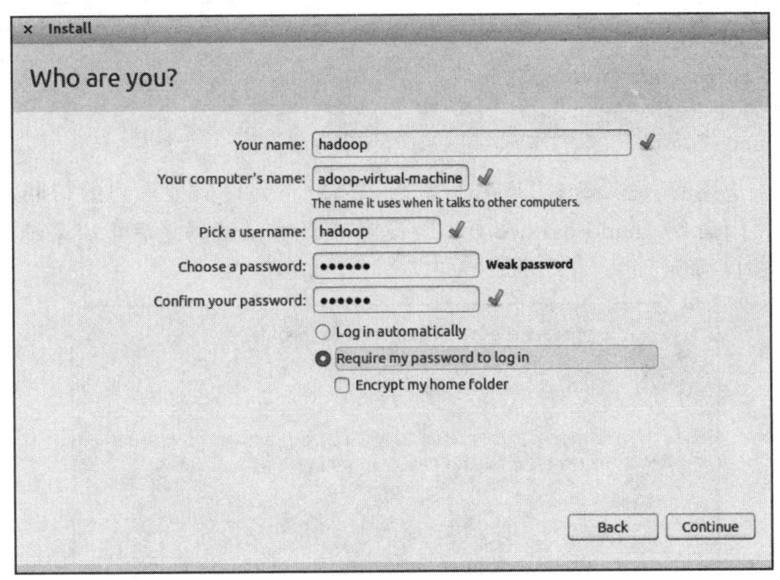

图 8-15 设置用户名密码

点击“continue”后开始加载数据写入磁盘空间，等待数据写入完成，会提示 restart 完成安装过程，重启后填写用户密码登录到 Ubuntu 系统图形化界面，然后点击浏览器查看网络是连接的状态，网络可以正常访问，至此 Ubuntu 系统安装完成。

为了使 Ubuntu 系统与本机之间更加方便地进行文件传输，虚拟机的分辨率也会自动跟随窗口调整而变化、拓展虚拟机的功能，就需要安装 VMware Tools 增强工具。

首先点击虚拟机工具栏中的“安装 WMware Tools 工具”，打开文件系统在找到 VMware Tool 文件后，复制到桌面上提取该复制的 VMware Tool 文件，然后使用“sudo -s”命令切换为 root 用户，之后在终端中输入命令“sudo. /vmware-install. pl”运行提取后的文件 vmware-install. pl，运行完成后表示安装完成，此时就可以实现本机与虚拟机之间的文件传输，Ubuntu 系统的分辨率随着窗口进行调整。

8.1.3 Hadoop 集群的安装和设置

安装好 Ubuntu 系统后，需要配置 root 用户登录，因为 Ubuntu 系统默认 root 是关闭状态，那么要切换到 root 用户的话，就需要使用命令“sudo -s”，然后输入当前用户的登录密码进行切换，同样从 root 用户切回普通登录用户可以使用三种方式。

① su username；

② 直接在通道中输入“exit”；

③ 使用快捷键 Ctrl+D。

要完成 Hadoop 集群的安装需要设置并安装以下内容。

(1) root 用户设置

对于初学者而言，设置 root 用户方便其聚焦于对 Spark 内容的学习，同时免去因为文件权限等带来的阻碍。root 用户的设置需要修改/etc/light 配置文件夹下的 lightdm. conf 文件，下面为 lightdm. conf 设置的具体内容。

```
#Setting user of root
[SeatDefaults]
user-session=ubuntu
greeter-session=unity-greeter
greeter-show-manual-login=true
allow-guest=false
```

主要设置了手工输入登录的用户名和密码，以及不允许其他用户的访问，设置好 root 用户后，需要使用命令“sudo passwd root”设置 root 登录密码之后重启系统，重启系统后会报错，如图 8-16 所示。

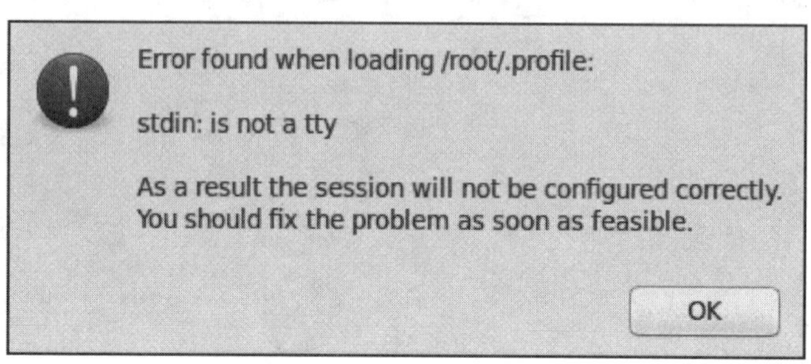

图 8-16　root 用户登录后报错信息

此时点击“OK”按钮，然后执行命令“nano/root/. profile”修改其配置文件的内容，“mesg n”前面添加“tty-s&&”，重启后即可使用 root 用户登录。

(2) 安装 SSH

SSH（Secure Shell 的缩写）是建立在 TCP/IP 协议的应用层和传输层基础上的安全协议。SSH 保障了远程登录会话和网络传输服务的安全性，起到防止信息泄露等作用，通过 SSH 可以对文件进行加密处理，SSH 也可以运行于多平台。

为了 Hadoop 集群的正常运行需要安装 SSH 安全协议，方便集群之间文件的安全传输，实际可以使用 scp 远程拷贝、sftp 安全文件传输等功能，当启动 SSH 服务后，会在默认的 22 端口进行监听且等待连接请求，当监听到请求信息后 SSH 守护进程会产生子进程，然后由子进程进行请求的连接。

① 安装 SSH

Hadoop 采用 SSH 进行通信，所以为了保证通信顺畅安装的时候需要设置成无密码登录（密码为空），即可在 Slaves 之间进行通信的时候免去输入密码的繁琐和低效。下面开始安装 SSH。

```
root@hadoop-virtual-machine:~/Desktop# apt-get install ssh
Reading package lists...Done
Building dependency tree
Reading state information...Done
ssh is already the newest version.
The following packages were automatically installed:and are no longer required:
diffstat gettext hardening-includes intltool-debian libapt-pkg-perl
libarchive-zip-perl libasprintf-dev libauthen-sasl-perl libautodie-perl
libclass-accessor-perl libclone-perl libdigest-hmac-perl libdpkg-perl
libemail-valid-perl libfile-fcntllock-perl libgettextpo-dev libio-pty-perl
```

```
libio-socket-inet6-perl libio-socket-ssl-perl libio-string-perl
libipc-run-perl libipc-system-simple-perl liblist-moreutils-perl
libmailtools-perl libnet-dns-perl libnet-domain-tld-perl libnet-ip-perl
libnet-libidn-perl libnet-smtp-ssl-perl libnet-ssleay-perl
libparse-debianchangelog-perl libperlio-gzip-perl libsocket6-perl
libsub-identify-perl libsub-name-perl libtext-levenshtein-perl
libtimedate-perl liburi-perl patchutils t1utils
Use'apt-get autoremove' to remove them.
0 upgraded,0 newly installed,0 to remove and 456 not upgraded.
root@hadoop-virtual-machine:~/Desktop# ps -e|grep ssh
1311 ?        00:00:00 sshd
root@hadoop-virtual-machine:~/Desktop#
```

可以看到这里已经安装好了 SSH 最新的版本，输入“ps-e | grep ssh”查看 SSH 进程发现 SSHD 进程已经启动。

② 设置 SSH 免密码登录。

```
root@hadoop-virtual-machine:~/Desktop# ssh-keygen -t rsa -P ""
Generating public/private rsa key pair.
Enter file in which to save the key (/root/.ssh/id_rsa):
Created directory '/root/.ssh'.
Your identification has been saved in/root/.ssh/id_rsa.
Your public key has been saved in/root/.ssh/id_rsa.pub.
The key fingerprint is:
e6:d6:28:11:e1:df:95:08:77:0a:0a:37:35:ce:ce:3e root@hadoop-virtual-machine
The key's randomart image is:
+--[ RSA 2048]----+
|   . +.= . .     |
|    + * = + .    |
|     + o o o     |
|      = . .      |
|     . S .       |
|      = o        |
|     . E .       |
|      o .        |
|                 |
+-----------------+
root@hadoop-virtual-machine:~/Desktop#
root@hadoop-virtual-machine:~/Desktop# cd ~/.ssh/
root@hadoop-virtual-machine:~/.ssh# ls
id_rsa  id_rsa.pub
root@hadoop-virtual-machine:~/.ssh# cat id_rsa.pub >> ~/.ssh/authorized_keys
```

上面的 ssh-keygen 用来生成 RSA 类型的密钥以及管理该密钥，参数“-t”用于指定要创建的 SSH 密钥的类型为 RSA。安装好 SSH 安全协议之后协议设置 SSH 免密码登录，生成私有密钥和共有密钥两个文件 id _ rsa 和 id _ rsa. pub。

其中 id _ rsa 为私有密钥、id _ rsa. pub 为共有密钥，然后将公钥追加到 authorized _ keys 中，因为 authorized _ keys 可以用来保存所有允许当前用户登录到 SSH 客户端用户的公钥内容，从而实现无密钥通信。下面最后验证 SSH 是否能够无密钥登录。

```
root@hadoop-virtual-machine:~/.ssh# ssh localhost
The authenticity of host 'localhost (127.0.0.1)' can't be established.
ECDSA key fingerprint is 45:54:a9:bb:5c:c8:d2:76:f5:5c:b3:f2:70:38:d4:4b.
Are you sure you want to continue connecting (yes/no)? yes
Warning: Permanently added 'localhost' (ECDSA) to the list of known hosts.
Welcome to Ubuntu 14.04.2 LTS (GNU/Linux 3.16.0-30-generic x86_64)
* Documentation:  https://help.ubuntu.com/
The programs included with the Ubuntu system are free software;
the exact distribution terms for each program are described in the
individual files in /usr/share/doc/*/copyright.
Ubuntu comes with ABSOLUTELY NO WARRANTY, to the extent permitted by
applicable law.
root@hadoop-virtual-machine:~#
```

（3）安装 rsync

rsync（remote synchronize）是一个工具，主要用于远程同步数据，也可以使用 Rsync 同步本地磁盘文件的目录，Rsync 支持多数的 Unix 系统。Rsync 可以带来较高的文件传输效率、镜像保存整个目录树和文件系统，也可以使用 SSH 等方式实现文件的传输。具体使用 Rsync 同步文件时可选用 SSH 的方式，由 SSH 验证用户。安装过程如下所述。

```
root@hadoop-virtual-machine:~/Desktop# rpm -qa|grep rsync
The program 'rpm' is currently not installed.You can install it by typing:
apt-get install rpm
root@hadoop-virtual-machine:~/Desktop# apt-get install rsync
Reading package lists...Done
Building dependency tree
Reading state information...Done
```

（4）安装 Java

本机上传或者将在线下载好的 Java 安装包拷贝到指定的目录，需要特别说明的是为了避免版本问题的困扰，学习者在下载 Java 版本的时候需要选择 Java7 以上的版本，然后解压 Java 安装包到指定的安装目录下，配置 Java 运行的环境变量，编辑进入环境配置文件“. bashrc”中，export Java 的环境变量，之后在终端中输入“java-version”验证安装是否正确，输入后终端中会显示 Java 的版本信息。Java 环境配置变量如下所示。

```
#Setting the env of java
export JAVA_HOME=/usr/java/jdk1.8.0_60
export JRE_HOME=${JAVA_HOME}/jre
export CLASS_PATH=.:${JAVA_HOME}/lib:${JRE_HOME}/lib
export PATH=${JAVA_HOME}/bin:$PATH
```

（5）安装 Hadoop

本机传入下载好的 Hadoop 安装包到指定的目录，或者在 Ubuntu 系统中直接下载 Hadoop 的安装包到指定目录下，然后解压 Hadoop 压缩文件到指定的目录下，进入解压的 Hadoop 目录下，如下所示。

```
root@hadoop-virtual-machine:/usr/local/hadoop# cd hadoop-2.6.0-src/
root@hadoop-virtual-machine:/usr/local/hadoop/hadoop-2.6.0-src# ls
BUILDING.txt          hadoop-hdfs-project        hadoop-tools
dev-support           hadoop-mapreduce-project   hadoop-yarn-project
hadoop-assemblies     hadoop-maven-plugins       LICENSE.txt
hadoop-client         hadoop-minicluster         NOTICE.txt
hadoop-common-project hadoop-project             pom.xml
hadoop-dist           hadoop-project-dist        README.txt
root@hadoop-virtual-machine:/usr/local/hadoop/hadoop-2.6.0-src#
```

解压好 Hadoop 安装文件后，进入 hadoop-env. sh 中添加 Java 安装目录信息，添加 Java 安装目录后按“ESC”键后输入“wq”保存退出，然后使用命令“source hadoop-env. sh”使更改生效。hadoop-env. sh 中添加 Java 的安装信息如下所示。

```
# The java implementation to use.
export JAVA_HOME=/usr/java/jdk1.8.0_60
```

此时 Hadoop 单机版本已经安装完成，但是如果不想在使用 Hadoop 的 bin 目录下的“hadoop”等命令的时候，必须到 Hadoop 的 bin 目录下使用的话，可以将 Hadoop 的 bin 目录配置到环境变量“. bashrc”中，这样只要用户登录系统后即可访问环境变量中的 Hadoop 安装目录下的 bin 目录，随时随地使用“hadoop”等命令。添加 Hadoop 的 bin 目录如下所示。

```
export PATH=${JAVA_HOME}/bin:/usr/local/hadoop/hadoop-2.6.0/bin:$PATH
之后验证Hadoop单机版本是否安装成功。如下所示：
root@hadoop-virtual-machine:/usr/local/hadoop/hadoop-2.6.0/bin# hadoop version
Hadoop2.6.0
Subversion https://git-wip-us.apache.org/repos/asf/hadoop.git
-r e3496499ecb8d220fba99dc5ed4c99c8f9e33bb1
Compiled by jenkins on 2014-11-13T21:10Z
Compiled with protoc 2.5.0
From source with checksum 18e43357c8f927c0695f1e9522859d6a
This command was run using
/usr/local/hadoop/hadoop-2.6.0/share/hadoop/common/hadoop-common-2.6.0.jar
root@hadoop-virtual-machine:/usr/local/hadoop/hadoop-2.6.0/bin#
```

可以看出 Hadoop 单机版本已经安装成功且显示 Hadoop 的版本为 2.6.0。

8.1.4 Hadoop 单机模式下运行 WordCount 示例

运行 Hadoop 自带的 WordCount 程序，需要在 Hadoop 的安装根目录下创建 input 文件夹，将 WordCount 运行的结果输出到 output 文件中。运行 WordCount 程序的命令如下所述。

```
root@hadoop-virtual-machine:/usr/local/hadoop/hadoop-2.6.0# hadoop jar share/hadoop/mapreduce
/hadoop-mapreduce-examples-2.6.0.jar wordcount input output
```

查看运行的过程如下所述。

```
16/01/2718:20:03 INFO mapred.LocalJobRunner: reduce task executor complete.
16/01/2718:20:04 INFO mapreduce.Job:  map 100% reduce 100%
16/01/2718:20:04 INFO mapreduce.Job:Job job_local1784484959_0001 completed successfully
16/01/2718:20:05 INFO mapreduce.Job:Counters:33
```

```
  FileSystemCounters
      FILE:Number of bytes read=547426
      FILE:Number of bytes written=1051740
      FILE:Number of read operations=0
      FILE:Number of large read operations=0
      FILE:Number of write operations=0
 Map-ReduceFramework
      Map input records=31
      Map output records=179
      Map output bytes=2055
      Map output materialized bytes=1836
      Input split bytes=117
      Combine input records=179
      Combine output records=131
      Reduce input groups=131
      Reduce shuffle bytes=1836
      Reduce input records=131
      Reduce output records=131
      SpilledRecords=262
      ShuffledMaps=1
      FailedShuffles=0
      MergedMap outputs=1
      GC time elapsed (ms)=224
      CPU time spent (ms)=0
      Physical memory (bytes) snapshot=0
      Virtual memory (bytes) snapshot=0
      Total committed heap usage (bytes)=254550016
 ShuffleErrors
      BAD_ID=0
      CONNECTION=0
      IO_ERROR=0
      WRONG_LENGTH=0
      WRONG_MAP=0
      WRONG_REDUCE=0
 FileInputFormatCounters
     BytesRead=1366
 FileOutputFormatCounters
     BytesWritten=1326
```

从上面运行的文件系统的统计可以看出，Hadoop 经过 mapper、Shuffle、reducer 阶段的执行细节，Map-Reduce Framework 阶段总包含了经过 Map 映射、Combine、Reduce、Shuffle 阶段的记录和输入输出字节数，以及物理 Memory 的使用，JVM 垃圾回收 GC 时间统计，还有 Shuffle Errors 的统计等。运行结果如下（只截取了少部分输出）。

```
functions     1
has           1
have          1
http://hadoop.apache.org/core/    1
http://wiki.apache.org/hadoop/    1
provides      1
```

```
re-export      2
regulations    1
reside         1
restrictions   1
security       1
see            1
software       2
software,      2
software.      2
software:      1
root@hadoop-virtual-machine:/usr/local/hadoop/hadoop-2.6.0#
```

至此 Hadoop 单机版安装并测试成功。

8.2 Hadoop 伪分布式环境

Hadoop 伪分布式环境顾名思义在单节点上模拟分布式的计算，也就是说 Hadoop 伪分布式的方式运行于单一节点之上。伪分布式模式是将单机模式节点分割为几个模块，其目的与单机模式相同都是用来进行 hadoop 程序开发和测试。配置好开发环境就可以方便 hadoop 重新开发了。

Hadoop 伪分布模式可看作是只有一个节点的集群，在这个集群中，这个节点既是 Master 也是 Slave，既是 NameNode 也是 DataNode，既是 JobTracker 也是 TaskTracker。

也就是说在该节点上 Namenode 的元数据操作交互完成对 HDFS 的访问，且有一个 Datanode 用来存储数据，将数据定期汇报给 Namenode，而 Namenoode 主要用于数据的存储，通常这些元数据中包括了文件名称、文件目录结构、文件属性等。JobTracker 主要负责集群和任务的资源调度，监控任务执行，跟踪 MapReduce 与作业的执行状态，TaskTracker 属于应用层，用来执行 Jobtracker 分发的任务，并向 Jobtracker 汇报任务执行完成情况，伪分布式模式中这些都发生在当前一台机器上。

8.2.1 Hadoop 伪分布式环境搭建

要完成 Hadoop 伪分布式环境的搭建，需要在 Hadoop 的安装目录下创建文件夹，用来保存输入的数据缓存等信息以及用来将原始数据上传到 HDFS 分布式文件系统上，如下所示。

```
root@Master:/usr/local/hadoop/hadoop-2.6.0# mkdir tmp
root@Master:/usr/local/hadoop/hadoop-2.6.0# mkdir -p hdfs/date
root@Master:/usr/local/hadoop/hadoop-2.6.0# mkdir -p hdfs/name
root@Master:/usr/local/hadoop/hadoop-2.6.0# ll hdfs/
total 20
drwxr-xr-x  5 root  root  40961月2807:10./
drwxr-xr-x 13200002000040961月2807:10../
drwxr-xr-x  2 root  root  40961月2807:10 date/
drwxr-xr-x  2 root  root  40961月2807:10 name/
drwxr-xr-x  2 root  root  40961月2807:10 namels/
```

创建完文件夹后，配置以下文件指定 NameNode 和 JobTracker 的位置和端口号，DataNode、TaskTracker 等的位置，具体配置如下所示。

（1）配置 hosts 文件。

```
127.0.0.1       localhost
192.168.1.10Master
```

配置 hostname，到 hostname 所在的目录下，使用 vim hostname 命令编辑 hostname 主机名，将主机名配置为 Master。

（2）配置 core-site. xml 文件。

```
        <!--Put site-specific property overrides in this file.-->

<configuration>
<property>
<name>fs.defaultFS</name>
<value>hdfs://Master:9000</value>
<description>The name of the default file system</description>
</property>

<property>
<name>hadoop.tmp.dir</name>
<value>/usr/local/hadoop/hadoop-2.6.0/tmp</value>
<description>The base for other temporary directories</description>

</property>
<property>
<name>io.file.buffer.size</name>
<value>4096</value>
</property>
</configuration>
```

从上面的配置信息可以看出这里指定了 NameNode 的位置和 9000 端口号等信息。

（3）配置 hsfs-site. xml（伪分布式模式下配置生成文件副本数为 1 个）。

```
<configuration>
<property>
<name>fs.replication</name>
<value>1</value>
<description>onefiles</description>
</property>

<property>
<name>dfs.namenode.name.dir</name>
<value>/usr/local/hadoop/hadoop-2.6.0/hdfs/name</value>
</property>

<property>
<name>dfs.datanode.data.dir</name>
<value>/usr/local/hadoop/hadoop-2.6.0/hdfs/data</value>
</property>
</configuration>
```

(4) 配置 mapred-site. xml（资源管理框架设置为 yarn）。

```
<configuration>
<property>
<name>mapreduce.framework.name</name>
<value>yarn</value>
</property>
</configuration>
```

(5) 配置 slaves：使用命令 vim slaves 将从节点设置为 Master。

```
Master
```

(6) 配置 yarn-site. xml（配置 yarn 的资源管理器 resourcemanager 为 Master 节点）。

```
<configuration>
<property>
<name>yarn.resourcemanager.hostname</name>
<value>Master</value>
</property>
<property>
<name>yarn.nodemanager.aux-services</name>
<value>mapreduce_shuffle</value>
</property>
</configuration>
```

8.2.2 Hadoop 伪分布式模式下运行 WordCount 示例

配置好 NameNode、TaskTracker 等位置及端口号后，需要格式化 NameNode。格式化 NameNode 如下所示（中间格式化过程省略）。

```
root@Master:/usr/local/hadoop/hadoop-2.6.0# hadoop namenode -format
DEPRECATED: Use of this script to execute hdfs command is deprecated.
Instead use the hdfs command for it.
16/02/28 07:38:04 INFO namenode.NameNode: STARTUP_MSG:
/************************************************************
     .
     .
     .
16/01/2807:38:17 INFO namenode.NameNode: SHUTDOWN_MSG:
/************************************************************
SHUTDOWN_MSG:Shutting down NameNode at Master/127.0.1.1
************************************************************/
```

进入 Hadoop 安装目录下的 sbin 目录下，使用命令“start-all. sh”来启动 Hadoop 伪分布式集群，启动完成后使用命令“jps”查看启动进程，如下所示。

```
4804NodeManager
4340DataNode
4692ResourceManager
4501SecondaryNameNode
4230NameNode
5885Jps
```

在浏览器中访问 Master：50070 端口，如图 8-17 所示。

DFS Used%:	0%
DFS Remaining%:	52.01%
Block Pool Used:	24 KB
Block Pool Used%:	0%
DataNodes usages% (Min/Median/Max/stdDev):	0.00% / 0.00% / 0.00% / 0.00%
Live Nodes	1 (Decommissioned: 0)

图 8-17　查看 50070 端口

然后在分布式文件系统中创建 input 文件夹，并将 Hadoop 安装目录下的 README. txt 文件上传到 input 文件夹中，如下所示。

```
root@Master:/usr/local/hadoop/hadoop-2.6.0/bin# hadoop fs -mkdir  /input
root@Master:/usr/local/hadoop/hadoop-2.6.0/bin# hadoop fs -ls /
Found 1 items
drwxr-xr-x   - root supergroup          0 2016-01-28 18:03 /input

root@Master:/usr/local/hadoop/hadoop-2.6.0# hadoop fs -put README.txt /input
root@Master:/usr/local/hadoop/hadoop-2.6.0# hadoop fs -ls /
Found 1 items
drwxr-xr-x   - root supergroup          0 2016-01-28 18:20 /input
root@Master:/usr/local/hadoop/hadoop-2.6.0# hadoop fs -ls /input/
Found 1 items
-rw-r--r--   3 root supergroup       1366 2016-01-28 18:20 /input/README.txt
```

至此伪分布式模式准备完成，此时开始运行伪分布式 WordCount 程序，运行 Hadoop 自带的/hadoop-mapreduce-examples-2. 6. 0. jar 文件中的 Wordcount 程序，并且将结果输出到 myWordCount 文件夹下，在终端中输入命令如下所示。

```
root@Master:/usr/local/hadoop/hadoop-2.6.0#        hadoop        jar share/hadoop/mapreduce/hadoop-mapreduce-examples-2.6.0.jar wordcount /input//output/myWordCount
```

查看程序运行结果（只截取部分结果），如下所示。

```
16/01/28 18:42:15 INFO mapreduce.Job:  map 0% reduce 0%
16/01/28 18:43:01 INFO mapreduce.Job:  map 100% reduce 0%
16/01/28 18:43:40 INFO mapreduce.Job:  map 100% reduce 100%
16/01/28 18:43:44 INFO mapreduce.Job: Job job_1458985846933_0001 completed successfully
16/01/28 18:43:46 INFO mapreduce.Job: Counters: 49

 File System Counters
            FILE: Number of bytes read=1836
            FILE: Number of bytes written=215081
            FILE: Number of read operations=0
            FILE: Number of large read operations=0
            FILE: Number of write operations=0
            HDFS: Number of bytes read=1466
            HDFS: Number of bytes written=1306
            HDFS: Number of read operations=6
```

```
            HDFS: Number of large read operations=0
            HDFS: Number of write operations=2
    Job Counters
            Launched map tasks=1
            Launched reduce tasks=1
            Data-local map tasks=1
```

通过 50070 查看 WordCount 程序的运行结果，如图 8-18 所示。

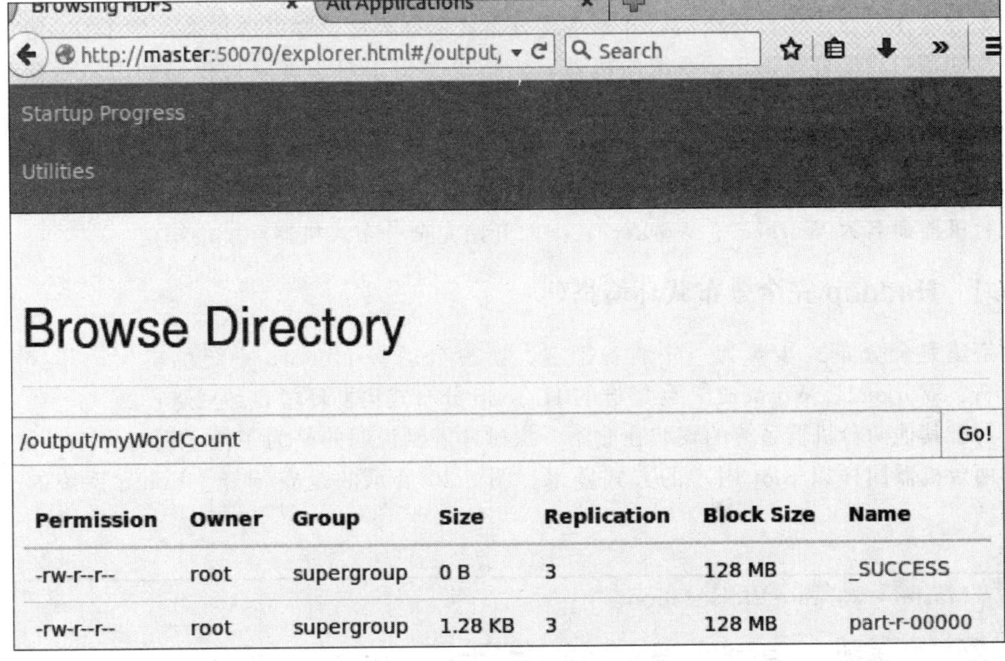

图 8-18　程序运行生成的结果文件

通过 yarn 资源管理器的访问端口 8088 查看程序运行结果，如图 8-19 所示。

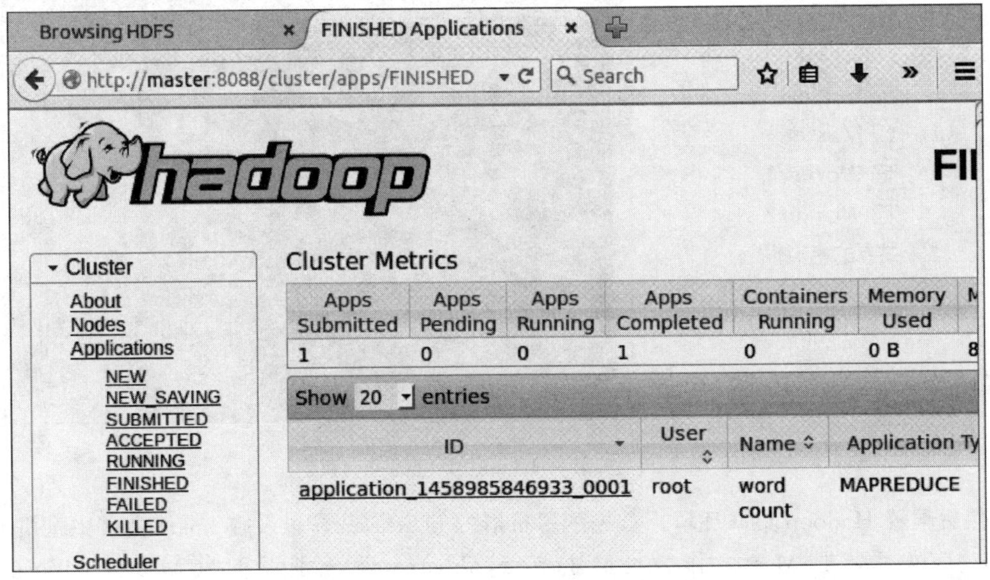

图 8-19　WordCount 程序运行结果

此时伪分布式模式搭建完成并成功运行了 Hadoop 自带的 Wordcount 程序。伪分布式同单机模式一样主要用来测试，搭建 Hadoop 伪分布式集群环境对于初学者来说是很有必要掌握的内容。

8.3 Hadoop 完全分布式环境

搭建完伪分布式集群环境后，开始搭建分布式集群环境，这里选择最小的 Hadoop 的集群，即使用三台同样 Ubuntu 系统的机器，当然也是出于对学者硬件机器的限制做出的选择，避免因为机器的内存或者磁盘空间的局限导致集群环境搭建失败。

这里可以使用跟已经搭建完成伪分布式模式的 Master 机器相同的安装方式，安装好 Ubuntu Kylin 操作系统在第二、第三台机器上，为了便于区分和模拟真实环境，将第二、第三台机器命名为 Worker1、Worker2。下面开始完全分布式机器环境的搭建。

8.3.1 Hadoop 完全分布式环境搭建

搭建完全分布式步骤为：先准备第二、第三台运行 Ubuntu 系统的机器，其次配置 Master、Worker1、Worker2 三台集群的 Hadoop 分布式集群环境，最后进行测试。

对于其他两台机器系统的安装在此不多探讨，需要说明的是为了便于说明将 Worker 节点的两台机器同样以 root 用户的方式登录。图 8-20 展示的是安装好 Ubuntu 系统的三台机器。

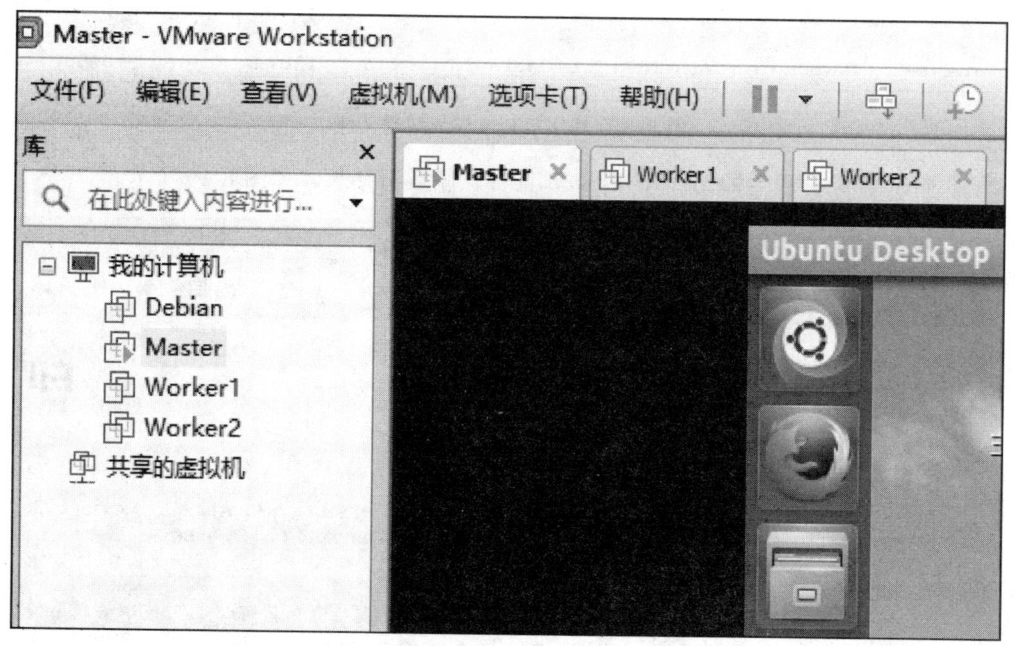

图 8-20　安装好 Ubuntu 系统的三台机器

然后配置 Hadoop 集群环境。首先配置机器的 hostname，以及在 hosts 文件中主机名和 IP 的对应关系，把 Master 作为集群主节点，hostname 分别设置为 Master、Worker1、Worker2，然后设置 hosts 文件。如下所示。

```
127.0.0.1       localhost.localdomain    localhost
192.168.1.10    Master.localdomain    Master
192.168.1.11    Worker1.localdomain   Worker1
192.168.1.13    Worker2.localdomain   Worker2

# The following lines are desirable for IPv6 capable hosts
::1     ip6-localhost ip6-loopback
fe00::0 ip6-localnet
ff00::0 ip6-mcastprefix
ff02::1 ip6-allnodes
ff02::2 ip6-allrouters

```

然后使用命令将 hosts 文件传到 Worker1 和 Worker2 节点上，如下所示。

```
root@Master:~/Desktop# vim /etc/host
host.conf    hostname     hosts        hosts.allow  hosts.deny
root@Master:~/Desktop# vim /etc/hosts
root@Master:~/Desktop# scp /etc/hosts root@Worker1:/etc/hosts
hosts                              100%  364     0.4KB/s   00:00
root@Master:~/Desktop# scp /etc/hosts root@Worker2:/etc/hosts
hosts          100%  364     0.4KB/s   00:00
root@Master:~/Desktop# ping Worker1
PING Worker1.localdomain (192.168.1.11) 56(84) bytes of data.
64 bytes from Worker1.localdomain (192.168.1.11): icmp_seq=1 ttl=64 time=0.372 ms
64 bytes from Worker1.localdomain (192.168.1.11): icmp_seq=2 ttl=64 time=0.349 ms
64 bytes from Worker1.localdomain (192.168.1.11): icmp_seq=3 ttl=64 time=0.488 ms
^C
---Worker1.localdomain ping statistics---
3 packets transmitted, 3 received, 0% packet loss, time 1998ms
rtt min/avg/max/mdev = 0.349/0.403/0.488/0.060 ms
root@Master:~/Desktop#
```

分别传给 Worker1 和 Worker2 后互相 ping 都能通，然后验证 ssh 协议，以 Master 连接 Worker 为例（首先三台机器都安装配置好了 ssh 安全通信协议），如下所示。

```
root@Master:~/Desktop# ssh Worker1
Welcome to Ubuntu 14.04.2 LTS (GNU/Linux 3.16.0-30-generic x86_64)

 * Documentation:  https://help.ubuntu.com/

490 packages can be updated.
0 updates are security updates.

Last login: Sun Mar 20 18:48:37 2016 from master.localdomain
root@Worker1:~#
```

再修改三台机器 Hadoop 安装目录下的 etc/hadoop 文件下的如下文件，首先修改 Master 机器的 core-site. xml 文件，如下所示。

```
<!-- Put site-specific property overrides in this file. -->

<configuration>
<property>
```

```
<name>fs.defaultFS</name>
<value>hdfs://Master:9000</value>
<description>The name of the default file system</description>
</property>

<property>
<name>hadoop.tmp.dir</name>
<value>/usr/local/hadoop/hadoop-2.6.0/tmp</value>
<description>The base for other temporary directories</description>

</property>
<property>
<name>io.file.buffer.size</name>
<value>4096</value>
</property>
</configuration>
```

同样的方法修改两个 Worker 节点，然后修改 mapred-site. xml 文件，如下所示。

```
<!-- Put site-specific property overrides in this file. -->

<configuration>
<property>
<name>mapreduce.framework.name</name>
<value>yarn</value>
</property>
</configuration>
```

然后以同样的方式修改 hdfs-site. xml 文件（副本数设置为 3 个），如下所示。

```
<!-- Put site-specific property overrides in this file. -->

<configuration>
<property>
<name>fs.replication</name>
<value>2</value>
<description>two same files</description>
</property>

<property>
<name>dfs.namenode.name.dir</name>
<value>/usr/local/hadoop/hadoop-2.6.0/hdfs/name</value>
</property>

<property>
<name>dfs.datanode.data.dir</name>
<value>/usr/local/hadoop/hadoop-2.6.0/hdfs/data</value>
</property>
</configuration>
```

然后修改 Master 机器的 masters 文件，如下所示。

```
Master
```

再修改 Master 节点的 slaves 文件，如下所示。

```
Master
Worker1
Worker2
```

然后将 Master 机器的 masters 和 slaves 文件使用命令 scp masters root@Worker1：/usr/hadoop/hadoop-2.6.0/etc/hadoop/拷贝到 Worker 节点上。

最后使用命令 ./hadoop namenode - format 对集群进行格式化，格式化完毕后启动集群，若如无问题即表明已经成功完成 Hadoop 完全分布式集群的部署。

8.3.2 Hadoop 完全分布式模式下运行 WordCount 示例

首先启动分布式文件系统和 yarn 资源管理器，如下所示。

```
root@Master:/usr/local/hadoop/hadoop-2.6.0/sbin# ./start-all.sh
This script is Deprecated. Instead use start-dfs.sh and start-yarn.sh
Starting namenodes on [Master]
Master: starting namenode, logging to /usr/local/hadoop/hadoop-2.6.0/logs/hadoop-root-namenode-Master.out
Worker1: starting datanode, logging to /usr/local/hadoop/hadoop-2.6.0/logs/hadoop-root-datanode-Worker1.out
Master: starting datanode, logging to /usr/local/hadoop/hadoop-2.6.0/logs/hadoop-root-datanode-Master.out
Starting secondary namenodes [0.0.0.0]
0.0.0.0: starting secondarynamenode, logging to /usr/local/hadoop/hadoop-2.6.0/logs/hadoop-root-secondarynamenode-Master.out
starting yarn daemons
starting resourcemanager, logging to /usr/local/hadoop/hadoop-2.6.0/logs/yarn-root-resourcemanager-Master.out
Worker1: starting nodemanager, logging to /usr/local/hadoop/hadoop-2.6.0/logs/yarn-root-nodemanager-Worker1.out
Master: starting nodemanager, logging to /usr/local/hadoop/hadoop-2.6.0/logs/yarn-root-nodemanager-Master.out

root@Master:/usr/local/hadoop/hadoop-2.6.0/sbin# jps
22058 ResourceManager
22182 NodeManager
21581 NameNode
21702 DataNode
22226 Jps
21859 SecondaryNameNode
```

可以看到 HDFS 的进程 NameNode、DataNode、SecondaryNameNode，yarn 产生的负责资源管理的 ResourceManager 进程和负责节点资源管理的进程 NodeManager 都已经启动，查看 Worker1 节点的进程都已经启动，如下所示。

```
root@Worker1:/usr/local/hadoop/hadoop-2.6.0/hdfs/name/current# jps
12849 DataNode
12984 NodeManager
13101 Jps
```

访问 50070 端口，查看 DataNode 节点，如图 8-21 所示。

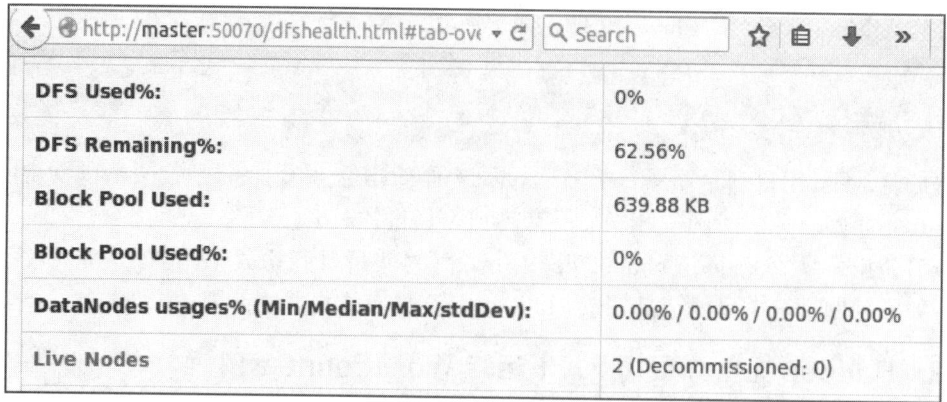

图 8-21　访问 50070 端口

下面在 hadoop 的安装主目录下创建一个文件夹 myTest，然后使用命令 touch mytest1 创建文本文件 mytest，往文件中输入，如图 8-22 所示。

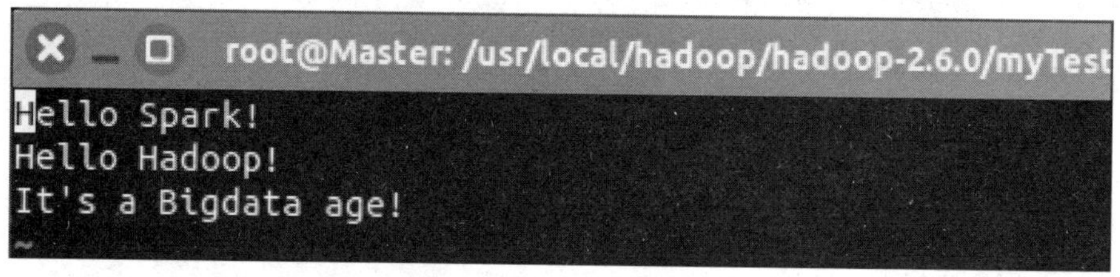

图 8-22　创建文本文件 mytest

保存文件，然后将该文件上传到 HDFS 分布式文件系统下，由于搭建环境的时候在配置文件 hdfs-site. xml 中设置为 2 个文件副本，因此上传的文件 mytest 是存储在 Master 和 Worker1 两个节点中的，保证数据的安全性。上传过程如下所示。

```
root@Master:/usr/local/hadoop/hadoop-2.6.0/myTest# ls
mytest1
root@Master:/usr/local/hadoop/hadoop-2.6.0/myTest# hadoop fs -put mytest1 /input/test
```

然后使用 hadoop 命令 hadoop fs - ls 查看上传的文件是否上传成功，如下所示。

```
root@Master:/usr/local/hadoop/hadoop-2.6.0/myTest# hadoop fs -ls /input/test
Found 2 items
-rw-r--r--   3 root supergroup       1366 2016-01-29 19:04 /input/test/README.txt
-rw-r--r--   3 root supergroup         47 2016-01-29 19:18 /input/test/mytest1
root@Master:/usr/local/hadoop/hadoop-2.6.0/myTest#
```

成功上传文件后，开始运行 WordCount 程序，这里将运行的结果文件存储到 KongResult 文件中，如下所示。

```
root@Master:/usr/local/hadoop/hadoop-2.6.0# hadoop jar share/hadoop/mapreduce/hadoop-mapreduce-examples-2.6.0.jar wordcount /input/test/mytest1 KongResult
16/01/29 19:21:14 INFO client.RMProxy: Connecting to ResourceManager at Master/192.168.1.10:8032
16/01/29 19:21:17 INFO input.FileInputFormat: Total input paths to process : 1
16/01/29 19:21:18 INFO mapreduce.JobSubmitter: number of splits:1
16/01/29 19:21:19 INFO mapreduce.JobSubmitter: Submitting tokens for job: job_1459163799198_0001
16/01/29 19:21:20 INFO impl.YarnClientImpl: Submitted application application_1459163799198_0001
16/01/29 19:21:21 INFO mapreduce.Job: The url to track the job:http://Master:8088/proxy/application_1459163799198_0001/
```

```
16/01/29 19:21:21 INFO mapreduce.Job: Running job: job_1459163799198_0001
16/01/29 19:21:54 INFO mapreduce.Job: Job job_1459163799198_0001 running in uber mode : false
16/01/29 19:21:54 INFO mapreduce.Job:  map 0% reduce 0%
16/01/29 19:22:54 INFO mapreduce.Job:  map 100% reduce 0%
16/01/29 19:23:16 INFO mapreduce.Job:  map 100% reduce 100%
```

成功运行程序后，通过 50070 端口查看运行 WordCount 程序的结果，如图 8-23 所示。

Utilities

Browse Directory

/user/root/KongResult Go!

Permission	Owner	Group	Size	Replication	Block Size	Name
-rw-r--r--	root	supergroup	0 B	3	128 MB	_SUCCESS
-rw-r--r--	root	supergroup	55 B	3	128 MB	part-r-00000

图 8-23 WordCount 运行结果文件

然后点击文件 part-r-00000，然后将其下载下来，保存该文件后，打开文件查看单词切分的结果，如图 8-24 所示。

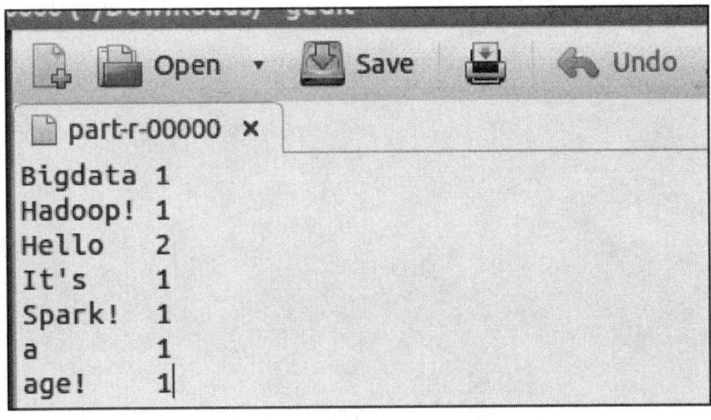

图 8-24 单词切分的结果

至此，hadoop 完全分布式环境搭建完成，并且实现了对单词的切分操作，只要细心操作，即可完全熟练掌握环境搭建，然后开启对 Hadoop 探索。并在此基础上部署 Spark 集群已游刃有余。

8.4 小结

Hadoop 的核心是可存储海量数据的分布式文件系统 HDFS 和分布式计算架构 MapReduce，HDFS 实现海量数据的存储和管理，将数据以多副本的方式存储于不同的节点上，保证数据的安全和冗余，同时可以增删数据，实时对数据进行处理，而 MapReduce 可将大规模数据块以任务的形式分割成若干小数据任务，之后通过分布式处理后重新加载到仓库。要学习 Hadoop 或者 Spark 应首先熟练掌握 Hadoop 集群环境的搭建。

本章对 Hadoop 单机模式搭建、Hadoop 伪分布式模式、完全分布式模式集群搭建进行了较为详细地讲解，并在搭建的集群环境上运行 Wordcount 程序。希望通过本章的讲解可以帮助初学者对于 Hadoop 环境搭建有更加深刻的认识。

第9章

精通Spark集群搭建与测试

本章将会手把手非常详细地介绍如何从零开始搭建 Spark 集群。对于 Spark 学习者而言构建 Spark 集群的重要性是不言而喻的，本章通过以下几步成功搭建集群。

① Spark 集群所需要的软件的安装；

② Spark 环境搭建；

③ Spark 集群测试。

本章是建立在前面已经对 Hadoop 的安装成功的基础上进行的，只需要按照上述步骤循序渐进就可以成功搭建集群，下面就开始这奇妙的旅程吧。

9.1 Spark 集群所需软件的安装

9.1.1 安装 JDK

① 请读者登录如下网站下载 JDK，本书安装的是 64 位的 1.8.60 版本的 JDK。http://www.oracle.com/technetwork/java/javase/downloads/java-archive-javase8-2177648.html#jdk-8u60-oth-JPR。如图 9-1 所示，选择 Accept License Agreement，选择 _jdk-8u60-linux-x64.tar.gz 下载。

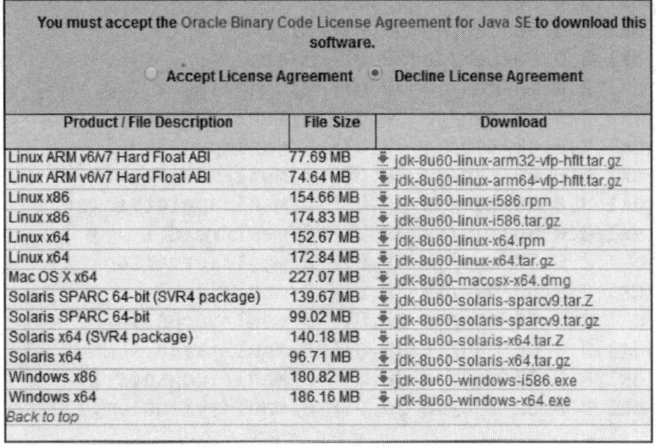

Java SE Development Kit 8u60

You must accept the Oracle Binary Code License Agreement for Java SE to download this software.

Accept License Agreement　Decline License Agreement

Product / File Description	File Size	Download
Linux ARM v6/v7 Hard Float ABI	77.69 MB	jdk-8u60-linux-arm32-vfp-hflt.tar.gz
Linux ARM v6/v7 Hard Float ABI	74.64 MB	jdk-8u60-linux-arm64-vfp-hflt.tar.gz
Linux x86	154.66 MB	jdk-8u60-linux-i586.rpm
Linux x86	174.83 MB	jdk-8u60-linux-i586.tar.gz
Linux x64	152.67 MB	jdk-8u60-linux-x64.rpm
Linux x64	172.84 MB	jdk-8u60-linux-x64.tar.gz
Mac OS X x64	227.07 MB	jdk-8u60-macosx-x64.dmg
Solaris SPARC 64-bit (SVR4 package)	139.67 MB	jdk-8u60-solaris-sparcv9.tar.Z
Solaris SPARC 64-bit	99.02 MB	jdk-8u60-solaris-sparcv9.tar.gz
Solaris x64 (SVR4 package)	140.18 MB	jdk-8u60-solaris-x64.tar.Z
Solaris x64	96.71 MB	jdk-8u60-solaris-x64.tar.gz
Windows x86	180.82 MB	jdk-8u60-windows-i586.exe
Windows x64	186.16 MB	jdk-8u60-windows-x64.exe

Back to top

图 9-1　JDK 下载

② 将下载下来的 jdk 拷贝到 Centos 里面，如图 9-2 所示。

图 9-2　虚拟机中的 JDK

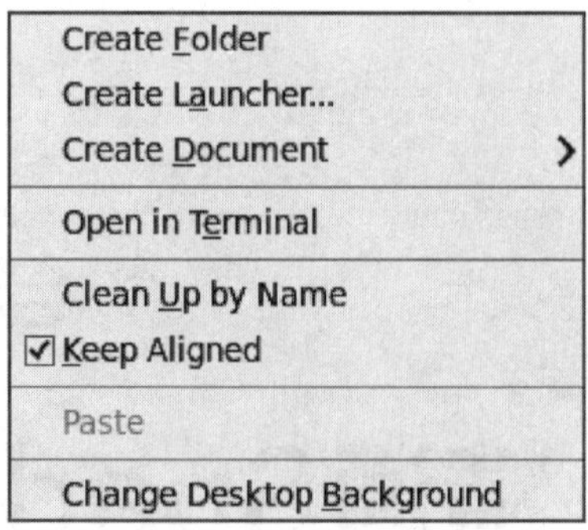

图 9-3　打开终端

③ 打开终端，右击鼠标，选择 Open in Terminal，如图 9-3 所示。

④ 创建 JDK 安装目录，使用 mkdir 命令。

```
[root@Master Desktop]# mkdir /usr/local/jdk
```

⑤ mv 命令将软件包移动到 jdk 下面，mv 相当于剪切。

```
[root@Master Desktop]# mv jdk-8u60-linux-x64.tar.gz /usr/local/jdk/
```

⑥ 用 tar 命令解压 jdk。

```
[root@Master jdk]# tar -zxvf jdk-8u60-linux-x64.tar.gz
```

⑦ 解压后，查看目录，解压成功，如图 9-4 所示。

```
jdk1.8.0_60/db/bin/NetworkServerControl
jdk1.8.0_60/db/bin/sysinfo.bat
jdk1.8.0_60/db/bin/dblook.bat
jdk1.8.0_60/db/bin/ij.bat
jdk1.8.0_60/db/bin/setEmbeddedCP
jdk1.8.0_60/db/bin/setNetworkServerCP.bat
jdk1.8.0_60/db/bin/derby_common.bat
jdk1.8.0_60/db/bin/sysinfo
jdk1.8.0_60/db/bin/setNetworkClientCP
jdk1.8.0_60/db/bin/stopNetworkServer
jdk1.8.0_60/db/bin/stopNetworkServer.bat
jdk1.8.0_60/src.zip
jdk1.8.0_60/javafx-src.zip
jdk1.8.0_60/THIRDPARTYLICENSEREADME.txt
jdk1.8.0_60/COPYRIGHT
jdk1.8.0_60/man/
jdk1.8.0_60/man/ja_JP.UTF-8/
jdk1.8.0_60/man/ja_JP.UTF-8/man1/
jdk1.8.0_60/man/ja_JP.UTF-8/man1/jps.1
jdk1.8.0_60/man/ja_JP.UTF-8/man1/javaws.1
jdk1.8.0_60/man/ja_JP.UTF-8/man1/pack200.1
jdk1.8.0_60/man/ja_JP.UTF-8/man1/extcheck.1
jdk1.8.0_60/man/ja_JP.UTF-8/man1/appletviewer.1
jdk1.8.0_60/man/ja_JP.UTF-8/man1/rmid.1
jdk1.8.0_60/man/ja_JP.UTF-8/man1/servertool.1
jdk1.8.0_60/man/ja_JP.UTF-8/man1/jar.1
jdk1.8.0_60/man/ja_JP.UTF-8/man1/jdeps.1
jdk1.8.0_60/man/ja_JP.UTF-8/man1/javah.1
jdk1.8.0_60/man/ja_JP.UTF-8/man1/jconsole.1
jdk1.8.0_60/man/ja_JP.UTF-8/man1/jvisualvm.1
```

图 9-4　jdk 解压过程

⑧ 在 bashrc 中配置，在 bashrc 中配置环境变量的目的，是为了可以在任何目录下使用 JDK。

```
[root@Master jdk]# vim ~/.bashrc
```

⑨ 进入配置文件。

```
# .bashrc
# User specific aliases and functions
alias rm='rm -i'
alias cp='cp -i'
alias mv='mv -i'
# Source global definitions
if [ -f /etc/bashrc ]; then
        . /etc/bashrc
fi
```

⑩ 按下 “i” 键，进入 INSERT 模式，配置 JAVA _ HOME 和 JRE _ HOME。

```
export JAVA_HOME=/usr/local/jdk/jdk1.8.0_60
export JRE_HOME=${JAVA_HOME}/jre
export PATH=${JAVA_HOME}/bin:$PATH
```

⑪ 完成配置之后，按下 Esc 按钮，然后 “Shift＋:” 之后左下角就会显示一个 “:”，这个时候输入 wq，保存并退出。

```
:wq
```

⑫ Source 一下 bashrc，使得在 bashrc 中配置的环境变量生效。

```
[root@Master jdk]# source ~/.bashrc
```

⑬ Java － version 显示 Java 的版本，JDK 安装成功。

```
[root@Master jdk]# java -version
java version "1.8.0_60"
Java(TM) SE Runtime Environment (build 1.8.0_60-b27)
Java HotSpot(TM) 64-Bit Server VM (build 25.60-b23, mixed mode)
```

⑭ 将 jdk 复制到另外两台机器上。

```
[root@Master jdk]# scp -r jdk1.8.0_60 root@Worker1:/usr/local/jdk/
[root@Master jdk]# scp -r jdk1.8.0_60 root@Worker2:/usr/local/jdk/
```

⑮ 然后再将 bashrc 的文件分别拷贝到 Worker1、Worker2 上，另外两台机器也要 source 一下 bashrc 文件。

```
[root@Master jdk]# scp ~/.bashrc root@Worker1:~/.bashrc
[root@Master jdk]# scp ~/.bashrc root@Worker2:~/.bashrc
```

⑯ 查看 Worker1 和 Worker2 上 jdk 安装均成功。

```
[root@Worker1 ~]# java -version
Javaversion "1.8.0_60"
Java(TM) SE Runtime Environment (build 1.8.0_60-b27)
Java HotSpot(TM) 64-Bit Server VM (build 25.60-b23, mixed mode)
[root@Worker2 ~]# java -version
Javaversion "1.8.0_60"
Java(TM) SE Runtime Environment (build 1.8.0_60-b27)
Java HotSpot(TM) 64-Bit Server VM (build 25.60-b23, mixed mode)
```

至此三台机器上面的 jdk 成功安装。

9.1.2 安装 Scala

① 官方要求 Scala 必须是 2.10x，所以这里的 Scala 版本一定要注意一下，笔者的版本是 2.10.4，具体的下载地址如下。

http：//www.scala-lang.org/download/all.html 下载 scala 下载 scala-2.10.4，如图 9-5所示。

Scala 2.10.5

Scala 2.10.4

图 9-5 Scala 版本

② 点击下载，如图 9-6 所示。

图 9-6 Scala 下载

③ 将下载的 Scala 安装包拷贝到 Centos 里面，如图 9-7 所示。

图 9-7 Scala 安装包

④ mkdir 命令创建 Scala 的安装目录。

```
[root@Master jdk]# mkdir /usr/local/scala/
```

⑤ 将 Scala 移动到安装目录下面。

```
[root@Master Desktop]# mv scala-2.10.4.tgz /usr/local/scala/
```

⑥ 同样，使用 tar 来解压文件。

```
[root@Master scala]# tar -zxvf  scala-2.10.4.tgz
```

⑦ 解压成功。

```
scala-2.10.4/examples/oneplacebuffer.scala
scala-2.10.4/examples/sort.scala
scala-2.10.4/examples/package.scala
scala-2.10.4/examples/actors/
scala-2.10.4/examples/actors/seq.scala
scala-2.10.4/examples/actors/producers.scala
scala-2.10.4/examples/actors/links.scala
scala-2.10.4/examples/actors/boundedbuffer.scala
```

```
scala-2.10.4/examples/actors/message.scala
scala-2.10.4/examples/actors/auction.scala
scala-2.10.4/examples/actors/channels.scala
scala-2.10.4/examples/actors/fringe.scala
scala-2.10.4/examples/actors/pingpong.scala
scala-2.10.4/examples/actors/looping.scala
scala-2.10.4/examples/xml/
scala-2.10.4/examples/xml/phonebook/
scala-2.10.4/examples/xml/phonebook/phonebook1.scala
scala-2.10.4/examples/xml/phonebook/phonebook2.scala
scala-2.10.4/examples/xml/phonebook/embeddedBook.scala
scala-2.10.4/examples/xml/phonebook/verboseBook.scala
scala-2.10.4/examples/xml/phonebook/phonebook.scala
scala-2.10.4/examples/xml/phonebook/phonebook3.scala
scala-2.10.4/examples/gadts.scala
scala-2.10.4/examples/maps.scala
scala-2.10.4/misc/
scala-2.10.4/misc/scala-devel/
scala-2.10.4/misc/scala-devel/plugins/
scala-2.10.4/misc/scala-devel/plugins/continuations.jar
scala-2.10.4/lib/
scala-2.10.4/lib/typesafe-config.jar
scala-2.10.4/lib/akka-actors.jar
scala-2.10.4/lib/scala-actors.jar
scala-2.10.4/lib/scala-compiler.jar
scala-2.10.4/lib/scala-reflect.jar
scala-2.10.4/lib/scala-library.jar
scala-2.10.4/lib/scala-swing.jar
scala-2.10.4/lib/jline.jar
scala-2.10.4/lib/scala-actors-migration.jar
scala-2.10.4/lib/scalap.jar
scala-2.10.4/bin/
scala-2.10.4/bin/scaladoc.bat
scala-2.10.4/bin/scala.bat
scala-2.10.4/bin/scalac.bat
scala-2.10.4/bin/scala
scala-2.10.4/bin/scaladoc
scala-2.10.4/bin/fsc.bat
scala-2.10.4/bin/fsc
scala-2.10.4/bin/scalac
scala-2.10.4/bin/scalap.bat
scala-2.10.4/bin/scalap
[root@Master scala]# ls
scala-2.10.4  scala-2.10.4.tgz
```

⑧ 按下“i”进入 INSERT 模式，在 bashrc 中配置 Scala 环境。

```
export SCALA_HOME=/usr/local/scala/scala-2.10.4
export PATH=${SCALA_HOME}/bin:${JAVA_HOME}/bin:$PATH
```

⑨ Source 一下文件。

```
[root@Master scala]# source ~/.bashrc
```

⑩ Scala 安装成功。

```
[root@Master scala]# scala-version
Scala code runner version 2.10.4 -- Copyright 2002-2013, LAMP/EPFL
```

⑪ 启动 Scala 交互式命令，对于 Scala 的开发和学习非常重要，因为它会显示很多内部的实现，例如下面的案例，算 1+1 的时候，交互式命令直接将结果输出。

```
[root@Master ~]# scala
Welcome to Scala version 2.10.4 (Java HotSpot(TM) 64-Bit Server VM, Java 1.8.0_60).
Type in expressions to have them evaluated.
Type :help for more information.

scala> 1 + 1
res0: Int = 2

scala>
将Scala的源文件拷贝给Worker1、Worker2。
[root@Master scala]# scp -r scala-2.10.4 root@Worker1:/usr/local/scala/
[root@Master scala]# scp -r scala-2.10.4 root@Worker2:/usr/local/scala/
查看Worker1和Worker2上面的Scala安装情况，均安装成功。
[root@Worker1 ~]# scala -version
Scala code runner version 2.10.4 -- Copyright 2002-2013, LAMP/EPFL

[root@Worker1 ~]# scala -version
Scala code runner version 2.10.4 -- Copyright 2002-2013, LAMP/EPFL
```

同样 bashrc 的配置文件也要拷贝给 Worker1、Worker2，然后再 source，这里不再演示，此时三台机器上的 Scala 成功安装。

9.2 Spark 环境搭建

请读者进入 Spark 的下载页面，下载 spark-1. 6. 0-bin-hadoop2. 6. tgz 版本：http：//www. apache. org/dyn/closer. lua/spark/spark-1. 6. 0/spark-1. 6. 0-bin-hadoop2. 6. tgz，保存到 Windows 本地目录。

9.2.1 Spark 单机与单机伪分布式环境

为了满足不同学习者的要求，先安装单机版，单机版的 Spark 可以满足对 Spark 的应用程序测试工作，对于初学者而言是非常有益的，而单机伪分布式是满足很多可能硬件不是太好的读者的需求，对于 Spark 集群的理解是非常有益的。

（1）单机版

安装 Spark 单机版，直接将 Spark 解压即可，不需要配置其他环境变量之类。

① Spark 版本的下载，进入 Spark 官网，下载 spark-1. 6. 0-bin-hadoop2. 6. tgz 版本，http：//www. apache. org/dyn/closer. lua/spark/spark-1. 6. 0/spark-1. 6. 0-bin-hadoop2. 6. tgz。

② 创建 Spark 安装目录。

```
[root@Master Desktop]# mkdir /usr/local/spark/
```

③ 将 spark 软件包复制到虚拟机中，如图 9-8 所示。

图 9-8 spark 拷贝到虚拟机

④ 将 spark 软件包移动到 spark 安装目录下。

```
[root@Master Desktop]# mv spark-1.6.0-bin-hadoop2.6.tgz /usr/local/spark
```

⑤ 解压 spark。

```
[root@Master spark]# tar -zxvf spark-1.6.0-bin-hadoop2.6.tgz
```

⑥ 解压的时候，显示很多文件，解压成功。

```
spark-1.6.0-bin-hadoop2.6/examples/src/main/python/als.py
spark-1.6.0-bin-hadoop2.6/examples/src/main/python/mllib/
spark-1.6.0-bin-hadoop2.6/examples/src/main/python/mllib/gradient_boosting_classification_example.py
spark-1.6.0-bin-hadoop2.6/examples/src/main/python/mllib/ranking_metrics_example.py
spark-1.6.0-bin-hadoop2.6/examples/src/main/python/mllib/multi_label_metrics_example.py
spark-1.6.0-bin-hadoop2.6/examples/src/main/python/mllib/fpgrowth_example.py
spark-1.6.0-bin-hadoop2.6/examples/src/main/python/mllib/isotonic_regression_example.py
spark-1.6.0-bin-hadoop2.6/examples/src/main/python/mllib/gaussian_mixture_model.py
spark-1.6.0-bin-hadoop2.6/examples/src/main/python/mllib/random_forest_classification_example.py
spark-1.6.0-bin-hadoop2.6/examples/src/main/python/mllib/multi_class_metrics_example.py
spark-1.6.0-bin-hadoop2.6/examples/src/main/python/mllib/recommendation_example.py
spark-1.6.0-bin-hadoop2.6/examples/src/main/python/mllib/correlations.py
spark-1.6.0-bin-hadoop2.6/examples/src/main/python/mllib/binary_classification_metrics_example.py
spark-1.6.0-bin-hadoop2.6/examples/src/main/python/mllib/kmeans.py
spark-1.6.0-bin-hadoop2.6/examples/src/main/python/mllib/logistic_regression.py
spark-1.6.0-bin-hadoop2.6/examples/src/main/python/mllib/regression_metrics_example.py
spark-1.6.0-bin-hadoop2.6/examples/src/main/python/mllib/naive_bayes_example.py
spark-1.6.0-bin-hadoop2.6/examples/src/main/python/mllib/decision_tree_classification_example.py
spark-1.6.0-bin-hadoop2.6/examples/src/main/python/mllib/decision_tree_regression_example.py
spark-1.6.0-bin-hadoop2.6/examples/src/main/python/mllib/sampled_rdds.py
spark-1.6.0-bin-hadoop2.6/examples/src/main/python/mllib/random_rdd_generation.py
spark-1.6.0-bin-hadoop2.6/examples/src/main/python/mllib/gradient_boosting_regression_example.py
spark-1.6.0-bin-hadoop2.6/examples/src/main/python/mllib/word2vec.py
spark-1.6.0-bin-hadoop2.6/examples/src/main/python/mllib/random_forest_regression_example.py
spark-1.6.0-bin-hadoop2.6/examples/src/main/r/
spark-1.6.0-bin-hadoop2.6/examples/src/main/r/data-manipulation.R
spark-1.6.0-bin-hadoop2.6/examples/src/main/r/dataframe.R
spark-1.6.0-bin-hadoop2.6/examples/src/main/r/ml.R
spark-1.6.0-bin-hadoop2.6/examples/src/main/resources/
```

```
spark-1.6.0-bin-hadoop2.6/examples/src/main/resources/kv1.txt
spark-1.6.0-bin-hadoop2.6/examples/src/main/resources/people.txt
spark-1.6.0-bin-hadoop2.6/examples/src/main/resources/full_user.avsc
spark-1.6.0-bin-hadoop2.6/examples/src/main/resources/users.parquet
spark-1.6.0-bin-hadoop2.6/examples/src/main/resources/users.avro
spark-1.6.0-bin-hadoop2.6/examples/src/main/resources/people.json
spark-1.6.0-bin-hadoop2.6/examples/src/main/resources/user.avsc
spark-1.6.0-bin-hadoop2.6/NOTICE
spark-1.6.0-bin-hadoop2.6/RELEASE

[root@Master spark]# ls
spark-1.6.0-bin-hadoop2.6  spark-1.6.0-bin-hadoop2.6.tgz
```

⑦ 运行案例，计算 SparkPi 2 是指两个并行度。

```
[root@Master spark]# ./bin/run-example SparkPi 2
```

⑧ 运行结果如下。

```
Pi is roughly 3.1445
```

(2) 单机伪分布式

Spark 单机伪分布式是在一台机器上既有 Master，又有 Worker 节点。

① 在 bashrc 中配置 spark 环境，完成后 source 一下。

```
export HADOOP_HOME=/usr/local/hadoop/hadoop-2.6
export
PATH=${SPARK_HOME}/bin:${SPARK_HOME}/sbin/sbin::${SCALA_HOME}/bin:${JAVA_HOME}/bin:${HADOOP_HOME}/bin:${HADOOP_HOME}/sbin:$PATH
```

② source bashrc 文件。

```
[root@Master bin]# source ~/.bashrc
```

③ 现在要配置 spark 中的 conf 文件。

```
[root@Master conf]# cp spark-env.sh.template spark-env.sh
```

④ 配置 spar-env. sh。

```
# Options for the daemons used in the standalone deploy mode
#- SPARK_MASTER_IP, to bind the master to a different IP address or hostname
# - SPARK_MASTER_PORT / SPARK_MASTER_WEBUI_PORT, to use non-default ports for the master
# - SPARK_MASTER_OPTS, to set config properties only for the master (e.g. "-Dx=y")
#- SPARK_WORKER_CORES, to set the number of cores to use on this machine
# - SPARK_WORKER_MEMORY, to set how much total memory workers have to give executors (e.g. 1000m, 2g)
# - SPARK_WORKER_PORT / SPARK_WORKER_WEBUI_PORT, to use non-default ports for the worker
#- SPARK_WORKER_INSTANCES, to set the number of worker processes per node
# - SPARK_WORKER_DIR, to set the working directory of worker processes
# - SPARK_WORKER_OPTS, to set config properties only for the worker (e.g. "-Dx=y")
#- SPARK_DAEMON_MEMORY, to allocate to the master, worker and history server themselves (default: 1g).
# - SPARK_HISTORY_OPTS, to set config properties only for the history server (e.g. "-Dx=y")
```

```
# - SPARK_SHUFFLE_OPTS, to set config properties only for the external shuffle service (e.g. "-Dx=y")
# - SPARK_DAEMON_JAVA_OPTS, to set config properties for all daemons (e.g. "-Dx=y")
# - SPARK_PUBLIC_DNS, to set the public dns name of the master or workers

# Generic options for the daemons used in the standalone deploy mode
#- SPARK_CONF_DIR      Alternate conf dir. (Default: ${SPARK_HOME}/conf)
# - SPARK_LOG_DIR      Where log files are stored.  (Default: ${SPARK_HOME}/logs)
#- SPARK_PID_DIR       Where the pid file is stored. (Default: /tmp)
#- SPARK_IDENT_STRING  A string representing this instance of spark. (Default: $USER)
#- SPARK_NICENESS      The scheduling priority for daemons. (Default: 0)

export JAVA_HOME=/usr/local/jdk/jdk1.8.0_60
export SCALA_HOME=/usr/local/scala/scala-2.10.4
export HADOOP_HOME=/usr/local/hadoop/hadoop-2.6.0
export HADOOP_CONF_DIR=/usr/local/hadoop/hadoop-2.6.0/etc/hadoop
export SPARK_MASTER_IP=Master
export SPARK_LOCAL_IP=Master
```

⑤ 目录切换到 sbin 目录下启动集群。

```
[root@master sbin]# ./start-all.sh
starting org.apache.spark.deploy.master.Master, logging to/usr/local/spark-1.6.0-bin-hadoop2.6/logs/
spark-root-org.apache.spark.deploy.master.Master-1-master.out
Master: starting org.apache.spark.deploy.worker.Worker, logging to/usr/local/spark-1.6.0-bin-hadoop
2.6/logs/spark-root-org.apache.spark.deploy.worker.Worker-1-master.out
```

⑥ jps 查看进程，启动成功。

```
[root@Master sbin]# jps
6528 Worker
6561 Jps
6471 Master
```

9.2.2 Spark Standalone 集群环境搭建与配置

Spark 除了支持 Mesos 和 YARN 集群模式外，还支持 Standalone 模式，本书使用 Spark 集群的模式就是 Standalone，Standalone 是 Master/Slave 模式，本书将构建两个 Worker 和一个 Master。

① 进入 Spark 的 conf 目录下，对 Spark 进行配置。

```
[root@Master spark-1.6.0-bin-hadoop2.6]# cd conf
```

② 创建 spark-env. sh。

```
[root@Master conf]# cp spark-env.sh.template  spark-env.sh
```

③ 创建成功。

```
[root@Master conf]# ls
docker.properties.template  hive-site.xml metastore.log spark-defaults.conf spark-env.sh
fairscheduler.xml.template  log4j.properties.template  metrics.properties.template  slaves.template
spark-defaults.conf.template  spark-env.sh.template
```

④ 配置 spark-env. sh。

```
export JAVA_HOME=/usr/local/jdk/jdk1.8.0_60
export SCALA_HOME=/usr/local/scala/scala-2.10.4
```

```
export HADOOP_HOME=/usr/local/hadoop/hadoop-2.6.0
export HADOOP_CONF_DIR=/usr/local/hadoop/hadoop-2.6.0/etc/hadoop
export SPARK_MASTER_IP=Master
export SPARK_WORKER_MEMORY=2g
export SPARK_EXECUTOR_MEMORY=2g
export SPARK_DRIVER_MEMORY=2G
export SPARK_WORKER_CORES=2
```

⑤ 配置 slaves 将节点内容设置为 Worker1、Worker2，然后保存退出。

```
[root@Master conf]# cp slaves.template slaves
[root@Master conf]# vim slaves
# A Spark Worker will be started on each of the machines listed below.
Worker1
Worker2
```

⑥ 配置 spark-defaults. conf，然后保存退出。

```
[root@Master conf]# cp spark-defaults.conf.template spark-defaults.conf
[root@Master conf]# vim spark-defaults.conf
# Example:
# spark.master                     spark://master:7077
# spark.eventLog.enabled           true
# spark.eventLog.dir               hdfs://namenode:8021/directory
# spark.serializer                 org.apache.spark.serializer.KryoSerializer
# spark.driver.memory              5g
# spark.executor.extraJavaOptions  -XX:+PrintGCDetails -Dkey=value -Dnumbers="one two three"
spark.executor.extraJavaOptions    -XX:+PrintGCDetails -Dkey=value -Dnumbers="one two three"
spark.eventLog.enabled             true
spark.eventLog.dir                 hdfs://Master:9000/historyserverforSpark
spark.yarn.historyServer.address   Master:18080
spark.history.fs.logDirectory      hdfs://Master:9000/historyserverforSpark
#spark.default.parallelism         100
```

⑦ 下面就要将 bashrc 的配置文件发送给 worker1、worker2。

```
[root@Master conf]# scp ~/.bashrc root@Worker1:~/.bashrc
[root@Master conf]# scp ~/.bashrc root@Worker2:~/.bashrc
```

⑧ 对于 slaves 配置，同样在 Worker1 和 Worker2 节点中的 slaves 文件需要与 Master 同样配置。

⑨ 将 Master 节点中的 spark 里面的 conf 配置文件发送给 worker1、worker2

```
[root@Master spark]# scp -r ./spark-1.6.0-bin-hadoop2.6/ root@Worker1:/usr/local/spark/
[root@Master spark]# scp -r ./spark-1.6.0-bin-hadoop2.6/ root@Worker2:/usr/local/spark/
```

至此就集群就安装好了，下面就对环境进行验证。

9.2.3 Spark Standalone 环境搭建的验证

① 将目录切换到 Spark 的 sbin 目录下。

```
[root@Master sbin]# ls
derby.log             start-slaves.sh
slaves.sh             start-thriftserver.sh
spark-config.sh       stop-all.sh
spark-daemon.sh       stop-history-server.sh
```

```
spark-daemons.sh              stop-master.sh
start-all.sh                  stop-mesos-dispatcher.sh
start-history-server.sh       stop-mesos-shuffle-service.sh
start-master.sh               stop-shuffle-service.sh
start-mesos-dispatcher.sh     stop-slave.sh
start-mesos-shuffle-service.sh stop-slaves.sh
start-shuffle-service.sh      stop-thriftserver.sh
start-slave.sh
```

② 使用 start-all. sh 命令启动 Spark 集群。

```
[root@Master sbin]# ./start-all.sh
/usr/local/spark/spark-1.6.0-bin-hadoop2.6/conf/spark-env.sh:
 line 80: /root: is a directory
 starting org.apache.spark.deploy.master.Master, logging to/usr/local/spark/spark-1.6.0-bin-
      hadoop2.6/logs/spark-root-org.apache.spark.deploy.master.Master-1-Master.out
 /usr/local/spark/spark-1.6.0-
 bin-hadoop2.6/conf/spark-env.sh: line 80: /root: is a directory
Worker1: starting org.apache.spark.deploy.worker.Worker, logging to/usr/local/spark/spark-1.6.0-bin-
      hadoop2.6/logs/spark-root-org.apache.spark.deploy.worker.Worker-1-Worker1.out
Worker2: starting org.apache.spark.deploy.worker.Worker, logging to/usr/local/spark/spark-1.6.0-bin-
      hadoop2.6/logs/spark-root-org.apache.spark.deploy.worker.Worker-1-Worker2.out
```

③ jps 一下查看进程。

```
[root@Master sbin]# jps
8053 Master
8103 Jps
```

④ worker1、worker2 节点上的进程如下所示。

```
[root@Worker2 Desktop]# jps
2628Worker
2667 Jps
[root@Worker2 Desktop]# jps
3939 Worker
3975 Jps
```

⑤ 在 web 端通过 8080 端口查看，集群启动成功，如图 9-9 所示。

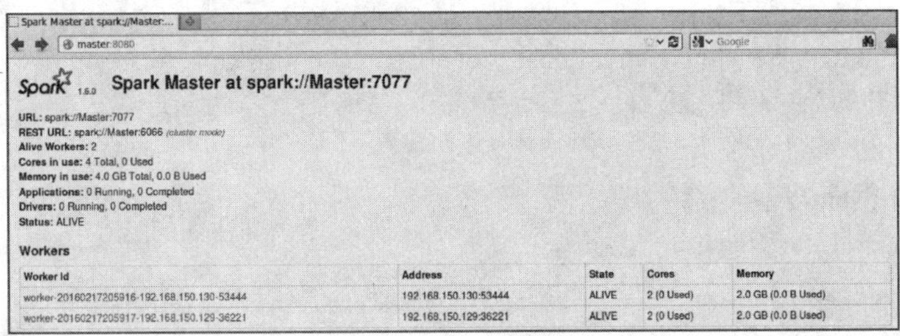

图 9-9　集群启动情况

9.3　Spark 集群的测试

9.3.1　通过 spark-shell 脚本进行测试

① 将目录切换到 hadoop 的 sbin 目录下。

```
[root@Master sbin]# ls
distribute-exclude.sh   start-all.cmd        stop-balancer.sh
hadoop-daemon.sh        start-all.sh         stop-dfs.cmd
hadoop-daemons.sh       start-balancer.sh    stop-dfs.sh
hdfs-config.cmd         start-dfs.cmd        stop-secure-dns.sh
hdfs-config.sh          start-dfs.sh         stop-yarn.cmd
httpfs.sh               start-secure-dns.sh  stop-yarn.sh
kms.sh                  start-yarn.cmd       yarn-daemon.sh
mr-jobhistory-daemon.sh start-yarn.sh        yarn-daemons.sh
refresh-namenodes.sh    stop-all.cmd
slaves.sh               stop-all.sh
```

② 使用 start-dfs. sh 启动 hdfs。

```
[root@Master sbin]# ./start-dfs.sh
Java HotSpot(TM) 64-Bit Server VM warning: You have loaded library/usr/local/hadoop/hadoop-
    2.6.0/lib/native/libhadoop.so which might have disabled stack guard. The VM will try to fix the
    stack guard now.
It's highly recommended that you fix the library with 'execstack -c <libfile>', or link it with '-z noexecstack'.
16/02/19 10:59:37 WARN util.NativeCodeLoader: Unable to load native-hadoop library for your
    platform...using builtin-java classes where applicable
Starting namenodes on [Master]
Master: starting namenode, logging to/usr/local/hadoop/hadoop-2.6.0/logs/hadoop-root-namenode-
    Master.out
Worker1: starting datanode, logging to/usr/local/hadoop/hadoop-2.6.0/logs/hadoop-root-datanode-
    Worker1.out
Worker2: starting datanode, logging to/usr/local/hadoop/hadoop-2.6.0/logs/hadoop-root-datanode-
    Worker2.out
Starting secondary namenodes [Master]
Master: starting secondarynamenode, logging to/usr/local/hadoop/hadoop-2.6.0/logs/hadoop-root-
    secondarynamenode-Master.out
Java HotSpot(TM) 64-Bit Server VM warning: You have loaded library/usr/local/hadoop/hadoop-
    2.6.0/lib/native/libhadoop.so which might have disabled stack guard. The VM will try to fix the
    stack guard now.
It's highly recommended that you fix the library with 'execstack -c <libfile>', or link it with '-z noexecstack'.
16/02/19 11:00:21 WARN util.NativeCodeLoader: Unable to load native-hadoop library for your
    platform..using builtin-java classes where applicable .
```

③ Jps 查看进程启动情况。

```
[root@Master sbin]# jps
8451 SecondaryNameNode
8053 Master
8299 NameNode
8620 Jps
```

④ Web 端通过 50070 端口显示没有问题，如图 9-10 所示。

⑤ 现在要从 HDFS 上读取数据，需要从本地将文件上传到 HDFS 上。

首先，在 HDFS 上创建 library 目录。然后再将 README. txt 上传到 HDFS 上。

```
[root@Master hadoop-2.6.0]# hadoop fs -mkdir /library
Java HotSpot(TM) 64-Bit Server VM warning: You have loaded library/usr/local/hadoop/hadoop-
    2.6.0/lib/native/libhadoop.so which might have disabled stack guard. The VM will try to fix
    the stack guard now.
```

```
It's highly recommended that you fix the library with 'execstack -c <libfile>', or link it with '-z noexecstack'.
16/02/19 11:13:50 WARN util.NativeCodeLoader: Unable to load native-hadoop library for your platform...
    using builtin-java classes where applicable
```

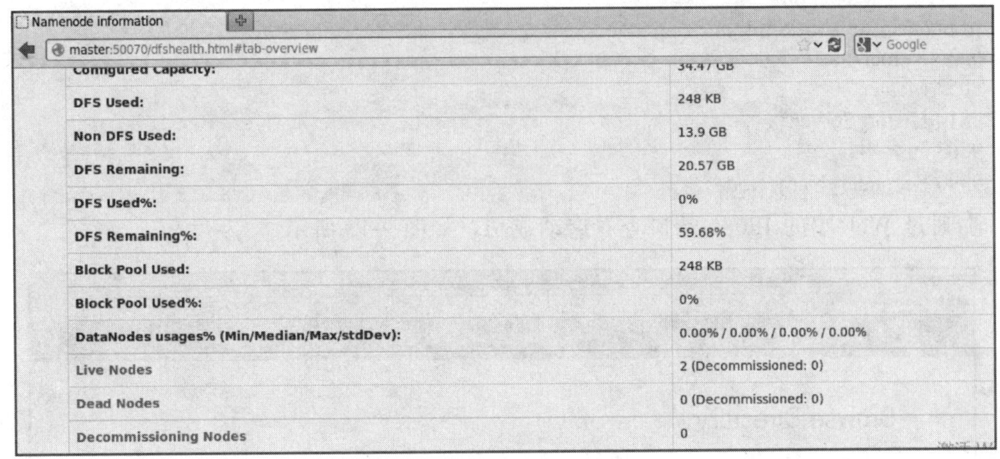

图 9-10　Hadoop 集群启动

⑥ Web 端查看，library 目录创建，如图 9-11 所示。

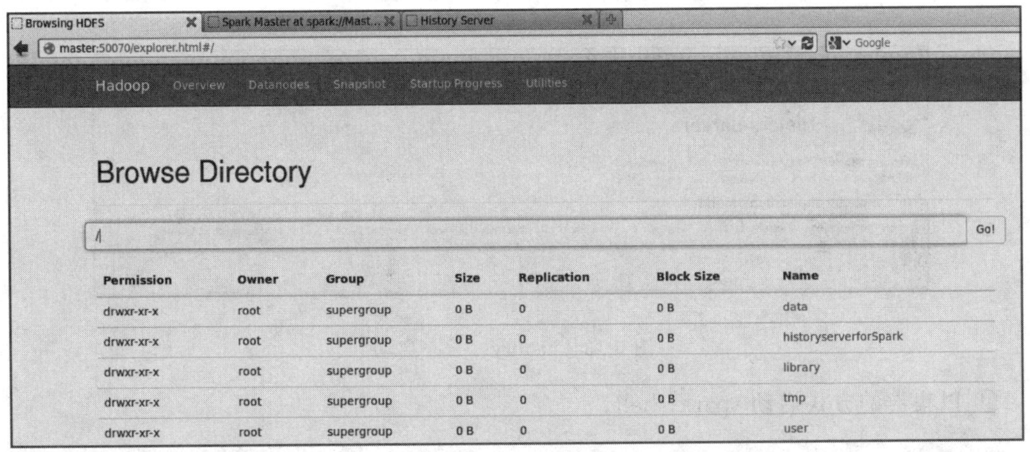

图 9-11　library 目录

⑦ 将本地的 README. txt 文件上传到 HDFS 上的 library 目录下。

```
[root@Master hadoop-2.6.0]# hadoop dfs -put README.txt /library/
DEPRECATED: Use of this script to execute hdfs command is deprecated.
Instead use the hdfs command for it.
16/04/24 13:04:51 WARN util.NativeCodeLoader: Unable to load native-hadoop library for your platform...
    using builtin-java classes where applicable
```

⑧ Web 端查看，文件上传成功，如图 9-12 所示。

⑨ 启动 Spark 的 start-history-server. sh，启动成功，history 可以记录集群工作的时候曾经运行的工程。

```
[root@Master sbin]# ./start-history-server.sh
/usr/local/spark/spark-1.6.0-bin-hadoop2.6/conf/spark-env.sh: line 80: /root: is a directory
starting org.apache.spark.deploy.history.HistoryServer, logging to
    /usr/local/spark/spark-1.6.0-bin-hadoop2.6/logs/spark-root-org.apache.spark.deploy.history.HistorySe
    rver-1-Master.out
[root@Master sbin]# jps
9236 NameNode
8053 Master
9814 HistoryServer
9863 Jps
9390 SecondaryNameNode
```

⑩ 通过 Web 端的 18080 端口查看启动成功，如图 9-13 所示。

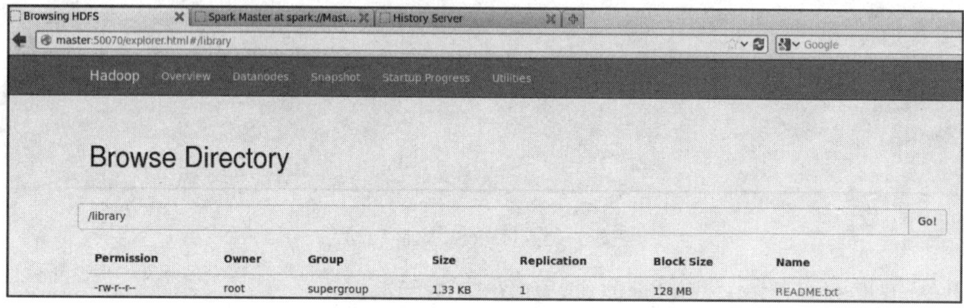

图 9-12　README. txt

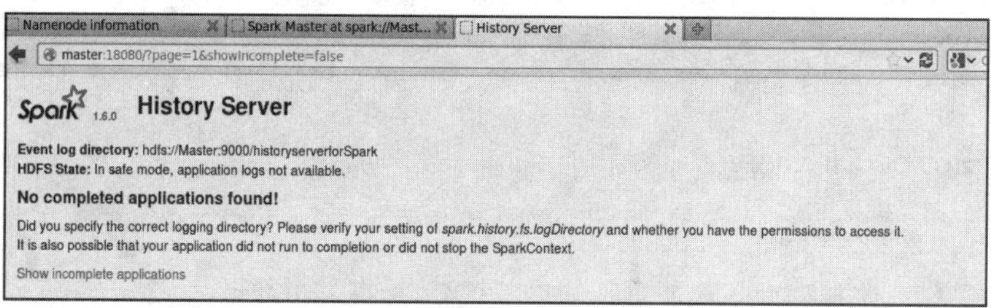

图 9-13　history-server 启动

⑪ 以集群的方式启动 spark-shell。

```
[root@Master bin]# ./spark-shell --master spark://Master:7077
16/02/19 11:20:25 INFO server.Server: jetty-8.y.z-SNAPSHOT
16/02/19 11:20:25 INFO server.AbstractConnector: Started SocketConnector@0.0.0.0:45830
16/02/19 11:20:25 INFO util.Utils: Successfully started service 'HTTP class server' on port 45830.
Welcome to
      ____              __
     / __/__  ___ _____/ /__
    _\ \/ _ \/ _ `/ __/  '_/
   /___/ .__/\_,_/_/ /_/\_\   version 1.6.0
      /_/

Using Scala version 2.10.5 (Java HotSpot(TM) 64-Bit Server VM, Java 1.8.0_60)
Type in expressions to have them evaluated.
16/04/24 12:56:12 INFO hive.metastore: Trying to connect to metastore with URI thrift://Master:9083
16/04/24 12:56:42 INFO hive.metastore: Connected to metastore.
```

```
16/04/24 12:57:01 INFO session.SessionState: Created local directory:
    /tmp/8e926832-fe26-4037-a7d8-186410bdd77a_resources
16/04/24 12:57:01 INFO session.SessionState: Created HDFS directory:
    /tmp/hive/root/8e926832-fe26-4037-a7d8-186410bdd77a
16/04/24 12:57:01 INFO session.SessionState: Created local directory:
    /tmp/root/8e926832-fe26-4037-a7d8-186410bdd77a
16/04/24 12:57:01 INFO session.SessionState: Created HDFS directory:
    /tmp/hive/root/8e926832-fe26-4037-a7d8-186410bdd77a/_tmp_space.db
16/04/24 12:57:02 INFO repl.SparkILoop: Created sql context (with Hive support)..
SQL context available as sqlContext.
scala>
```

⑫ Web 端查看 shell 启动情况，如图 9-14 所示。

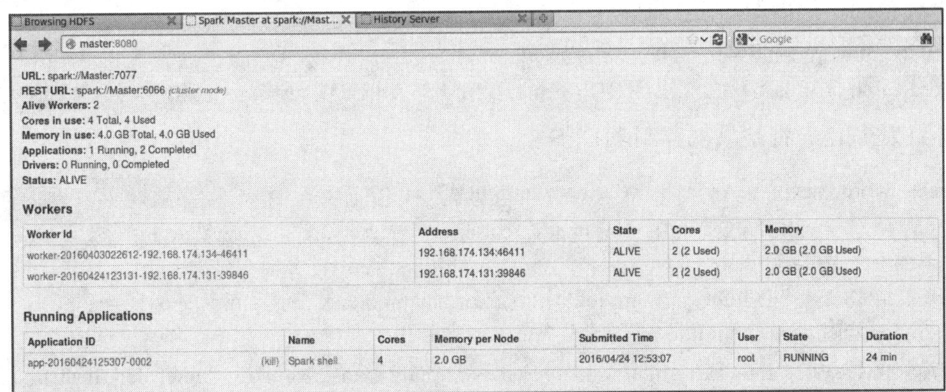

图 9-14　shell 启动

⑬ 通过 4040 端口可以查看更多 Job 详细信息，如图 9-15 所示。

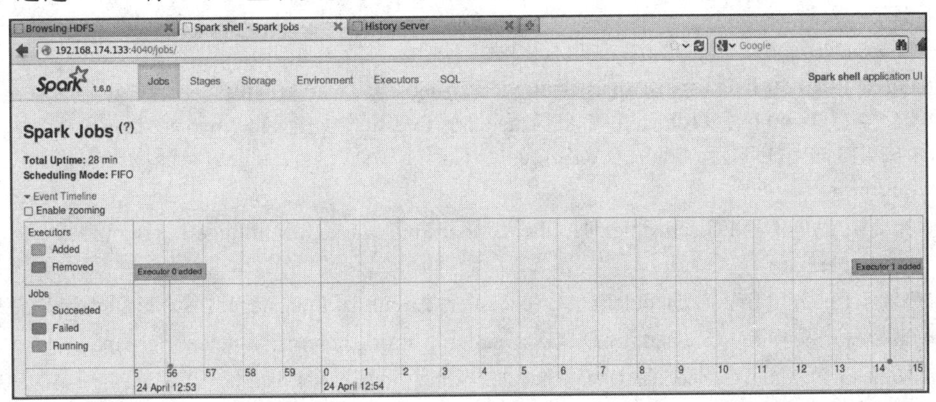

图 9-15　4040 端口

⑭ 从 HDFS 中读取数据进行 wordCount 统计案例，首先读取数据。

```
scala> val lines = sc.textFile("/library/README.txt")
16/04/24 13:26:32 INFO storage.MemoryStore: Block broadcast_0 stored as values in memory
    (estimated size 213.2 KB, free 213.2 KB)
16/04/24 13:26:32 INFO storage.MemoryStore: Block broadcast_0_piece0 stored as bytes in memory
    (estimated size 19.6 KB, free 232.8 KB)
16/04/24 13:26:32 INFO storage.BlockManagerInfo: Added broadcast_0_piece0 in memory on
    192.168.174.133:54788 (size: 19.6 KB, free: 1259.8 MB)
```

```
16/04/24 13:26:32 INFO spark.SparkContext: Created broadcast 0 from textFile at <console>:27
lines: org.apache.spark.rdd.RDD[String] = MapPartitionsRDD[1] at textFile at <console>:27
scala> sc.textFile("/library/README.txt").flatMap(line => line.split(" ")).map(word => (word ,
    1)).reduceByKey(_+_).saveAsTextFile("/library/output1")
```

⑮ 以空格的方式，对单词进行切分。

```
scala>  val words = lines.flatMap(line => line.split(" "))
words: org.apache.spark.rdd.RDD[String] = MapPartitionsRDD[4] at flatMap at <console>:29
```

⑯ 对每个单词计数为 1。

```
scala> val pairs = words.map(word => (word,1))
pairs: org.apache.spark.rdd.RDD[(String, Int)] = MapPartitionsRDD[5] at map at <console>:31
```

⑰ 对相同单词的数量进行总计。

```
scala> val wordcount = pairs.reduceByKey(_+_)
wordcount: org.apache.spark.rdd.RDD[(String, Int)] = ShuffledRDD[6] at reduceByKey at <console>:33
```

⑱ 触发 action 将结果保存到 HDFS 上。

```
scala> wordcount.saveAsTextFile("/library/output1")
16/04/24 13:39:38 INFO Configuration.deprecation: mapred.tip.id is deprecated. Instead, use
    mapreduce.task.id
16/04/24 13:39:38 INFO Configuration.deprecation: mapred.task.id is deprecated. Instead, use
    mapreduce.task.attempt.id
16/04/24 13:39:38 INFO Configuration.deprecation: mapred.task.is.map is deprecated. Instead, use
    mapreduce.task.ismap
16/04/24 13:39:38 INFO Configuration.deprecation: mapred.task.partition is deprecated. Instead, use
    mapreduce.task.partition
16/04/24 13:39:38 INFO Configuration.deprecation: mapred.job.id is deprecated. Instead, use
    mapreduce.job.id
16/04/24 13:39:39 INFO spark.SparkContext: Starting job: saveAsTextFile at <console>:36
16/04/24 13:39:40 INFO scheduler.DAGScheduler: Registering RDD 5 (map at <console>:31)
16/04/24 13:39:40 INFO scheduler.DAGScheduler: Got job 0 (saveAsTextFile at <console>:36) with 2 output
    partitions
16/04/24 13:39:40 INFO scheduler.DAGScheduler: Final stage: ResultStage 1 (saveAsTextFile at
    <console>:36)
16/04/24 13:39:40 INFO scheduler.DAGScheduler: Parents of final stage: List(ShuffleMapStage 0)
16/04/24 13:39:40 INFO scheduler.DAGScheduler: Missing parents: List(ShuffleMapStage 0)
16/04/24 13:39:40 INFO scheduler.DAGScheduler: Submitting ShuffleMapStage 0 (MapPartitionsRDD[5] at
    map at <console>:31), which has no missing parents
16/04/24 13:39:41 INFO storage.MemoryStore: Block broadcast_1 stored as values in memory (estimated
    size 4.1 KB, free 237.0 KB)
16/04/24 13:39:41 INFO storage.MemoryStore: Block broadcast_1_piece0 stored as bytes in memory
    (estimated size 2.3 KB, free 239.2 KB)
16/04/24 13:39:41 INFO storage.BlockManagerInfo: Added broadcast_1_piece0 in memory on
    192.168.174.133:54788 (size: 2.3 KB, free: 1259.8 MB)
16/04/24 13:39:41 INFO spark.SparkContext: Created broadcast 1 from broadcast at DAGScheduler.
    scala:1006
16/04/24 13:39:41 INFO scheduler.DAGScheduler: Submitting 2 missing tasks from ShuffleMapStage 0
```

```
    (MapPartitionsRDD[5] at map at <console>:31)
16/04/24 13:39:41 INFO scheduler.TaskSchedulerImpl: Adding task set 0.0 with 2 tasks
16/04/24 13:39:42 INFO scheduler.TaskSetManager: Starting task 0.0 in stage 0.0 (TID 0, Worker1, partition
    0,NODE_LOCAL, 2128 bytes)
16/04/24 13:39:42 INFO scheduler.TaskSetManager: Starting task 1.0 in stage 0.0 (TID 1, Worker1,
    partition 1,NODE_LOCAL, 2128 bytes)
16/04/24 13:39:53 INFO storage.BlockManagerInfo: Added broadcast_1_piece0 in memory on
    Worker1:39129 (size: 2.3 KB, free: 1259.8 MB)
16/04/24 13:39:56 INFO storage.BlockManagerInfo: Added broadcast_0_piece0 in memory on
    Worker1:39129 (size: 19.6 KB, free: 1259.8 MB)
16/04/24 13:40:54 INFO scheduler.TaskSetManager: Finished task 0.0 in stage 0.0 (TID 0) in
    72360 ms on Worker1 (1/2)
16/04/24 13:40:54 INFO scheduler.TaskSetManager: Finished task 1.0 in stage 0.0 (TID 1) in
    72223 ms on Worker1 (2/2)
16/04/24 13:40:54 INFO scheduler.TaskSchedulerImpl: Removed TaskSet 0.0, whose tasks have all
    completed, from pool
16/04/24 13:40:54 INFO scheduler.DAGScheduler: ShuffleMapStage 0 (map at <console>:31)
    finished in 72.777 s
16/04/24 13:40:54 INFO scheduler.DAGScheduler: looking for newly runnable stages
16/04/24 13:40:54 INFO scheduler.DAGScheduler: running: Set()
16/04/24 13:40:54 INFO scheduler.DAGScheduler: waiting: Set(ResultStage 1)
16/04/24 13:40:54 INFO scheduler.DAGScheduler: failed: Set()
16/04/24 13:40:55 INFO scheduler.DAGScheduler: Submitting ResultStage 1 (MapPartitionsRDD[7]
    at saveAsTextFile at <console>:36), which has no missing parents
16/04/24 13:40:55 INFO storage.MemoryStore: Block broadcast_2 stored as values in memory (estimated
    size 64.7 KB, free 304.0 KB)
16/04/24 13:40:55 INFO storage.MemoryStore: Block broadcast_2_piece0 stored as bytes in memory
    (estimated size 22.6 KB, free 326.6 KB)
16/04/2413:40:55 INFO storage.BlockManagerInfo: Added broadcast_2_piece0 in memory on
    192.168.174.133:54788 (size: 22.6 KB, free: 1259.8 MB)
16/04/24 13:40:55 INFO spark.SparkContext: Created broadcast 2 from broadcast at
    DAGScheduler.scala:1006
16/04/24 13:40:55INFO scheduler.DAGScheduler: Submitting 2 missing tasks from ResultStage 1
    (MapPartitionsRDD[7] at saveAsTextFile at <console>:36)
16/04/24 13:40:55 INFO scheduler.TaskSchedulerImpl: Adding task set 1.0 with 2 tasks
16/04/24 13:40:55 INFO scheduler.TaskSetManager: Starting task 0.0 in stage 1.0 (TID 2, Worker1,
    partition 0,NODE_LOCAL, 1894 bytes)
16/04/24 13:40:55 INFO scheduler.TaskSetManager: Starting task 1.0 in stage 1.0 (TID 3, Worker1,
    partition 1,NODE_LOCAL, 1894 bytes)
16/04/24 13:40:55 INFO storage.BlockManagerInfo: Added broadcast_2_piece0 in memory on
    Worker1:39129 (size: 22.6 KB, free: 1259.8 MB)
16/04/24 13:40:57 INFO spark.MapOutputTrackerMasterEndpoint: Asked to send map output locations
    for shuffle 0 to Worker1:42999
16/04/24 13:40:57 INFO spark.MapOutputTrackerMaster: Size of output statuses for shuffle 0 is 145 bytes
16/04/24 13:41:00 INFO scheduler.TaskSetManager: Finished task 1.0 in stage 1.0 (TID 3) in 4660
```

```
    ms on Worker1 (1/2)
16/04/24 13:41:00 INFO scheduler.TaskSetManager: Finished task 0.0 in stage 1.0 (TID 2) in 4857 ms
    on Worker1 (2/2)
16/04/24 13:41:00 INFO scheduler.DAGScheduler: ResultStage 1 (saveAsTextFile at <console>:36)
    finished in 4.857 s
16/04/24 13:41:00 INFO scheduler.TaskSchedulerImpl: Removed TaskSet 1.0, whose tasks have all
    completed, from pool
16/04/24 13:41:00 INFO scheduler.DAGScheduler: Job 0 finished: saveAsTextFile at <console>:36,
    took 80.467680 s
```

⑲ 通过 web 端查看运行成功，如图 9-16 所示。

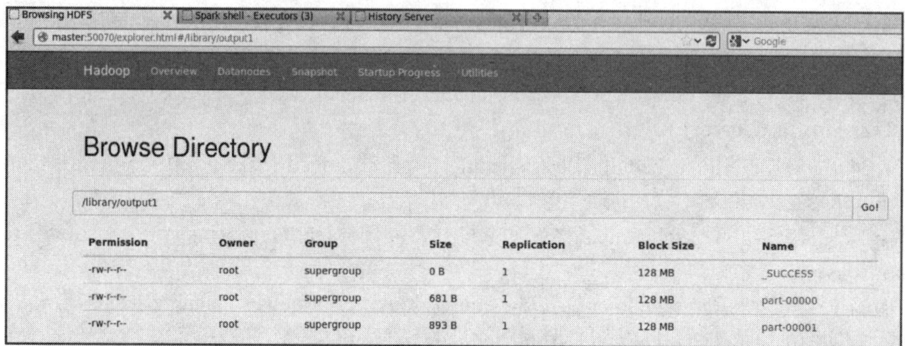

图 9-16　output1

⑳ 4040 端口查看，job 成功运行，如图 9-17 所示。

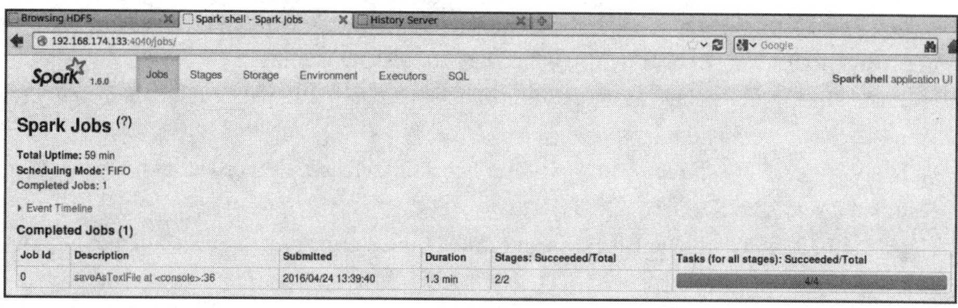

图 9-17　job

㉑ 成功生成 DAG，分为两个 stage，而 stage 的划分是 Shuffle，如图 9-18 所示。

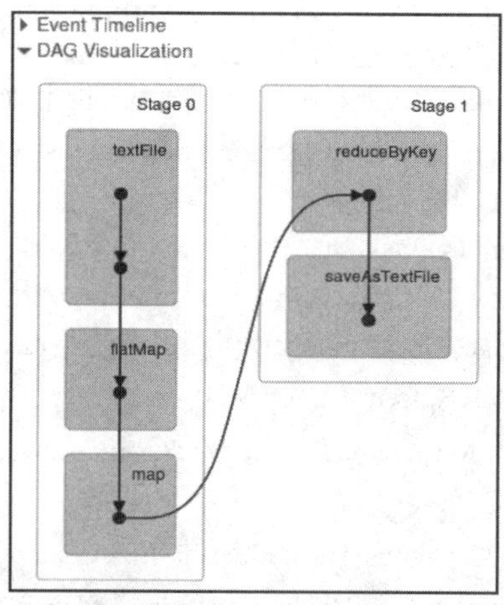

图 9-18　Stage

至此，Spark-shell 运行 wordcount 案例成功。

9.3.2 通过 spark-submit 脚本进行测试

spark-submit 是在命令行中提交 jar 运行程序，是用来向集群提交所编写的程序，根据参数配置，指定是本地还是集群进行。

计算圆周率。

--master MASTER_URL　　　　　　　集群的路径

--class CLASS_NAME　　　应用程序包的要运行的class

```
[root@Master bin]# ./spark-submit --class org.apache.spark.examples.SparkPi --master
spark://Master:7077 ../lib/spark-examples-1.6.0-hadoop2.6.0.jar 10
```

9.4 小结

本章主要介绍了 Linux 环境下的 Spark 集群安装。本章的内容非常重要，对于实战方面非常重要，Spark 集群的构建涉及很多细节方面，希望读者可以耐心地搭建。

第10章 Scala IDE开发Spark程序实战解析

本章主要通过图文的方式，讲解如何让零基础的人在 Windows 操作系统与 Ubuntu 操作系统上下载、安装、调试 Scala IDE 软件，并在此基础上完成基于 ScalaIDE 开发的基本工作（添加依赖导入 Jar 包，制作和调试 WordCount 程序等操作）。附录内容讲解如何在 ScalaIDE 中导入 Spark 源码，便于后续阅读源代码学习。

10.1 Scala IDE 安装

10.1.1 Ubuntu 系统下安装

① 下载最新版的 Scala for Eclipse 版本，选择 Linux 64 位，下载网址：http://scala-ide.org/download/sdk.html。如图 10-1 所示。

Download IDE Windows - 64 bit		
Windows	**Mac**	**Linux**
Windows 64 bit	Mac OS X Cocoa 64 bit	Linux GTK 64 bit
Windows 32 bit	Mac OS X Cocoa 32 bit	Linux GTK 32 bit

图 10-1 下载网页截图

② 将下载后的压缩包放入 Ubuntu 系统下的/usr/local/scalasdk 文件夹（如果没有此文件夹，创建一个相同名字的文件夹，方便获取环境配置代码的直接复制粘贴）。如图 10-2 所示。

③ 复制完毕后，进入命令行工具（快捷键：ctrl＋alt＋T），输入跳转命令 cd/usr/local/scalasdk，然后输入解压命令 tar - xzfscala-SDK-4.3.0-vfinal-2.11-linux.gtk.x86_64.tar.gz 进行解压。如图 10-3 所示。

图 10-2　文件系统

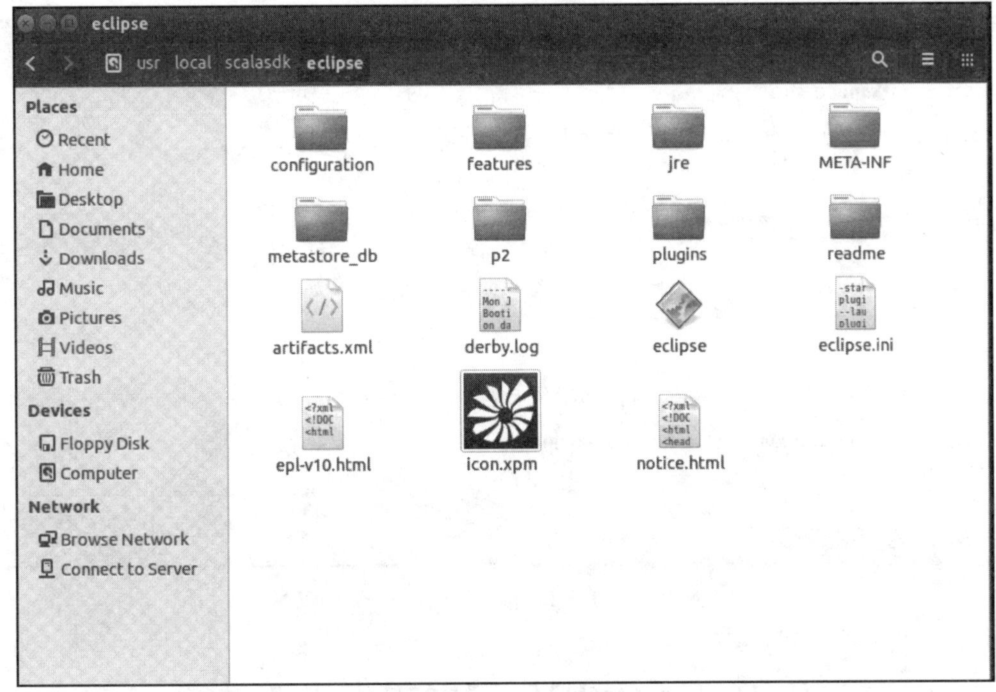

图 10-3　Eclipse 文件夹

④ 解压完成后会在文件夹内出现 eclipse 文件，进入文件夹，双击 eclipse 就可以打开程序，至此软件安装完成。如图 10-4 所示。

图 10-4　Eclipse 启动界面

10.1.2　Windows 系统下安装

① 下载最新版的 ScalaforEclipse 版本，选择 Windows64 位，下载网址：http://scala-ide.org/download/sdk.html。如图 10-5 所示。

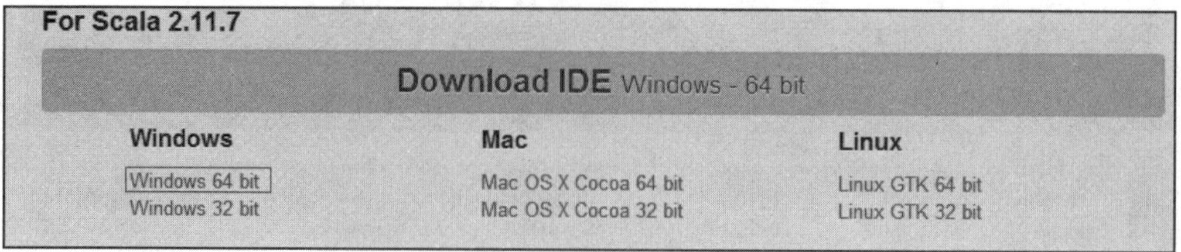

图 10-5 下载 Windows 64 bit IDE

② 文件下载完毕后，对文件进行解压，解压完毕后，运行程序并选择工作空间。如图 10-6 所示。

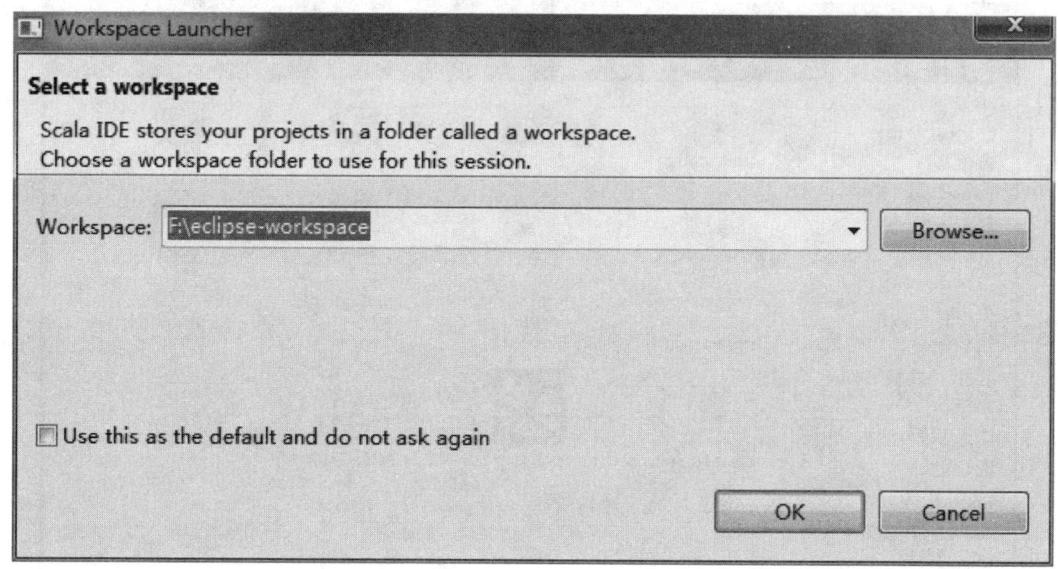

图 10-6 打开 Eclipse 后选择 Workspace

10.2 ScalaIDE 开发重点步骤详解

(1) 环境配置

① 打开上文安装完毕的程序，选择工作空间（通常选择在一个比较好寻找的文件夹，便于后序导出包之类的操作）。如图 10-7 所示。

② 完成工作空间创建后，需要完成 scala 环境的配置，右击项目选择 Properties，在对话框中选择 ScalaCompiler，在右面页签中勾选 Use Project Settings 和 Sca laInstallation 点击 ok，保存配置，操作如图 10-8 和图 10-9 所示。

(2) 项目创建

在完成 Scala 环境搭建后，开始创建测试项目。

① 首先，在 src 下创建 spark 工程包，并创建入口类。先选择项目 New->Package 创建 com. imf. spark 包。如图 10-10 所示。

② 选择 com. imf. spark 包名，修改 Kind，创建 ScalaObject。如图 10-11 所示。

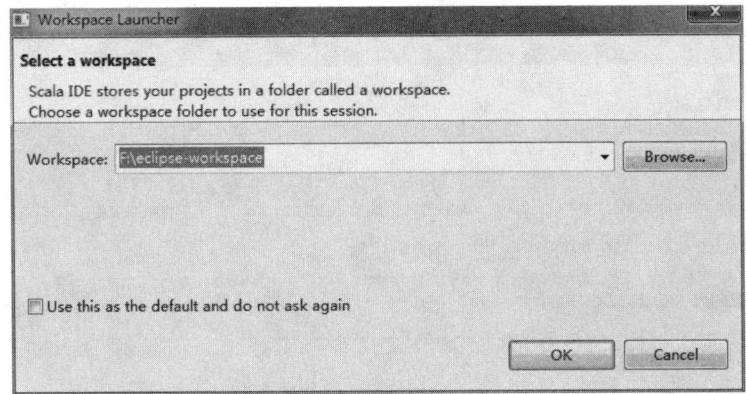

图 10-7　配置 workspace

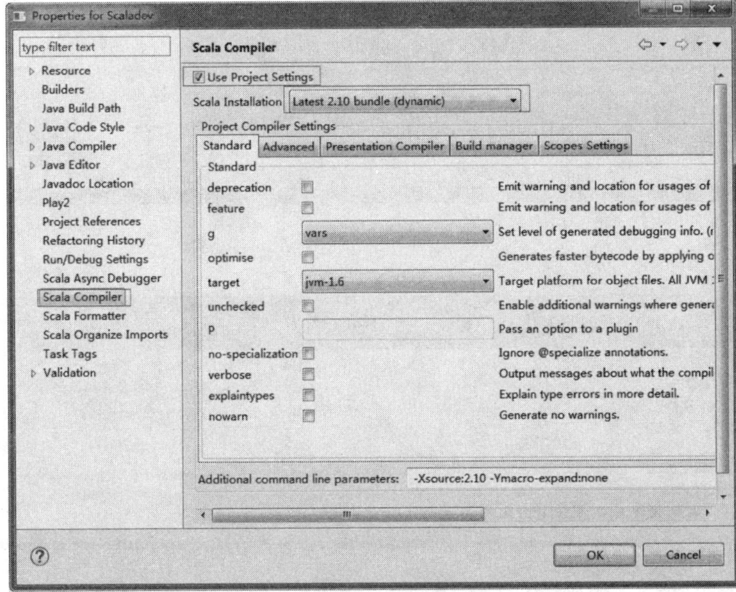

图 10-8　配置 Scala 环境（1）

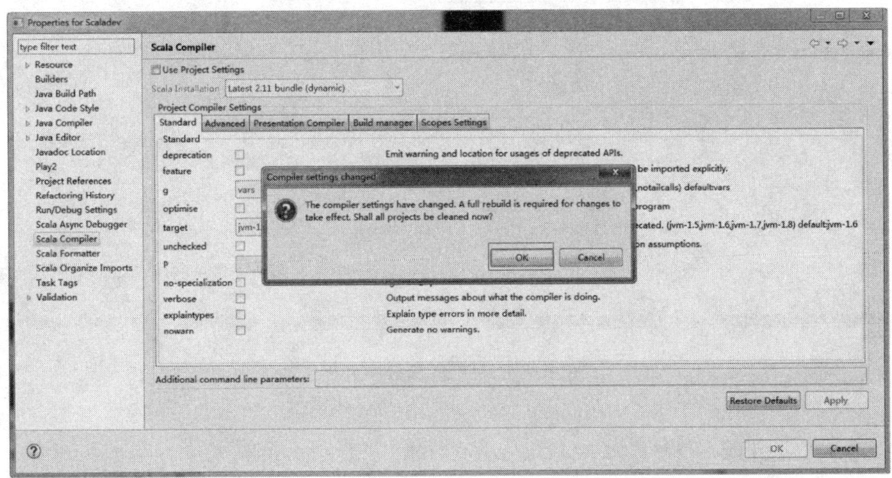

图 10-9　配置 Scala 环境（2）

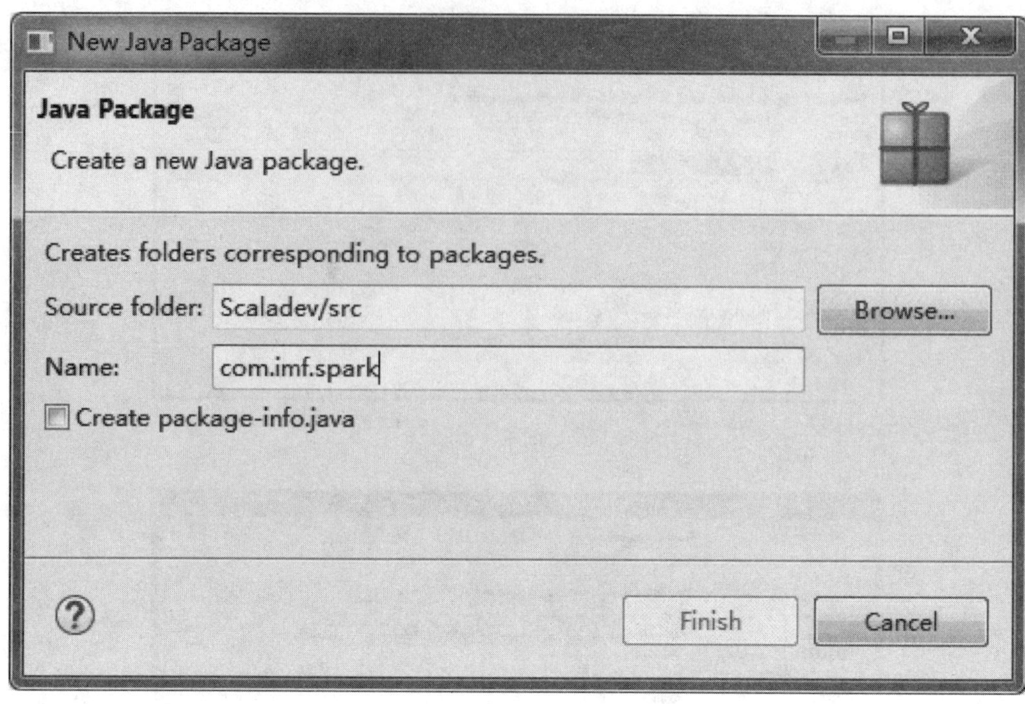

图 10-10　创建 jar 包

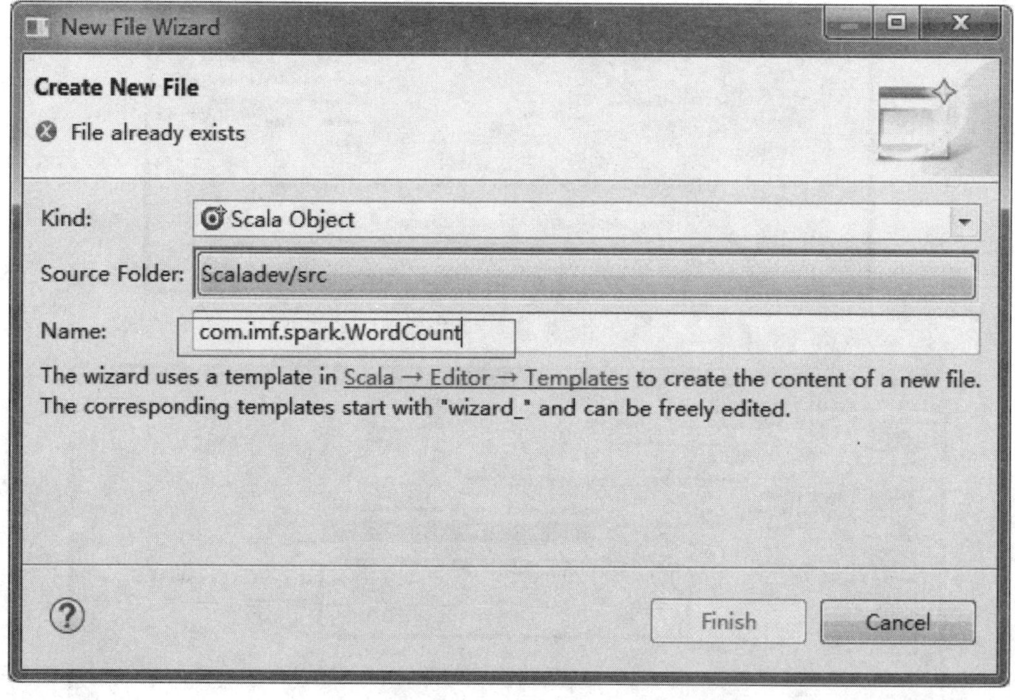

图 10-11　创建 Scala Object

(3) Spark 包引入

在完成项目创建后，还需要在项目中添加 spark1.6.0 的 jar 文件依赖，只有导入了 Jar

依赖，才可以使用和开发 Spark。Jar 依赖包名字为 spark-assembly-1.6.0-hadoop2.6.0.jar，位于 spark-1.6.0-bin-hadoop2.6.tgz 包中的 lib 下面（位于虚拟机中的 Spark 文件夹下 lib 文件夹中，如图 10-12 所示）。下文将对此步骤进行图文讲解。

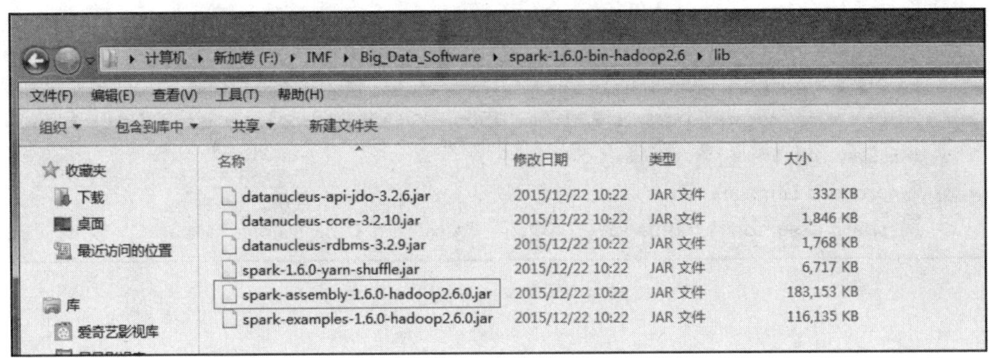

图 10-12 SparkJar 包地址

首先，右击 ScalaDev 项目选择 Build Path->Configure Build Path，然后如图 10-12 所示，点击 Java Bulid Path，选择右侧 Libraries 后，点击 Add Extermal JARs，然后在弹出的对话框中选择 spark-assembly-1.6.0-hadoop2.6.0.jar 文件并确定，导入成功后，结果如图 10-13 所示，项目中出现 spark-assembly-1.6.0-hadoop2.6.0.jar 内容。

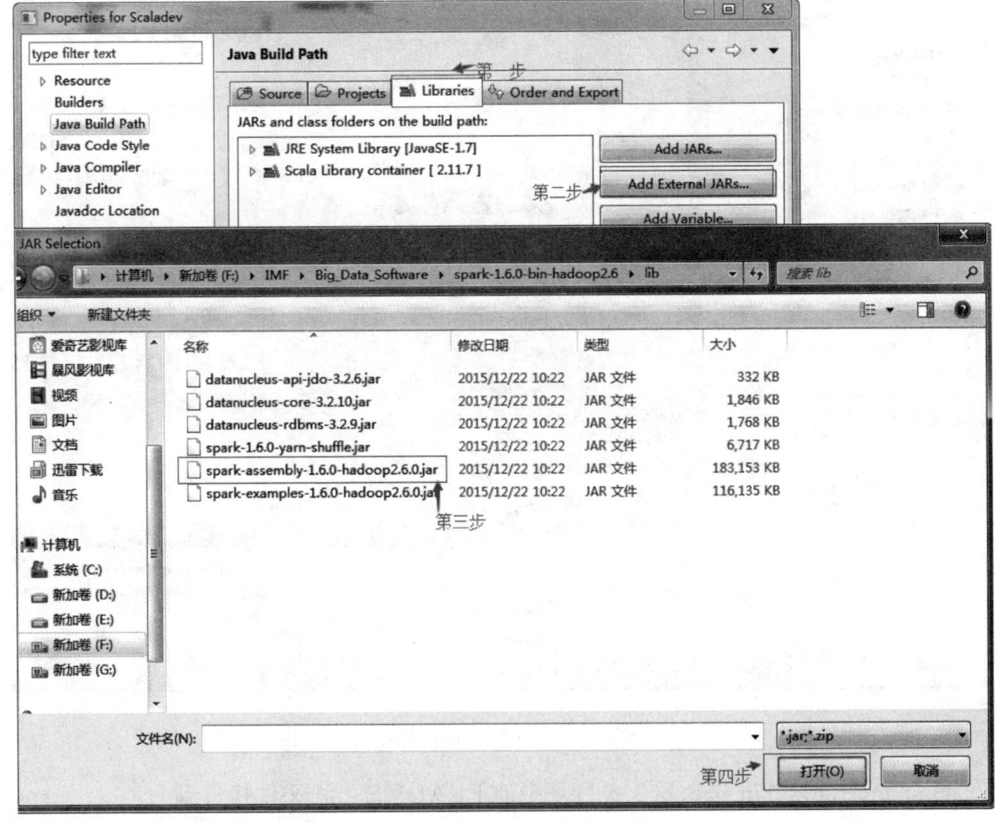

图 10-13 导入 SparkJar 包

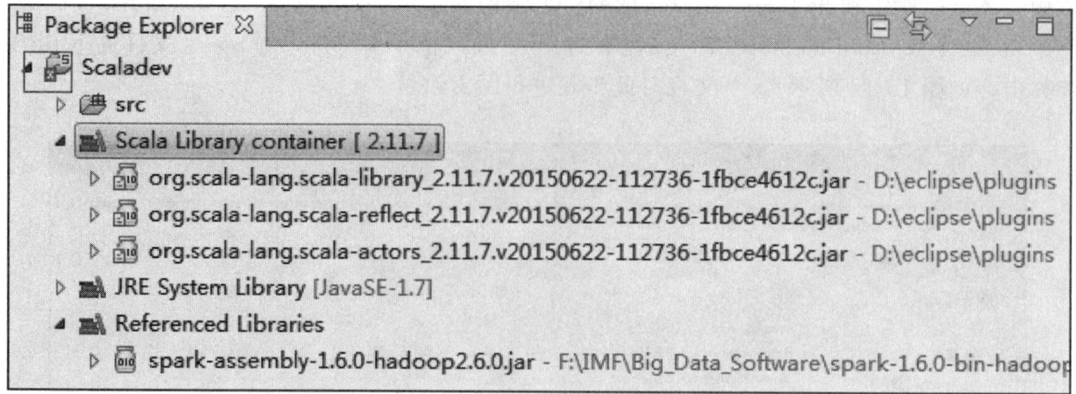

图 10-14　成功导入 Jar 包

10.3　Wordcount 创建实战

本小节主要通过在上文已经创建的项目基础上实战创建和调试 wordcount 程序。在创建程序前，需要完成以下准备工作。

① 打开相应的 WordCount 包。如图 10-15 所示。

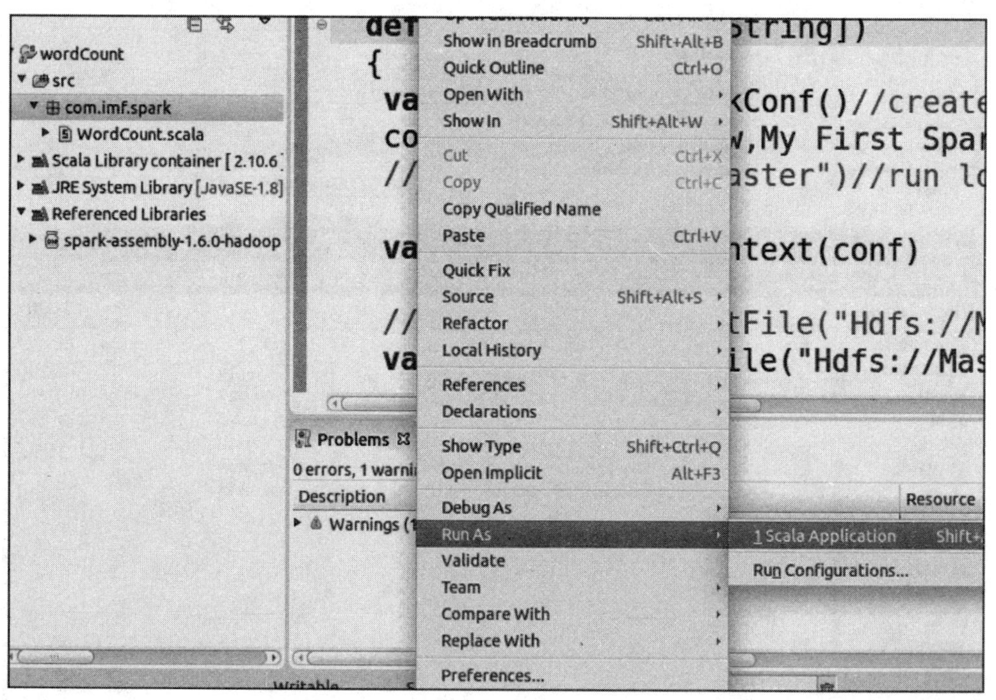

图 10-15　开始程序调试

② 将 spark-1.6.0-bin-hadoop2.6 目录中的 README.md 文件拷贝到 D://testspark// 目录下（其他目录也可，但是需要在程序中修改相关文件夹代码）。

③ 完成上述内容后，将下列代码复制到 WordCount.scala 中。

```
package com.imf.spark
import org.apache.spark.SparkConf
import org.apache.spark.SparkContext
object WordCount {
  def main(args: Array[String]): Unit = {
    /**
     * 1.创建Spark的配置对象SparkConf，设置Spark程序的运行时的配置信息，
     *  例如：通过setMaster来设置程序要链接的Spark集群的Master的URL,如果设置为local,则代表Spark程序在本地运行，特别适合于机器配置条件非常差的情况。
     */
    //创建SparkConf对象
    val conf = new SparkConf()
    //设置应用程序名称，在程序运行的监控界面可以看到名称
    conf.setAppName("My First Spark App!")
    //设置local使程序在本地运行，不需要安装Spark集群
    conf.setMaster("local")
    /**
     * 2.创建SparkContext对象
     * SparkContext是spark程序所有功能的唯一入口，无论是采用Scala,java,python,R等都必须有一个SprakContext
     * SparkContext核心作用：初始化spark应用程序运行所需要的核心组件，包括DAGScheduler,TaskScheduler,SchedulerBackend
     *  同时还会负责Spark程序往Master注册程序等；
     * SparkContext是整个应用程序中最为至关重要的一个对象；
     */
    //通过创建SparkContext对象，通过传入SparkConf实例定制Spark运行的具体参数和配置信息
    val sc = new SparkContext(conf)

    /**
     * 3.根据具体数据的来源（HDFS,HBase,Local,FS,DB,S3等）通过SparkContext来创建RDD；
     * RDD的创建基本有三种方式：根据外部的数据来源（例如HDFS）、根据Scala集合、由其他的RDD操作；
     *  数据会被RDD划分成为一系列的Partitions,分配到每个Partition的数据属于一个Task的处理范畴；
     */
    //读取本地文件，并设置一个partition
    val lines = sc.textFile("D://testspark//README.md",1)

    /**
     * 4. 对初始的RDD进行Transformation级别的处理，例如map,filter等高阶函数的变成，来进行具体的数据计算
     * 4.1将每一行的字符串拆分成单个单词
     */
    //对每一行的字符串进行拆分并把所有行的拆分结果通过flat合并成一个大的集合
     val words = lines.flatMap { line => line.split(" ") }
    /**
     * 4.2.在单词拆分的基础上对每个单词实例计数为1，也就是word => (word,1)
     */
```

```
    val pairs = words.map{word =>(word,1)}

    /**
     * 4.3. 在每个单词实例计数为1基础上统计每个单词在文件中出现的总次数
     */
    // 对相同的key进行value的累积（包括Local和Reducer级别同时Reduce）
    val wordCounts = pairs.reduceByKey(_+_)
    //打印输出
    wordCounts.foreach(pair => println(pair._1+":"+pair._2))
    sc.stop()
  }
}
```

④ 将代码输入完毕后右键代码所在区域（如图 10-15 所示），点击 RunAs-Scala Application 进行程序调试。

10.4 Spark 源码导入 Scala IDE

Windows 下源码环境搭建

本附录讲解如何通过 Egit 插件将 Spark 源码导入到 Eclipse 中。

导入源码前需要完成以下工作。

① 在 https：//github. com 上注册账号；

② 获得 Spark 下载的 git 地址；

③ 安装 egit 插件；

④ 开始导入源码。

（1）如何在 https://github. com 上注册账号。

① 进入网站 https：//github. com，如图 10-16 所示。

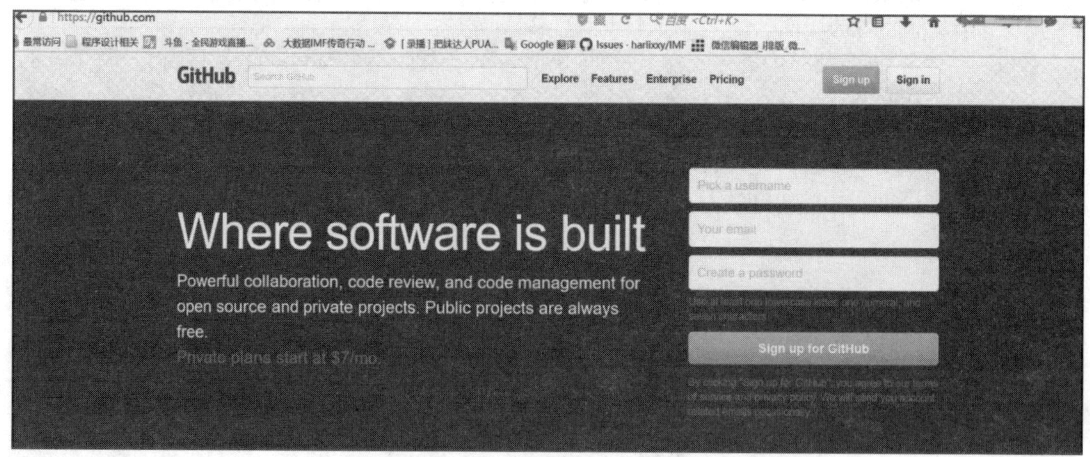

图 10-16　GitHub 官方网站

② 输入注册的信息账号，邮箱，密码等信息后，点击 Sign up for GitHub。如图 10-17 所示。

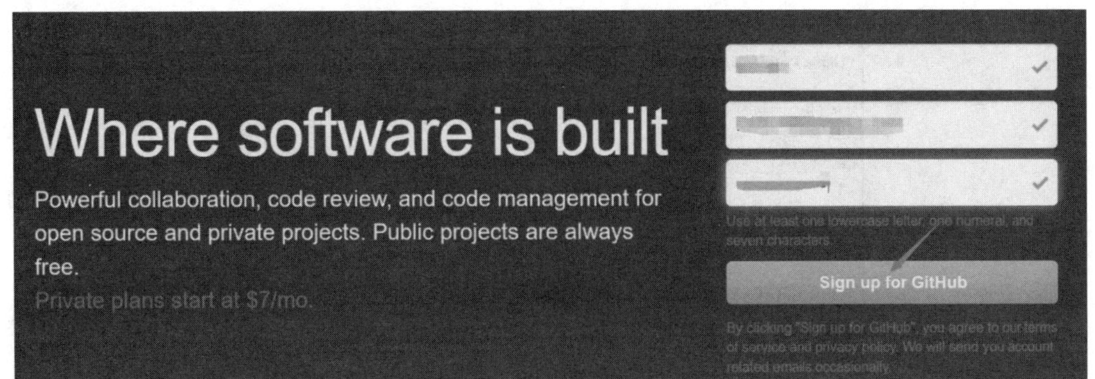

图 10-17　点击注册

（2）获得 Spark 的下载 Https 地址。

注册完毕后，需要去获取 Spark 下载的 Https 地址，此时登陆 ApacheSpark 网站的 download 地址 http：//spark. apache. org/downloads. html，并找到源码的 git 地址（git clonegit://github. com/apache/spark. git），如图 10-18 所示。

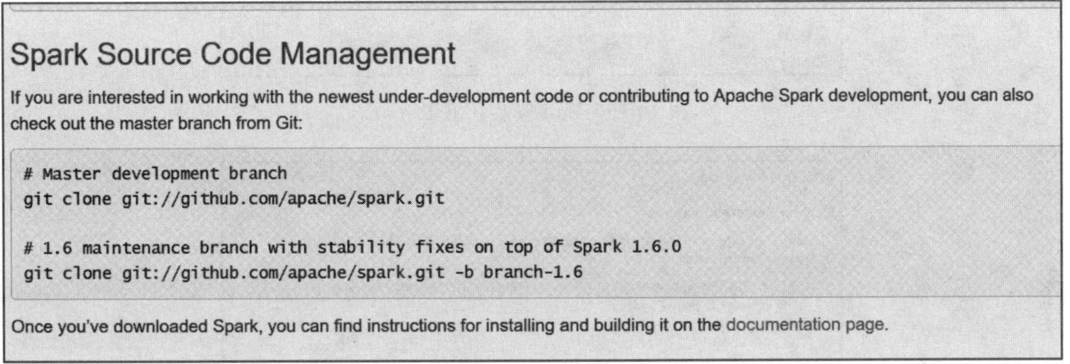

图 10-18　SparkGit 下载地址

（3）安装 Egit。

获取 Spark 下载地址以后，需要通过 Egit 插件进入 Eclipse 中下载源码。

① 先在 Eclipse 中选择 help->EclipseMarketplace，在弹出框中 search 栏 Find 中输入 egit，找到后安装即可，如图 10-19～图 10-21 所示。

点击 Finish 后重启 eclipse，完成 Egit 的安装。

② 完成 Egit 的安装后，开始导入 Spark 源码，打开 eclipse 中的 file->import，弹出框如图 10-22 所示。

选择上图中的 Git->projects from git 之后再点击“Next”，如图 10-23～图 10-28 所示。

当点击完最后一步的“Finish”之后，就会发现 eclipse 的左边会出现所导入的项目了。但是如果出现 Finish 之后出现莫名的错误，导致边上的项目没有显示（或者源码已经下载完毕）可以通过下述的方式导入。

打开 eclipse 中的 file -> import，弹出框如图 10-29～图 10-33 所示。

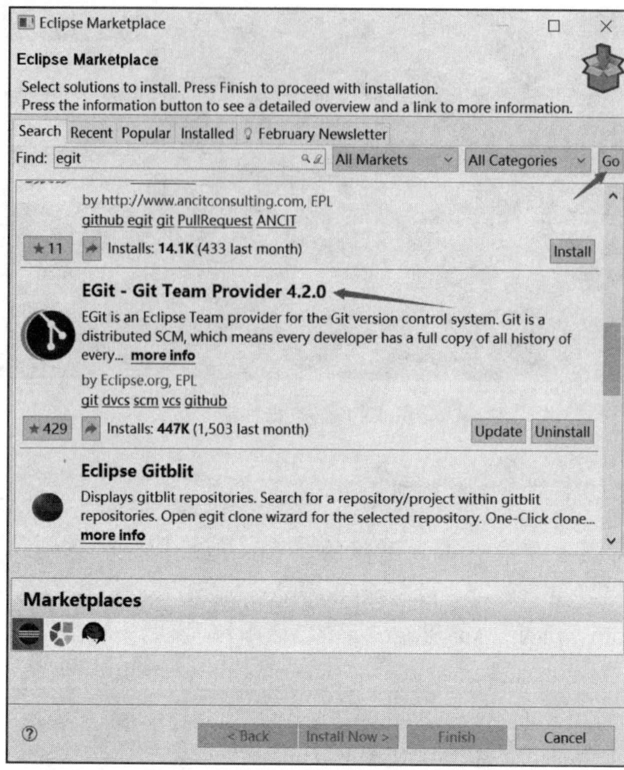

图 10-19　Egit 安装示意图（1）

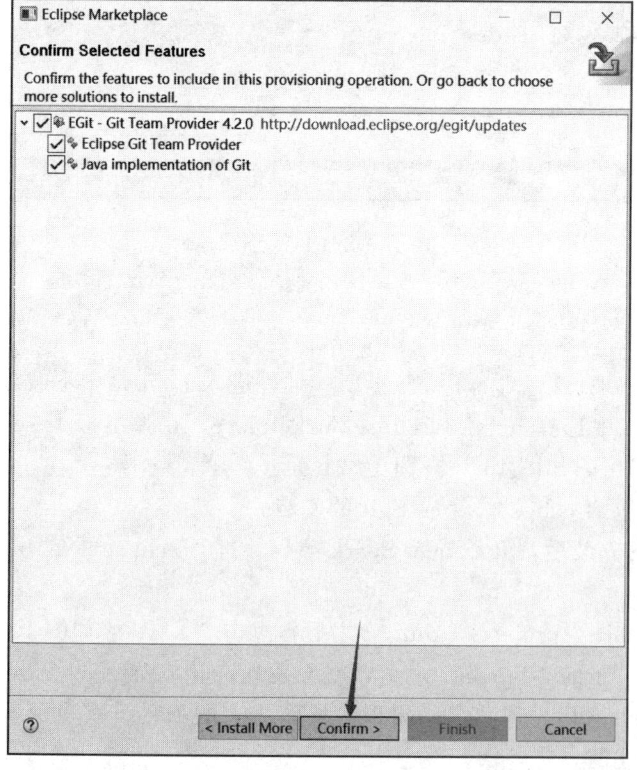

图 10-20　Egit 安装示意图（2）

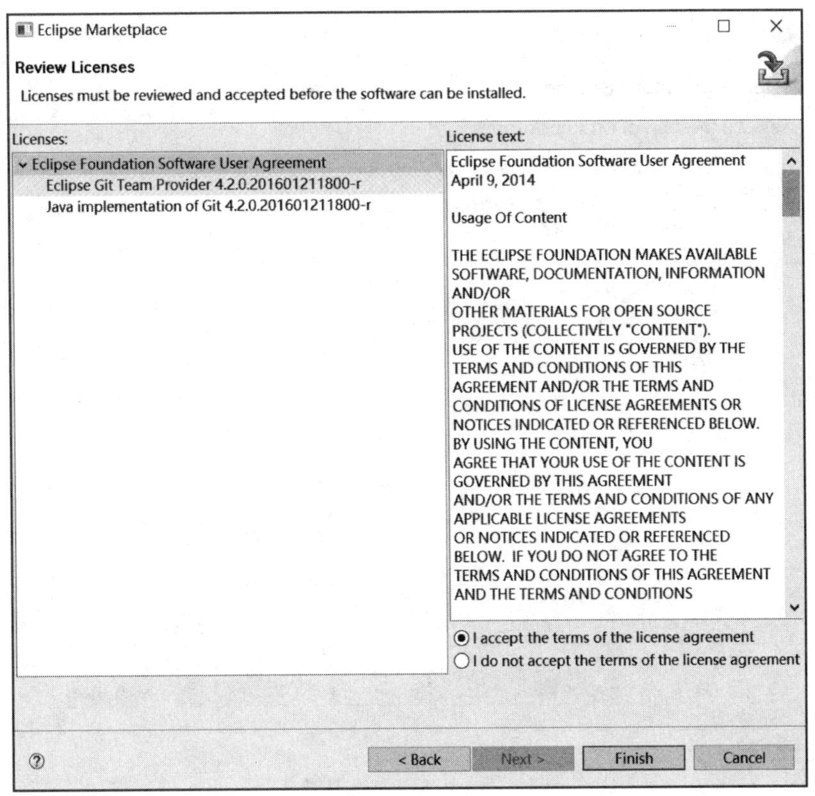

图 10-21　Egit 安装示意图（3）

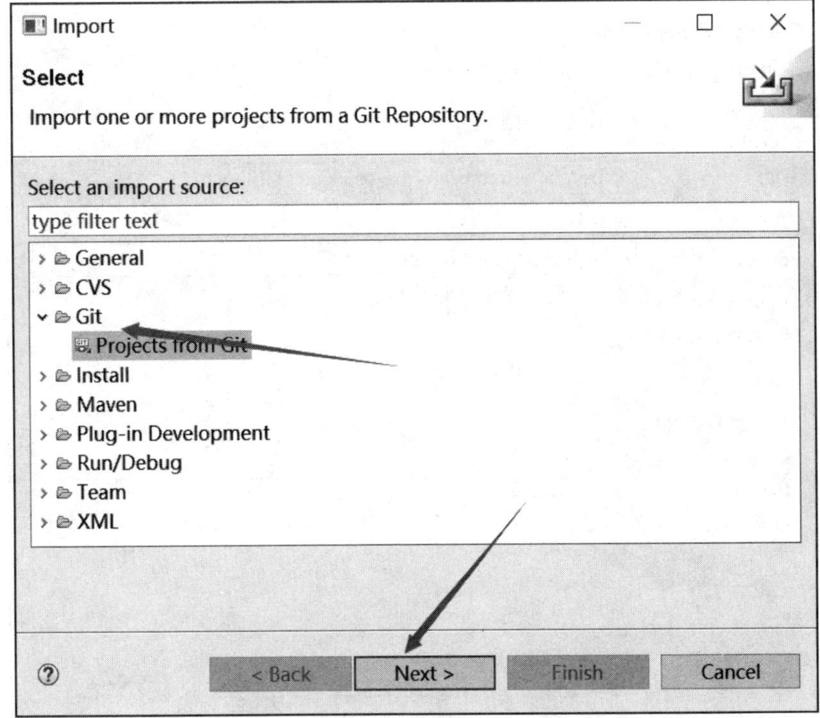

图 10-22　从 Git 导入源码（1）

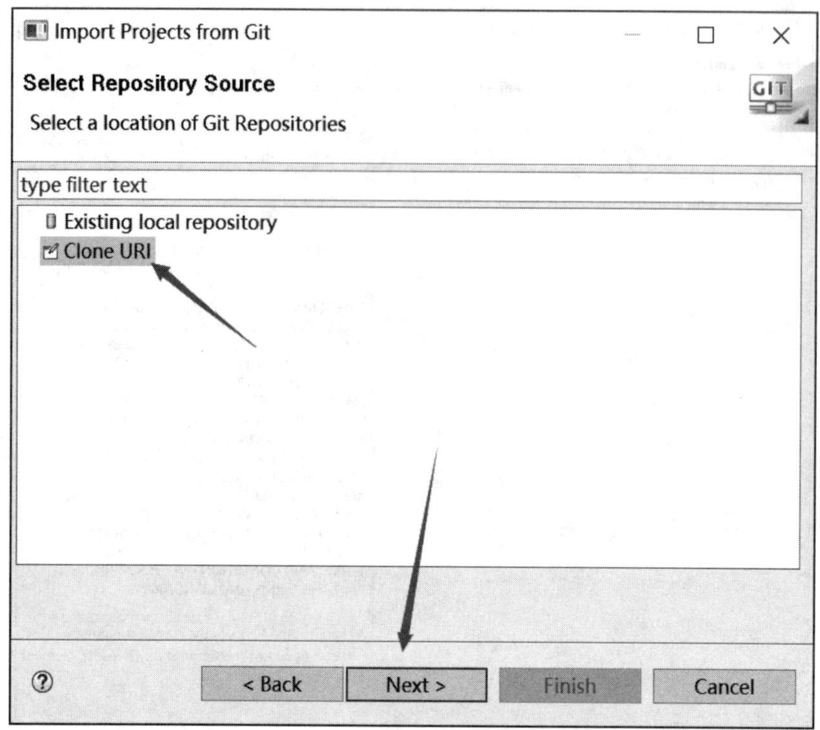

图 10-23　从 Git 导入源码（2）

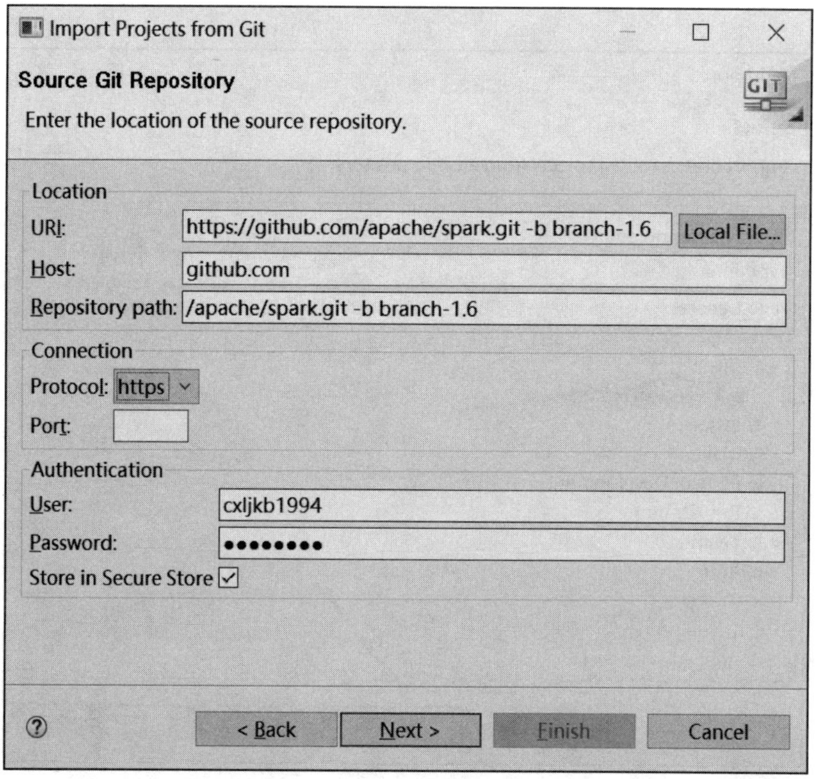

图 10-24　从 Git 导入源码（3）

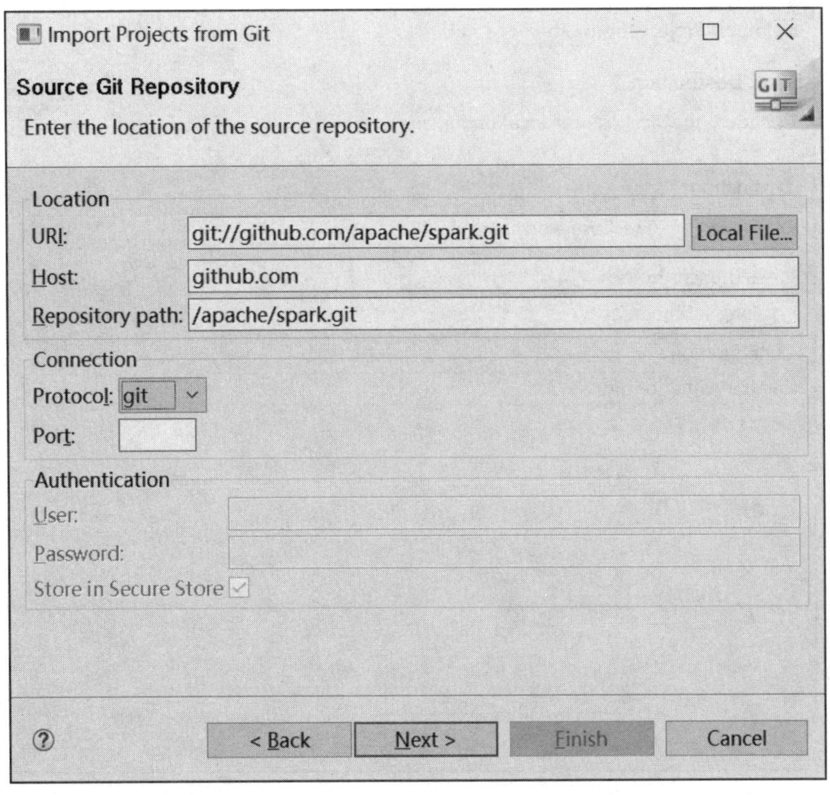

图 10-25　从 Git 导入源码（通过 Git 连接导入）（4）

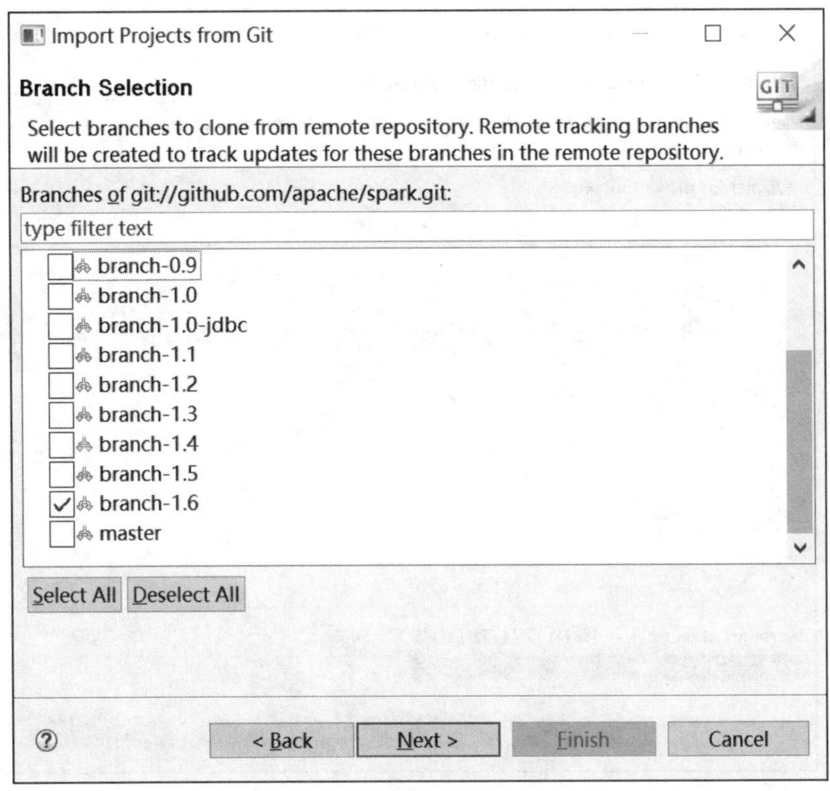

图 10-26　选择源码版本

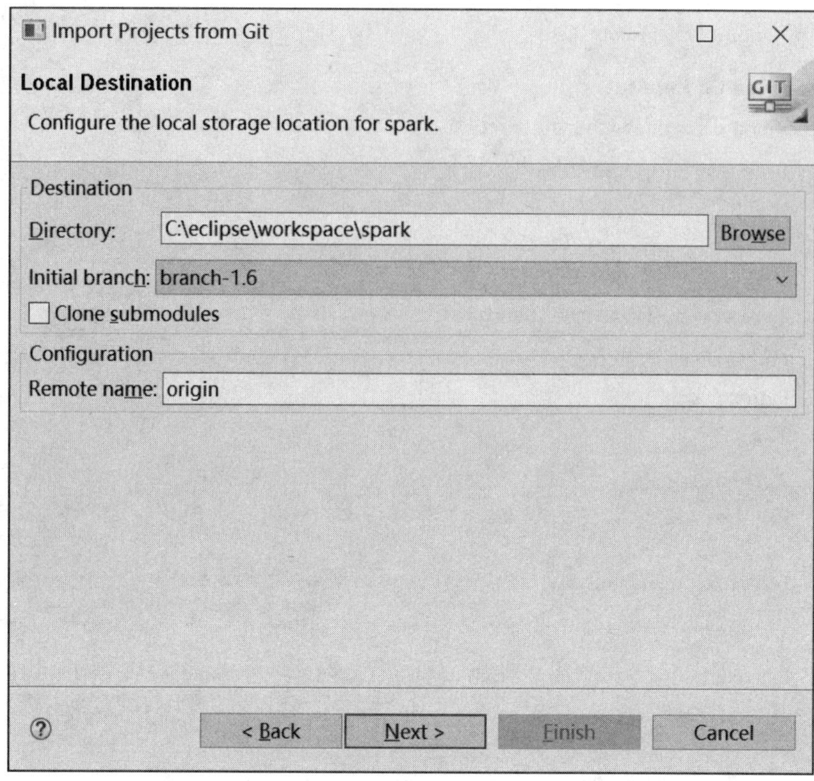

图 10-27　选择存放文件夹

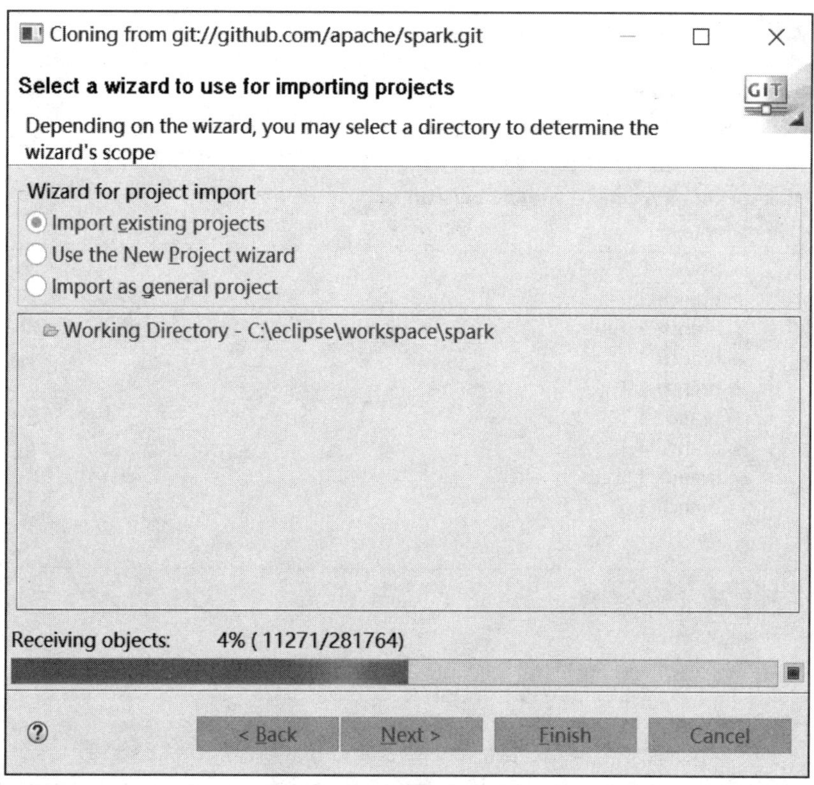

图 10-28　等待下载界面

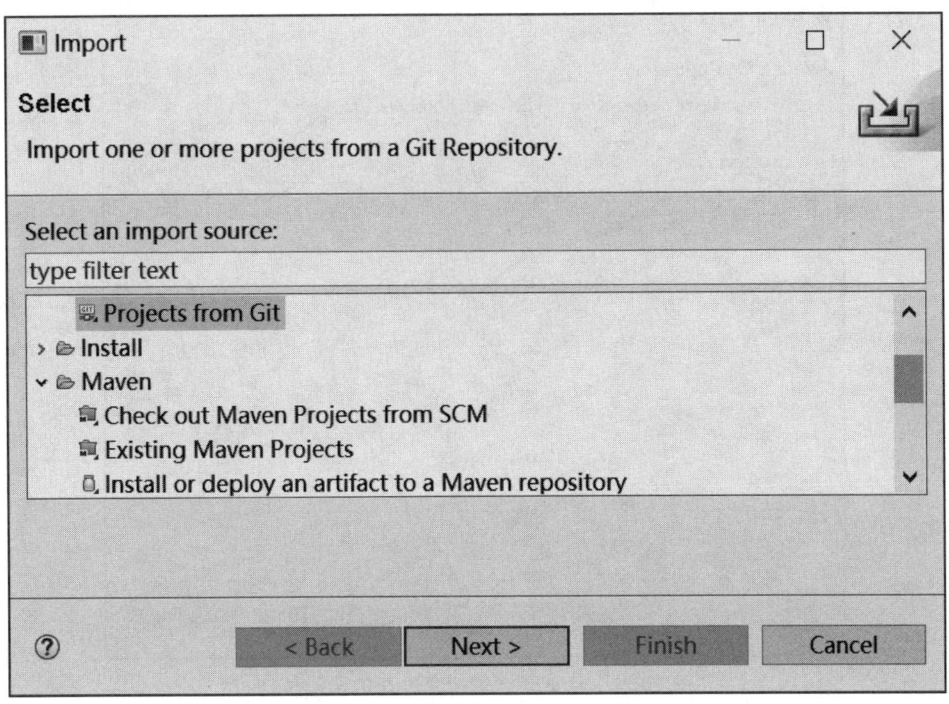

图 10-29 从下载完毕的源码直接导入（1）

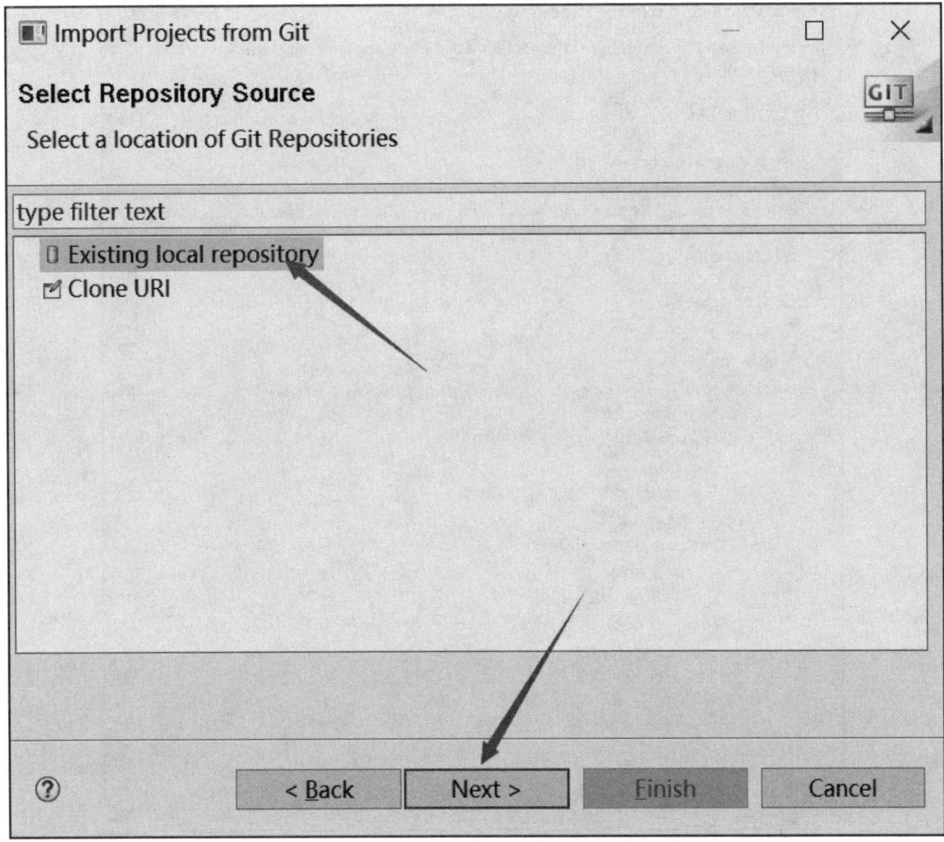

图 10-30 从下载完毕的源码直接导入（2）

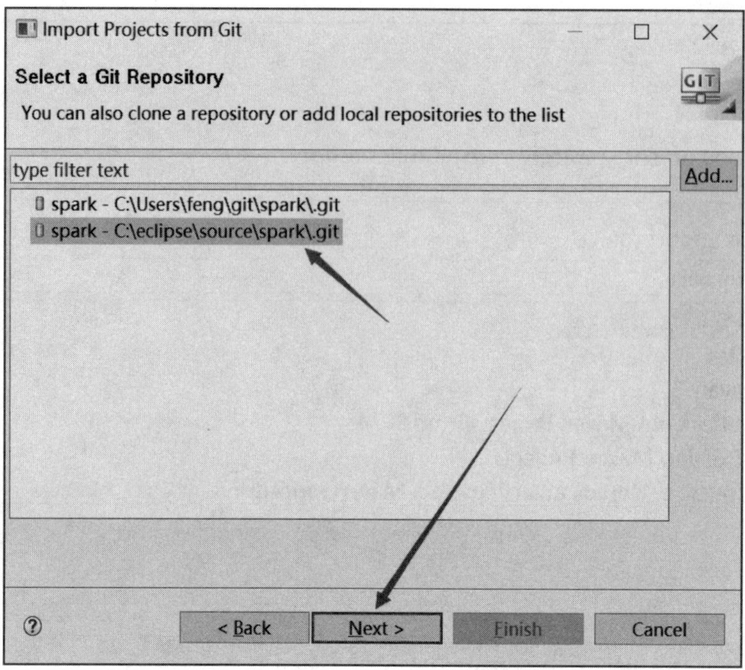

图 10-31 从下载完毕的源码直接导入（3）

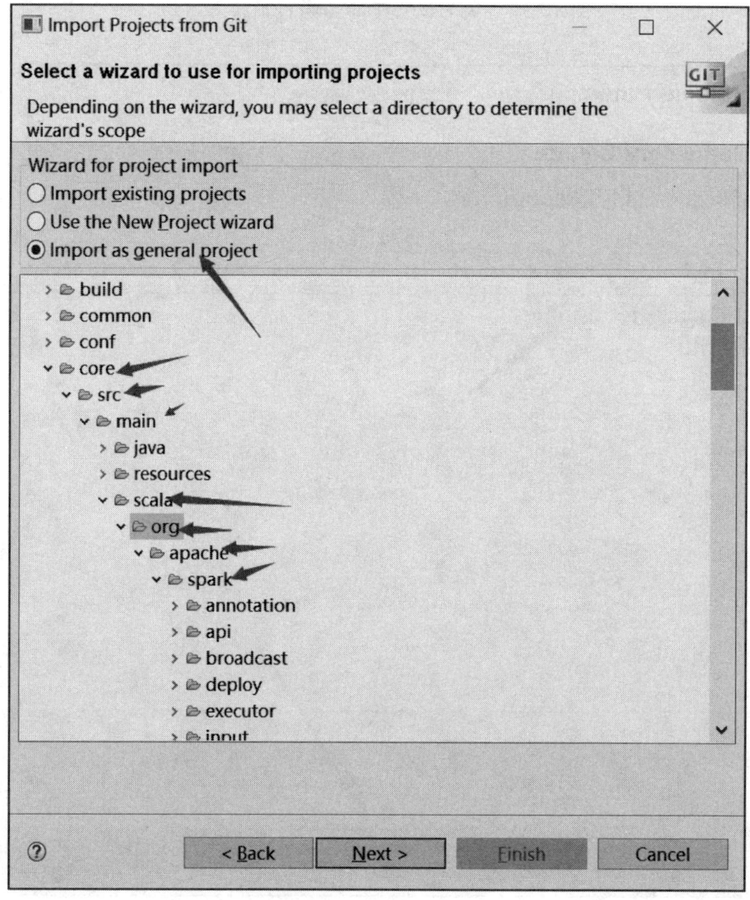

图 10-32 从下载完毕的源码直接导入（4）

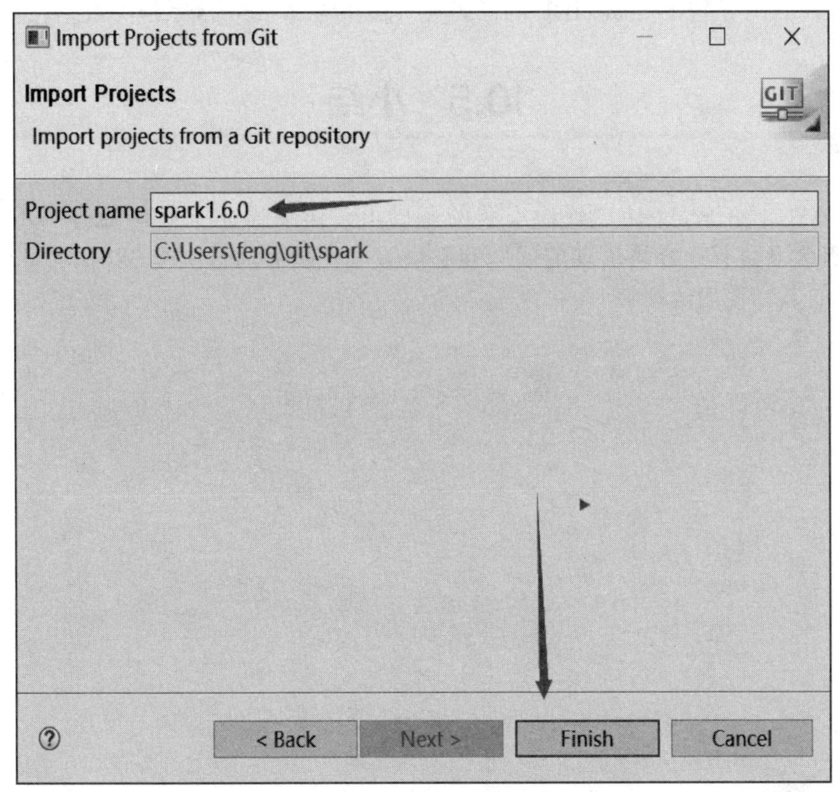

图 10-33　从下载完毕的源码直接导入（5）

点击 Finish，完成源码导入（如图 10-34 所示）。

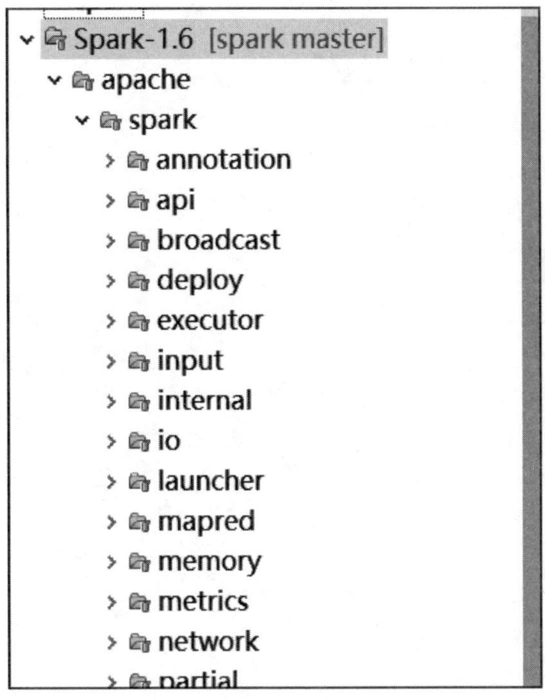

图 10-34　从下载完毕的源码直接导入（6）

注意：Ubuntu 系统下 Scala IDE 的源码导入操作与 Windows 系统一致。

10.5 小结

本章内容在介绍 Scala IDE 软件安装的基础上，详细描述了如何通过 IDE 软件进行 Spark 程序开发及过程中容易出现的问题和解决方法，最后通过一个 WordCount 实例让大家加深对 IDE 软件使用的理解。

第11章 实战详解IntelliJ IDEA下的Spark程序开发

本章节主要通过图文的方式，讲解如何让零基础的人在 Windows 操作系统与 Ubuntu 操作系统上下载、安装、调试 idea 软件，并在此基础上完成基于 idea 开发的基本工作（添加依赖导入 Jar 包，制作和调试 WordCount 程序等操作）。附录内容讲解如何在 idea 中导入 Spark 源码，便于后续阅读源代码学习。

11.1 IDEA 安装

11.1.1 Ubuntu 系统下安装

① 登录 IDEA 网站（https：//www.jetbrains.com/idea/download/#section=linux）下载 IDEA 文件包并放置在 linux 系统中，将下载后的压缩包放入 Ubuntu 系统下的/usr/local/idea 文件夹（如果没有此文件夹，请创建一个相同名字的文件夹，方便或许环境配置代码的直接复制粘贴）。如图 11-1 所示。

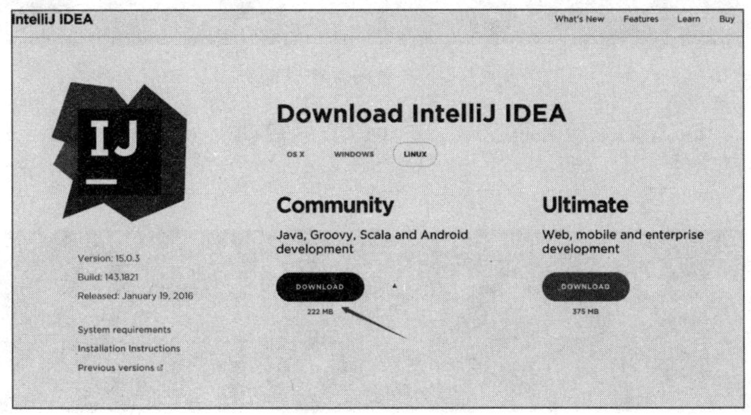

图 11-1 下载网页截图

② 复制完毕后，进入命令行工具（快捷键：ctrl＋alt＋T），输入跳转命令 cd/usr/local，创建“/usr/local/idea“目录。如图 11-2 所示。

```
root@Master: /usr/local/idea
root@Master:~# mkdir /usr/local/idea
root@Master:~# cd /usr/local/idea
root@Master:/usr/local/idea# ls
root@Master:/usr/local/idea#
```

图 11-2　创建 idea 文件夹

③ 然后输入解压命令 tar - xzf idea ＊进行解压。如图 11-3～图 11-4 所示。

```
root@lins:/usr/local/idea# tar -xzf ideaIC-15.0.2.tar.gz
```

图 11-3　解压 idea 压缩包

```
root@lins:/usr/local/idea# ls
idea-IC-143.1184.17  ideaIC-15.0.2.tar.gz
```

图 11-4　解压结果

④ 解压完成后需要配置系统环境变量，为了方便使用其 bin 目录下的命令，我们把它配置在”～/. bashrc”，输入 export IDEA _ HOME＝/usr/local/idea/idea-ID-143. 1184. 17（此处为你解压后的文件夹的名字）。如图 11-5 所示。

```
export SCALA_HOME=/usr/lib/scala/scala-2.10.4

export IDEA_HOME=/usr/local/idea/idea-IC-143.1184.17

export SPARK_HOME=/usr/local/spark/spark-1.6.0-bin-hadoop2.6

export CLASS_PATH=.:${JAVA_HOME}/lib:${JRE_HOME}/lib

export PATH=${IDEA_HOME}/bin:${HADOOP_HOME}/sbin:${HADOOP_HOME}/bin:${SPARK_HOM
E}/sbin:${SPARK_HOME}/bin:${SCALA_HOME}/bin:${JAVA_HOME}/bin:$PATH
                                                          116,1         Bot
```

图 11-5　系统环境变量配置

⑤ 配置完成 IDEA 的系统环境变量后，可以直接启动 idea，先进入 idea 的 bin 目录。如图 11-6 所示。

```
root@lins:/usr/local/idea/idea-IC-143.1184.17# cd bin
root@lins:/usr/local/idea/idea-IC-143.1184.17/bin# ls
appletviewer.policy  idea64.vmoptions  idea.vmoptions      log.xml
fsnotifier           idea.png          inspect.sh
fsnotifier64         idea.properties   libbreakgen64.so
fsnotifier-arm       idea.sh           libbreakgen.so
```

图 11-6　idea 的 bin 目录

⑥ 通过运行 bin 目录下的 idea. sh 可以运行 idea 应用程序。如图 11-7 所示。

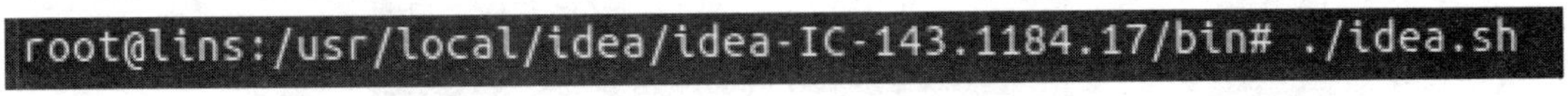

```
root@lins:/usr/local/idea/idea-IC-143.1184.17/bin# ./idea.sh
```

图 11-7　idea 的启动命令

11.1.2　Windows 系统下安装

① 登录 IDEA 网站（https://www.jetbrains.com/idea/download/#section=windows）下载 IDEA 文件包并放置在 windows 系统中。如图 11-8 所示。

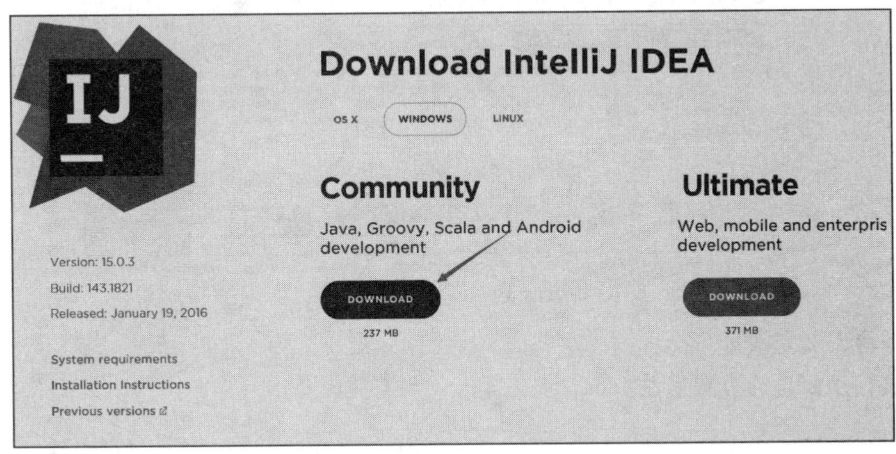

图 11-8　idea 下载页面

② 等软件下载完成后，运行 exe 安装包完成安装即可。如图 11-9～图 11-10 所示。

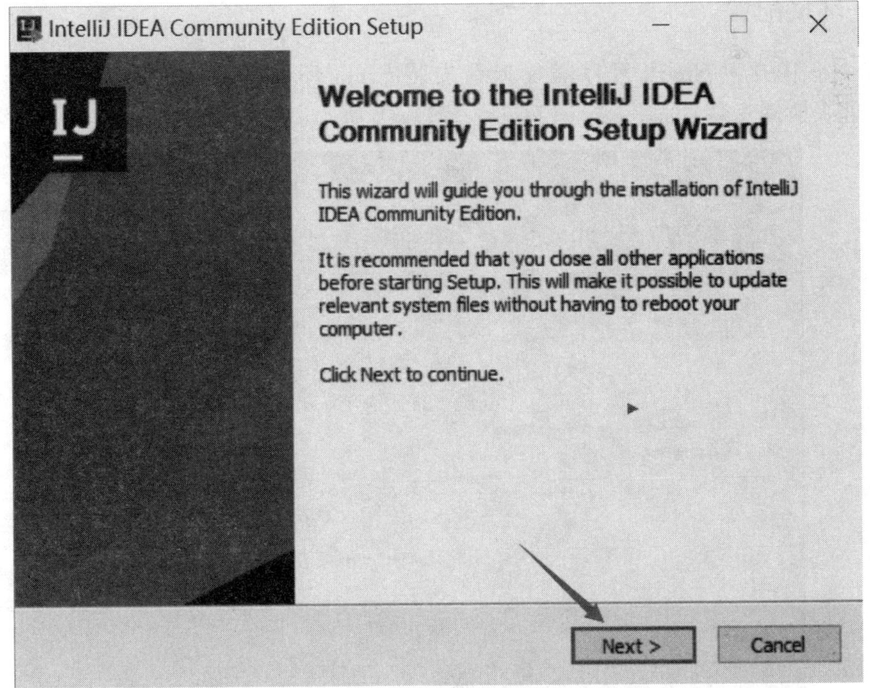

图 11-9　idea 安装界面

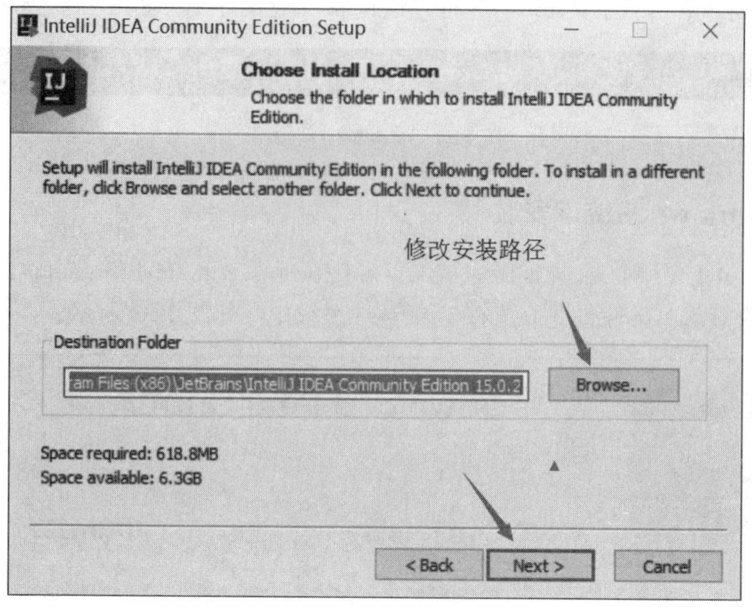

图 11-10 idea 安装界面

③ 等待安装完成即可。

④ 运行桌面快捷方式启动 idea。

11.2 IDEA 开发重点步骤详解

11.2.1 环境配置

① 在 idea 中开发 Spark 程序需要提前安装 Scala 环境，在首次开启 idea 页面，点击 plugins，进入插件安装页面。如图 11-11～图 11-12 所示。

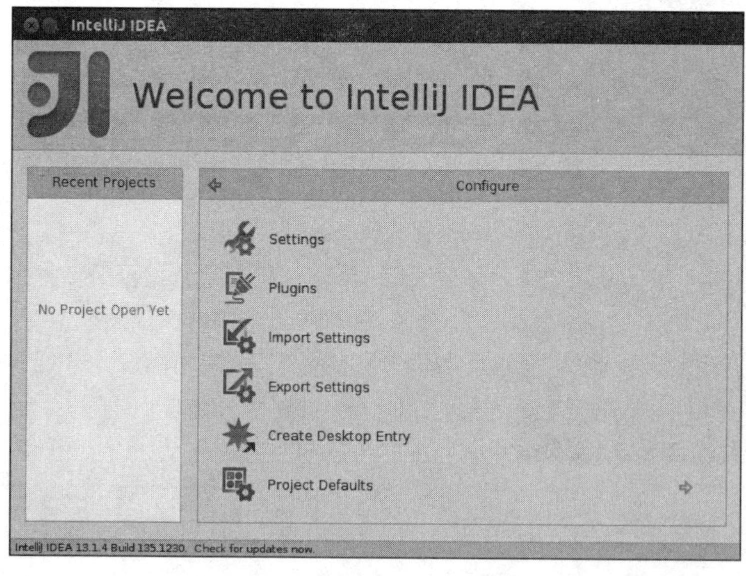

图 11-11 idea 首次启动界面

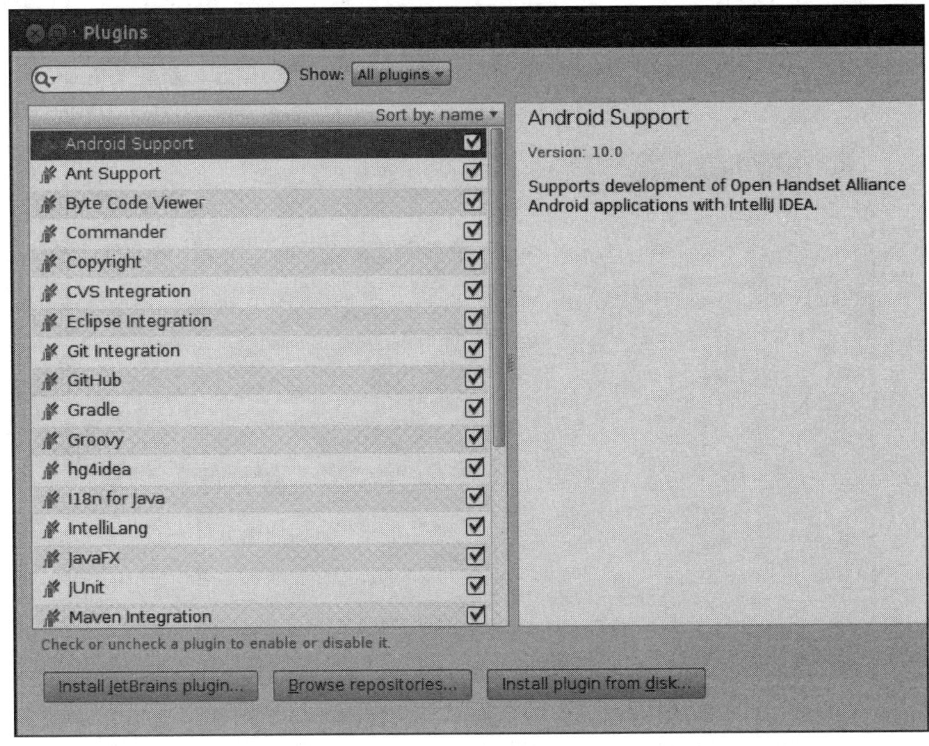

图 11-12　插件安装界面

② 进入界面后，点击左下角 Install JetBrains plugin 选项进入如下界面，如图 11-13 所示。

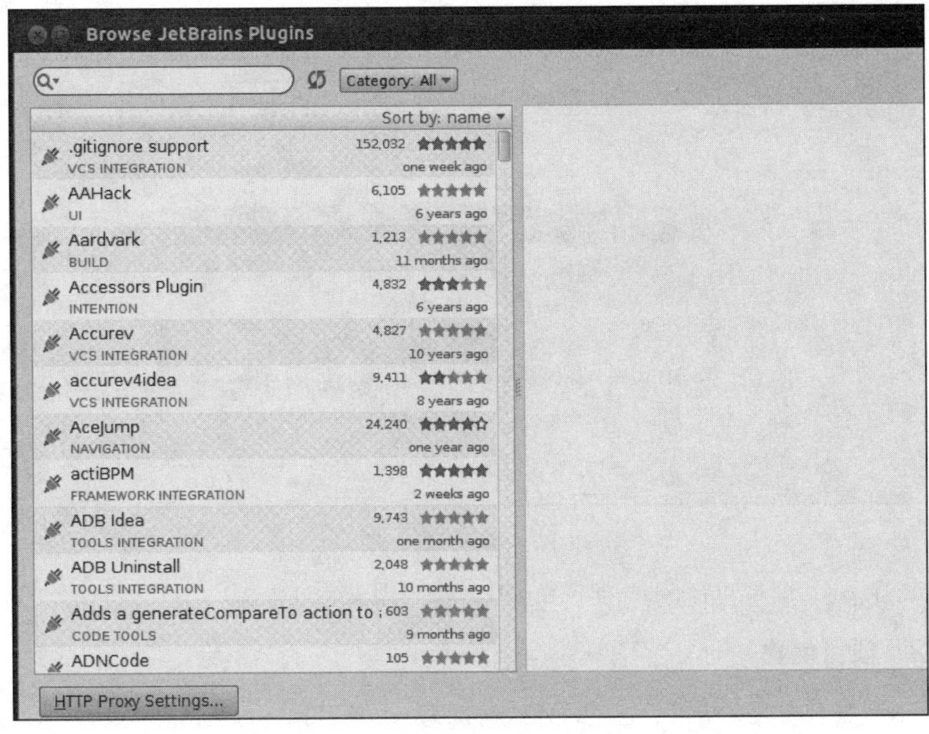

图 11-13　插件搜索界面

③ 在左上方输入框中输入 scala，查询 scala 插件。如图 11-14 所示。

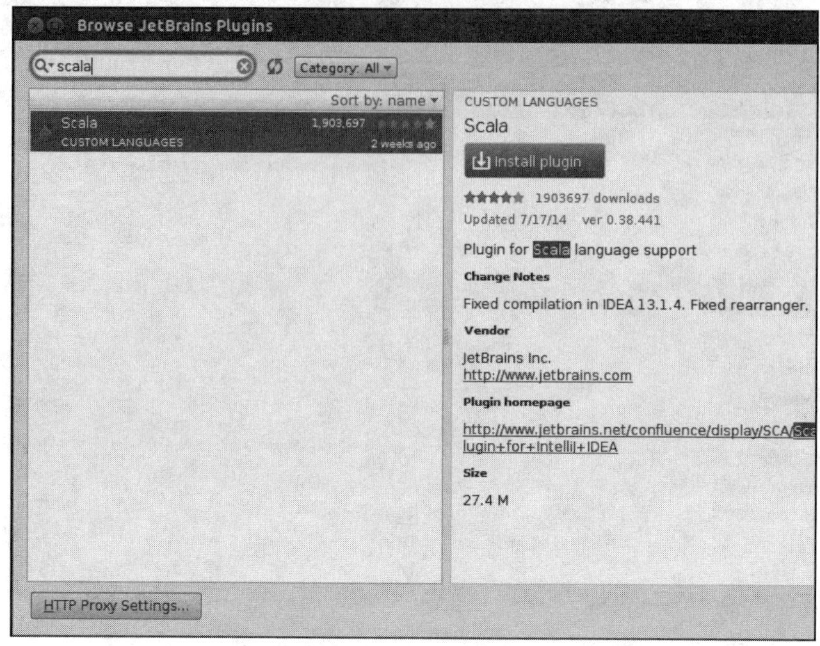

图 11-14 Scala 插件安装（1）

④ 此时点击右侧的 Install plugin，安装 scala 插件。如图 11-15 所示。

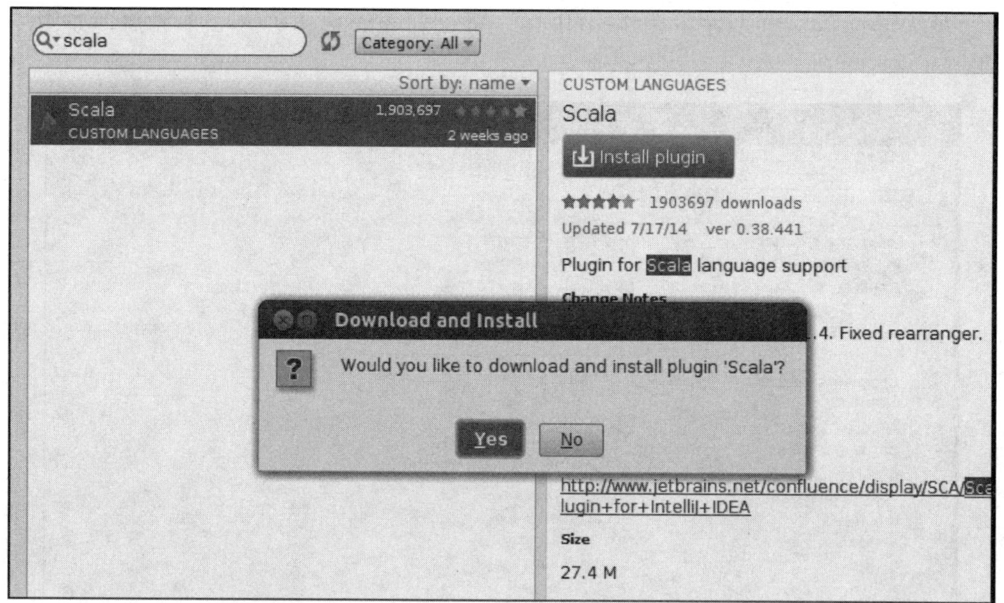

图 11-15 Scala 插件安装（2）

⑤ 选择 yes，即可开启 Scala 插件在 IDEA 中自动安装过程。

11.2.2 项目创建

① Scala 插件安装完毕以后，可以开始创建 Spark 项目，进行测试。双击 idea 程序，启动程序，如图 11-16 所示。

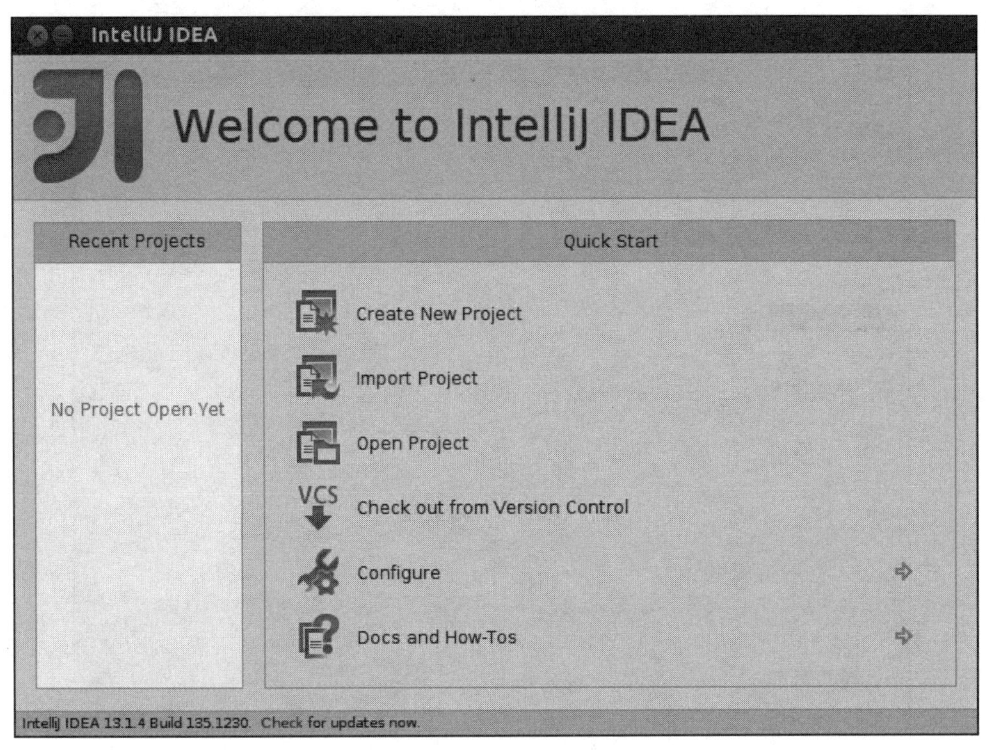

图 11-16　Scala 启动界面

② 启动应用程序后在进入界面选择 Create New Project，创建新应用程序，如图 11-17 所示。

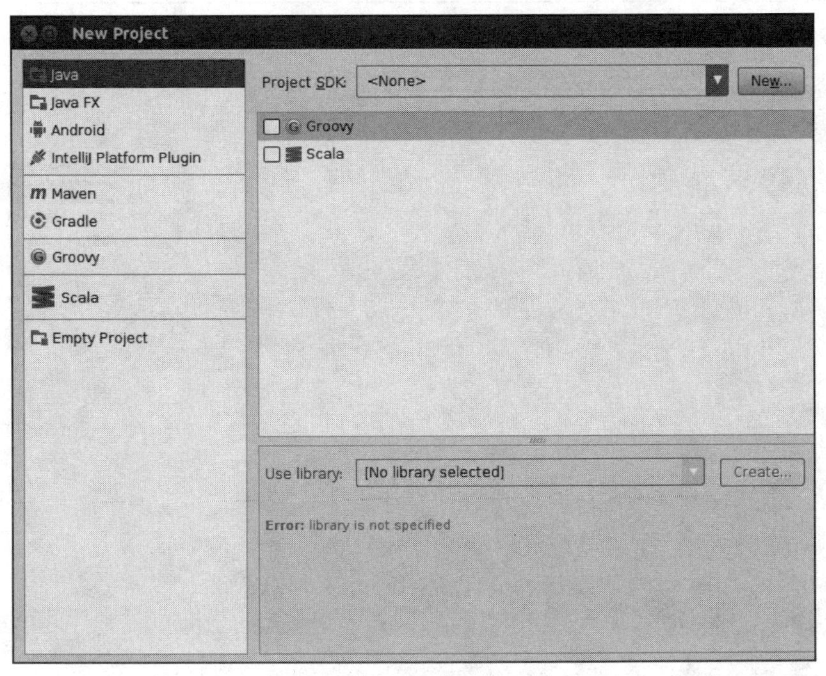

图 11-17　创建 Scala 项目

③ 选择左侧 Scala 选项并选择右侧 Scala 选项，确定创建 Scala 项目，如图 11-18 所示。

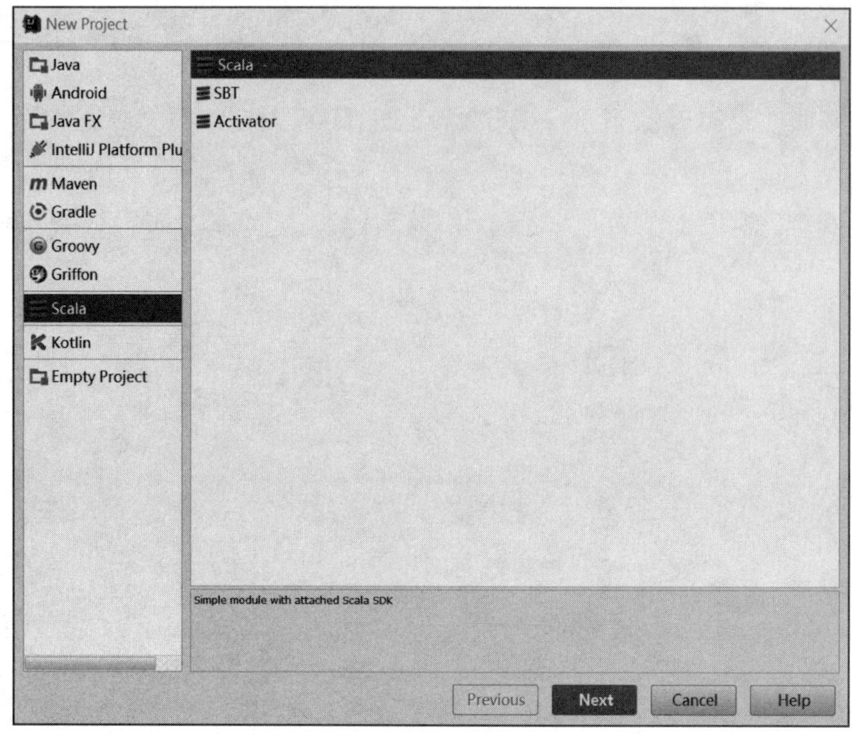

图 11-18　创建 Scala 项目

④ 点击 Next 进入下一步窗口，进行工程名字和目录的设置，同时也需要设置好 javasdk 的引用。如图 11-19～图 11-21 所示。

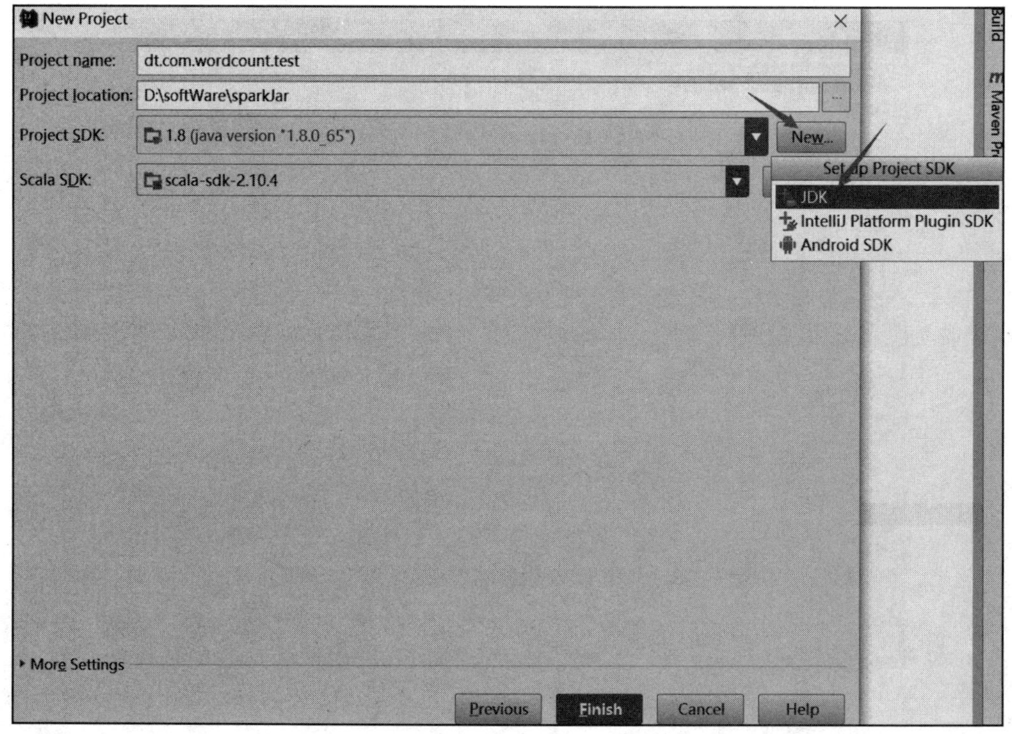

图 11-19　设置 jdk（1）

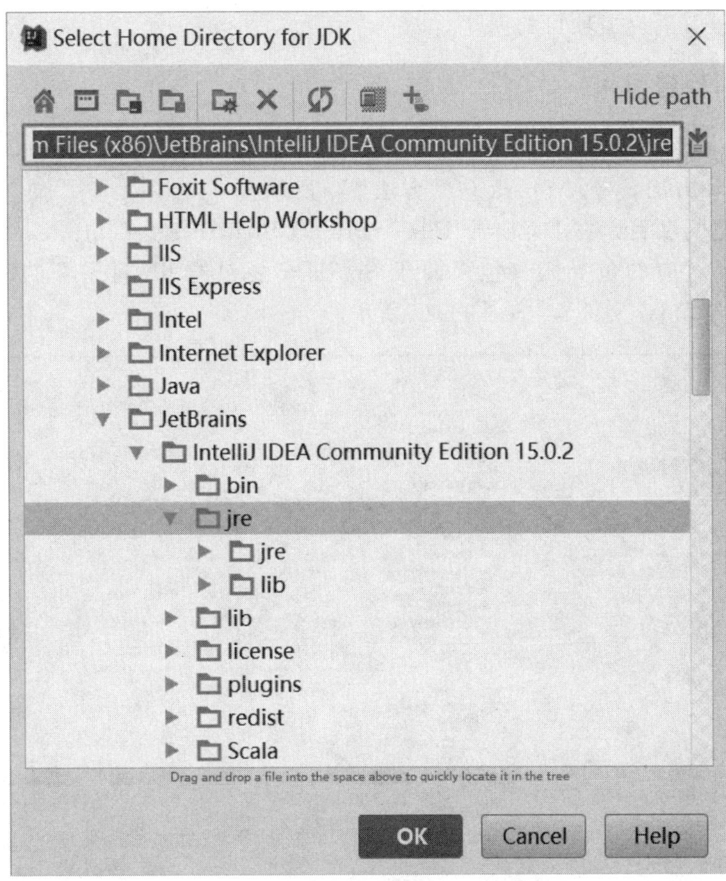

图 11-20 设置 jdk（2）

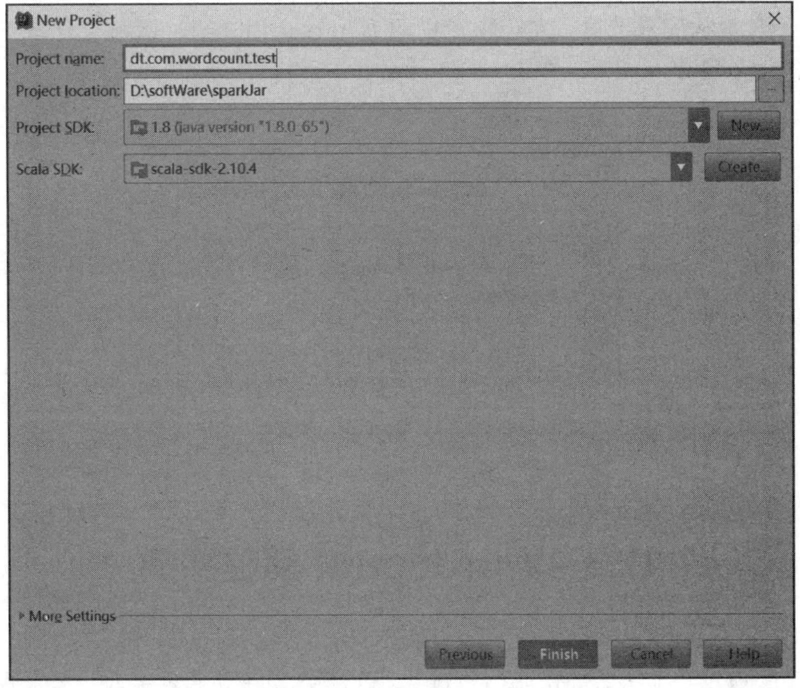

图 11-21 设置项目

⑤ 点击 Finish 后完成项目的创建。

11.2.3 Spark 包引入

在完成 Scala 环境配置以后，还需要导入 Spark 的编译后的代码包，以便可以使用 Spark 的各种类和方法，导入 Spark 包的方法如下所示。

① 打开创建的项目，右上角选择 File->Project Structure->Libraries，选择+->java，将 spark-hadoop 对应的包导入（包来源于 Spark 安装包的 lib 目录下的 spark-assembly-1.6.0-hadoop2.6.0.jar 文件），如图 11-22 所示。

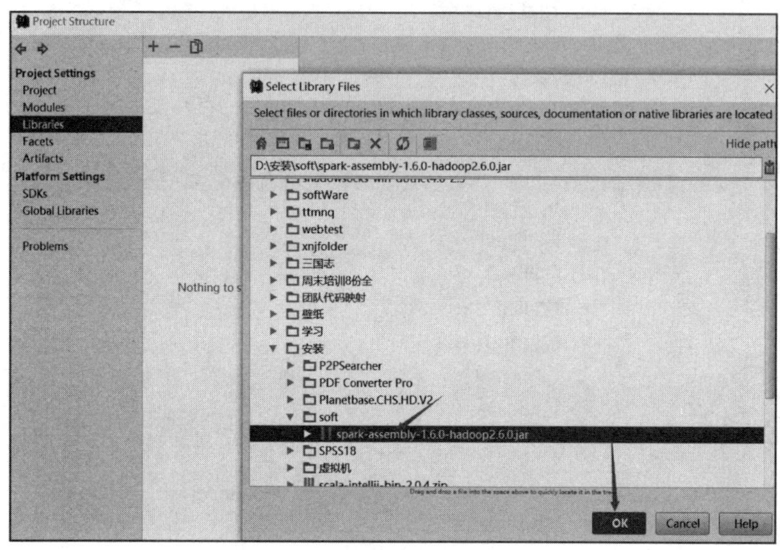

图 11-22 选择 jar 包

② 点击 OK 导入完成之后，左边项目栏的 External Libraries 区域会出现 Spark 的依赖包，标志着 Spark 包导入成功。如图 11-23 所示。

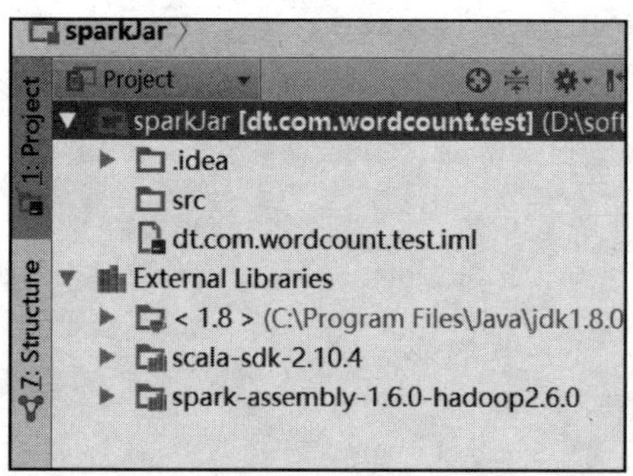

图 11-23 导入成功

11.3 Wordcount 创建实战

本小节主要通过在上文已经创建的项目基础上实战创建和调试 wordcount 程序。

在创建程序前，需要完成以下准备工作。

① 打开相应的 WordCount 包。

② 将 spark-1. 6. 0-bin-hadoop2. 6 目录中的 README. md 文件拷贝到 D：//testspark// 目录下（其他目录也可，但是需要在程序中修改相关文件夹代码）。

③ 完成上述内容后，创建一个新的 WordCount. scala 文件，创建方式如图 11-24 所示，并将下列代码复制入 WordCount. scala 中。

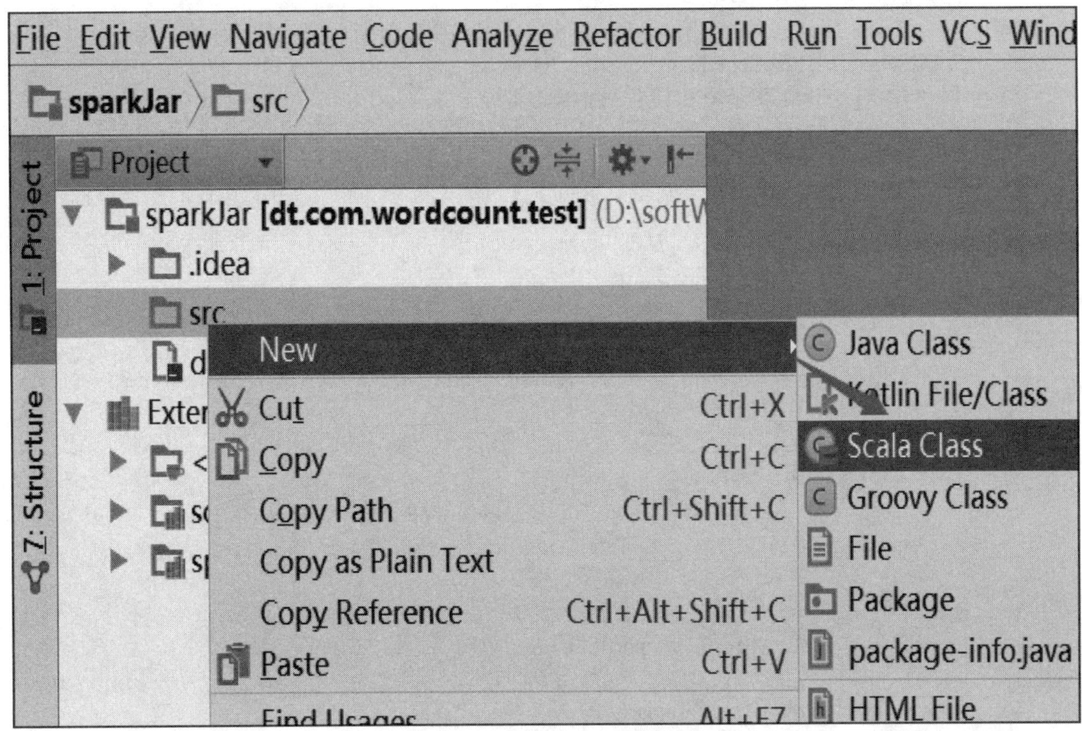

图 11-24　创建 Scala 文件夹

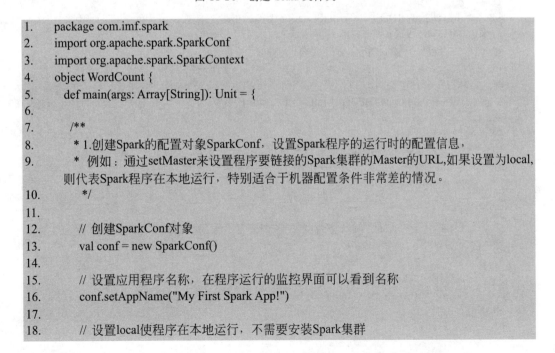

```
package com.imf.spark
import org.apache.spark.SparkConf
import org.apache.spark.SparkContext
object WordCount {
  def main(args: Array[String]): Unit = {

    /**
     * 1.创建Spark的配置对象SparkConf，设置Spark程序的运行时的配置信息，
     * 例如：通过setMaster来设置程序要链接的Spark集群的Master的URL,如果设置为local,
   则代表Spark程序在本地运行，特别适合于机器配置条件非常差的情况。
     */

    // 创建SparkConf对象
    val conf = new SparkConf()

    // 设置应用程序名称，在程序运行的监控界面可以看到名称
    conf.setAppName("My First Spark App!")

    // 设置local使程序在本地运行，不需要安装Spark集群
```

```
    conf.setMaster("local")

    /**
     * 2.创建SparkContext对象
     * SparkContext是spark程序所有功能的唯一入口，无论是采用Scala,java,python,R等都必须有一个SprakContext
     * SparkContext核心作用：初始化spark应用程序运行所需要的核心组件，包括DAGScheduler,TaskScheduler,SchedulerBackend
     * 同时还会负责Spark程序往Master注册程序等；
     * SparkContext是整个应用程序中最为至关重要的一个对象；
     */
    // 通过创建SparkContext对象，通过传入SparkConf实例定制Spark运行的具体参数和配置信息

    val sc = new SparkContext(conf)

    /**
     * 3.根据具体数据的来源（HDFS,HBase,Local,FS,DB,S3等）通过SparkContext来创建RDD；
     * RDD的创建基本有三种方式：根据外部的数据来源（例如HDFS）、根据Scala集合、 由其他的RDD操作；
     * 数据会被RDD划分成为一系列的Partitions,分配到每个Partition的数据属于一个Task的处理范畴；
     */

  //读取本地文件，并设置一个partition

val lines = sc.textFile("D://testspark//README.md",1)

/**
 * 4.对初始的RDD进行Transformation级别的处理，例如map,filter等高阶函数的变成，来进行具体的数据计算
 * 4.1.将每一行的字符串拆分成单个单词
 */
//对每一行的字符串进行拆分并把所有行的拆分结果通过flat合并成一个大的集合
  val words = lines.flatMap { line => line.split(" ") }

  /**
   * 4.2.在单词拆分的基础上对每个单词实例计数为1，也就是word => (word,1)
   */
  val pairs = words.map{word =>(word,1)}

  /**
   * 4.3.在每个单词实例计数为1基础上统计每个单词在文件中出现的总次数
   */

   /对相同的key进行value的累积（包括Local和Reducer级别同时Reduce）
  val wordCounts = pairs.reduceByKey(_+_)

   // 打印输出
```

```
    wordCounts.foreach(pair => println(pair._1+":"+pair._2))
    sc.stop()
  }
}
```

④ 创建完成之后右键点击 Run wordcount 就可以运行了，如图 11-25 所示。

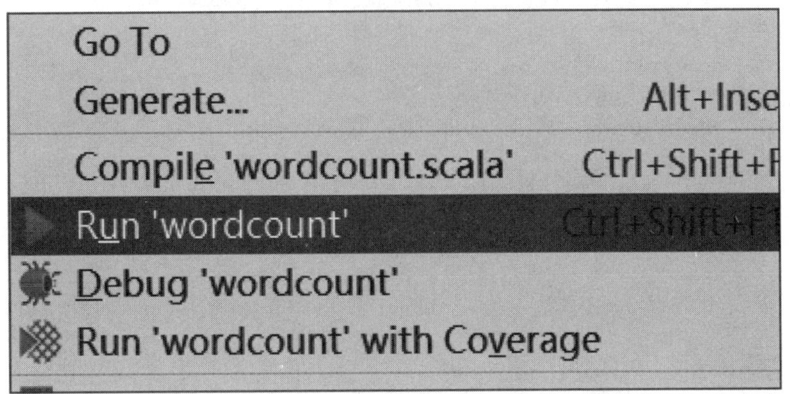

图 11-25 调试 wordcount

11.4 IDEA 导入 Spark 源码

Windows 源码环境搭建

此方法可分为以下三步骤。

① 配置 Maven。先配置 Maven 的 setting，使用国内镜像。可以先创建一个 txt 文件，将下面内容编辑进去，然后重命名为 setting. xml。

```
<mirror>
 <id>nexus-osc</id>
<mirrorOf>*</mirrorOf>
<name>Nexus osc</name>
<url>http://maven.oschina.net/content/groups/public/</url>
</mirror>
    <mirror>
<id>nexus-osc-thirdparty</id>
<mirrorOf>thirdparty</mirrorOf>
<name>Nexus osc thirdparty</name>
<url>http://maven.oschina.net/content/repositories/thirdparty/</url>
    </mirror>
```

在创建完毕 setting. xml 后，需要进入 idea 中，将修改后的 sitting. xml 放入如图 11-26 所示的 settingFile 中。

② 配置完成 Maven 后要准备下载 Spark 源码进行导入，进入 Spark 官网，下载最新的源代码，然后放入指定的目录。

③ 下载完成后，便可以在 idea 中通过导入额外代码来源的方式导入，流程如图 11-28～图 11-37 所示。

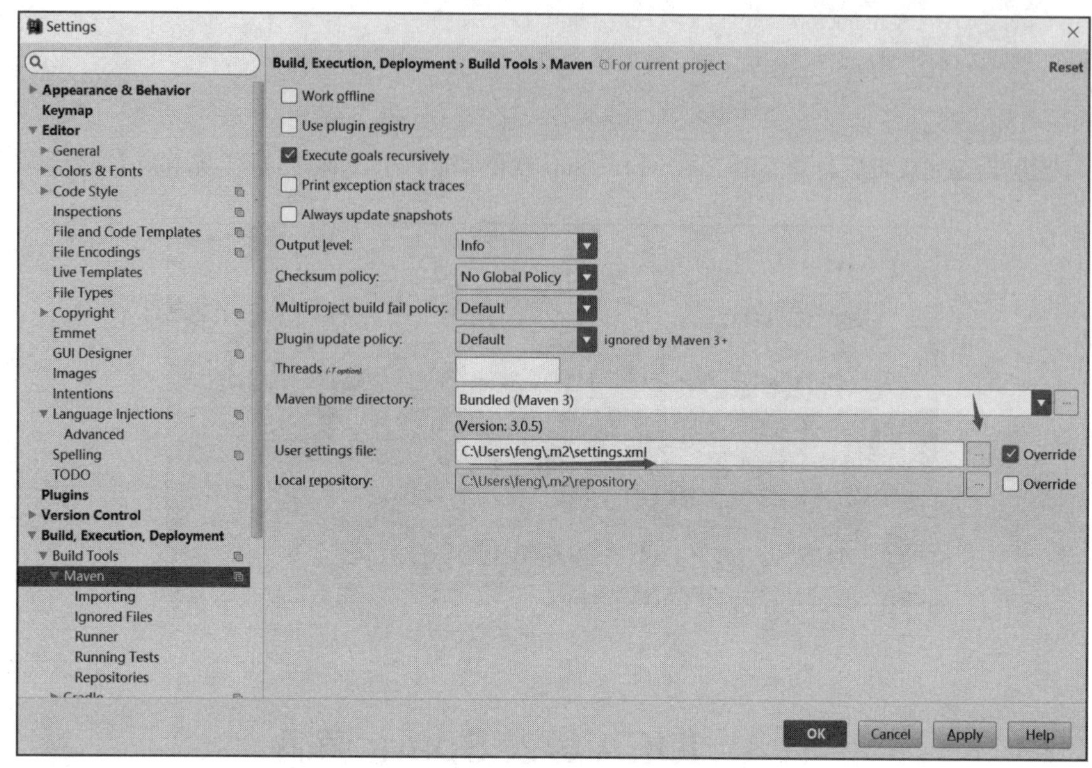

图 11-26 修改 sitting 文件

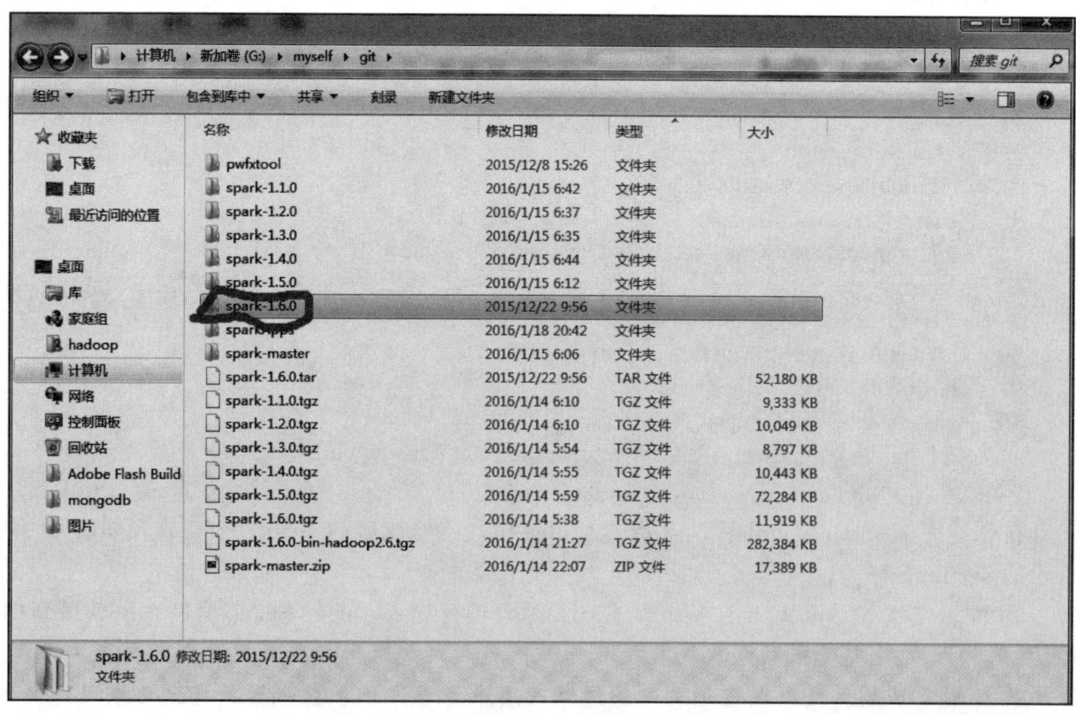

图 11-27 下载完毕的源码

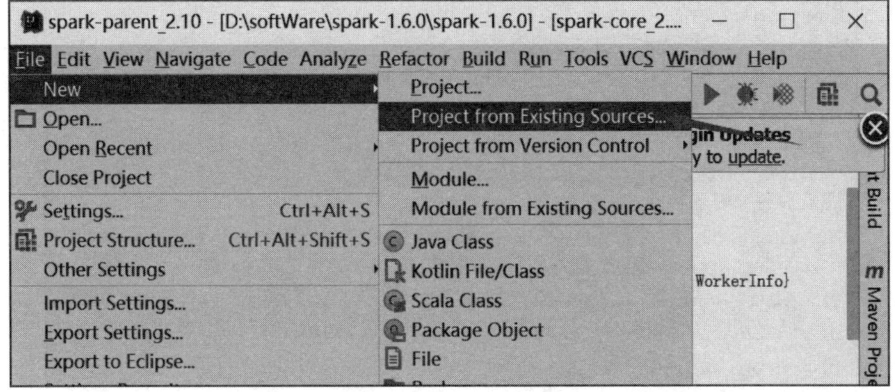

图 11-28 导入源码文件

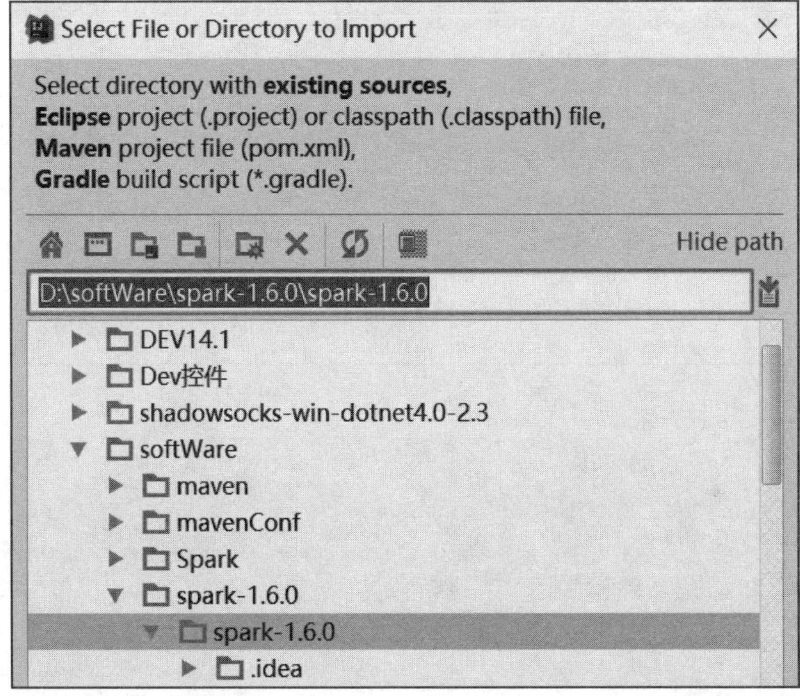

图 11-29 选择源码文件

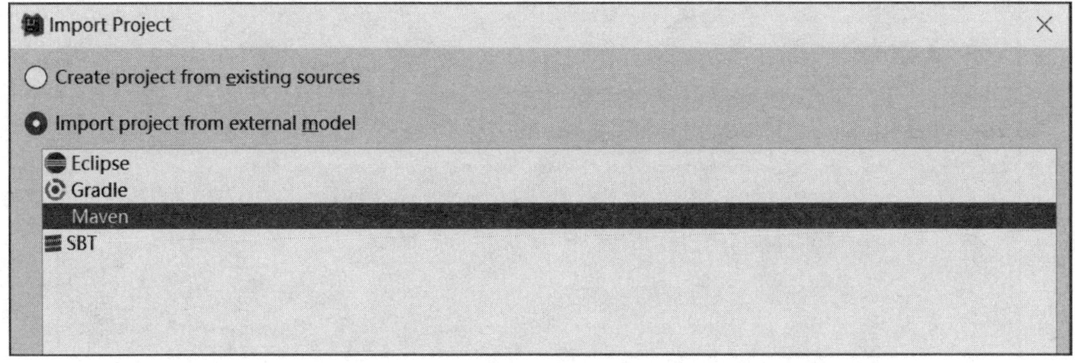

图 11-30 选择 maven 的方式导入包

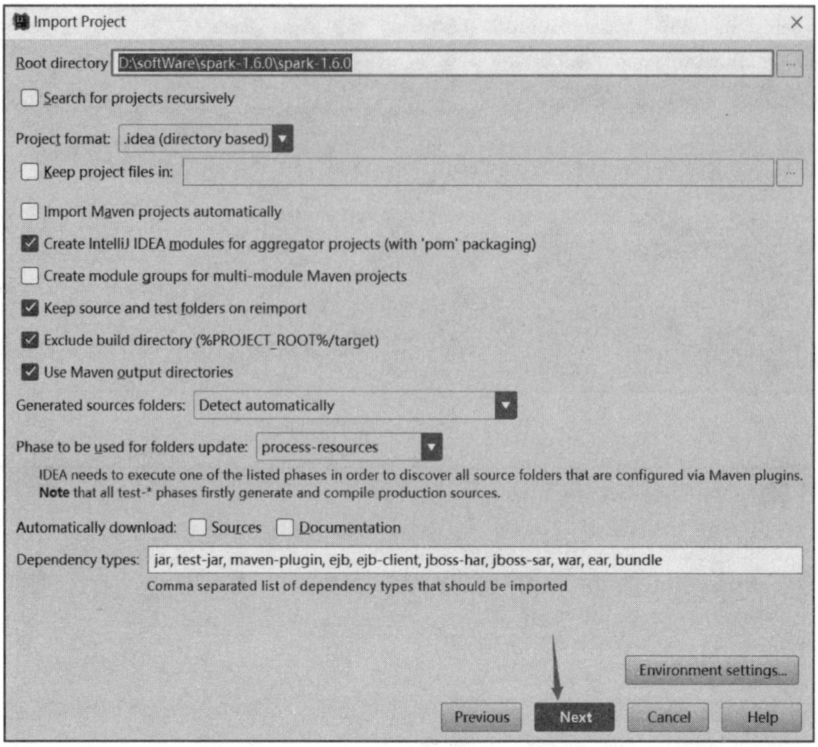

Import Project

Root directory D:\softWare\spark-1.6.0\spark-1.6.0

Search for projects recursively

Project format: .idea (directory based)

Keep project files in:

Import Maven projects automatically

Create IntelliJ IDEA modules for aggregator projects (with 'pom' packaging)

Create module groups for multi-module Maven projects

Keep source and test folders on reimport

Exclude build directory (%PROJECT_ROOT%/target)

Use Maven output directories

Generated sources folders: Detect automatically

Phase to be used for folders update: process-resources

IDEA needs to execute one of the listed phases in order to discover all source folders that are configured via Maven plugins. **Note** that all test-* phases firstly generate and compile production sources.

Automatically download: Sources Documentation

Dependency types: jar, test-jar, maven-plugin, ejb, ejb-client, jboss-har, jboss-sar, war, ear, bundle

Comma separated list of dependency types that should be imported

Environment settings...

Previous Next Cancel Help

图 11-31 项目设置界面

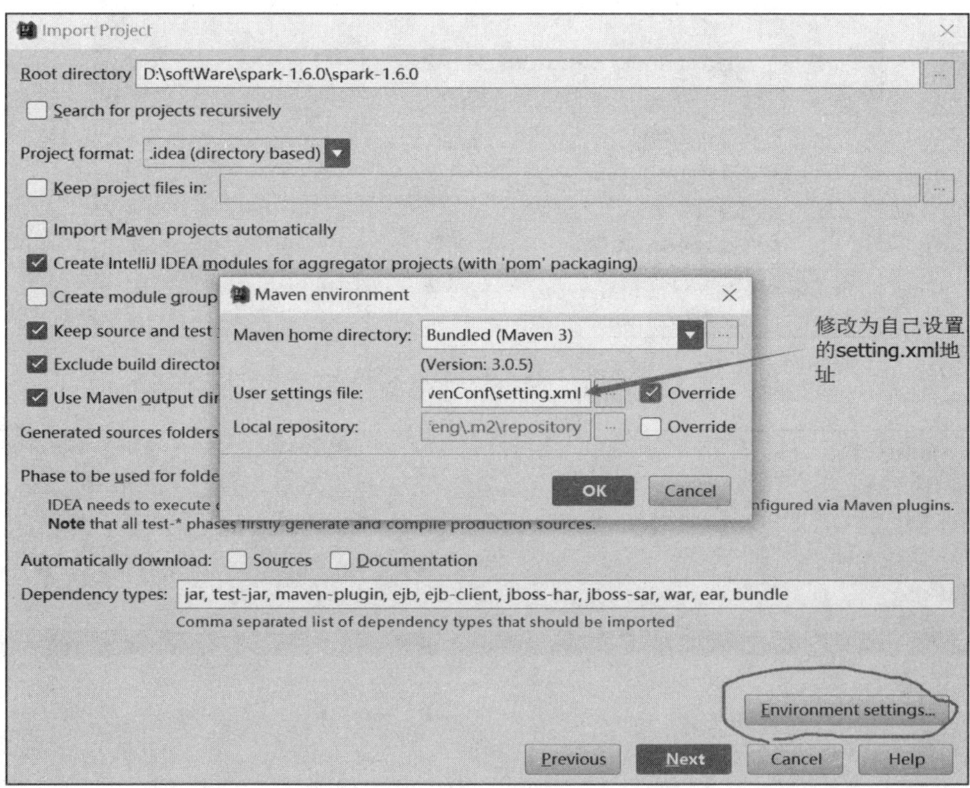

图 11-32 设置 setting. xml 内容

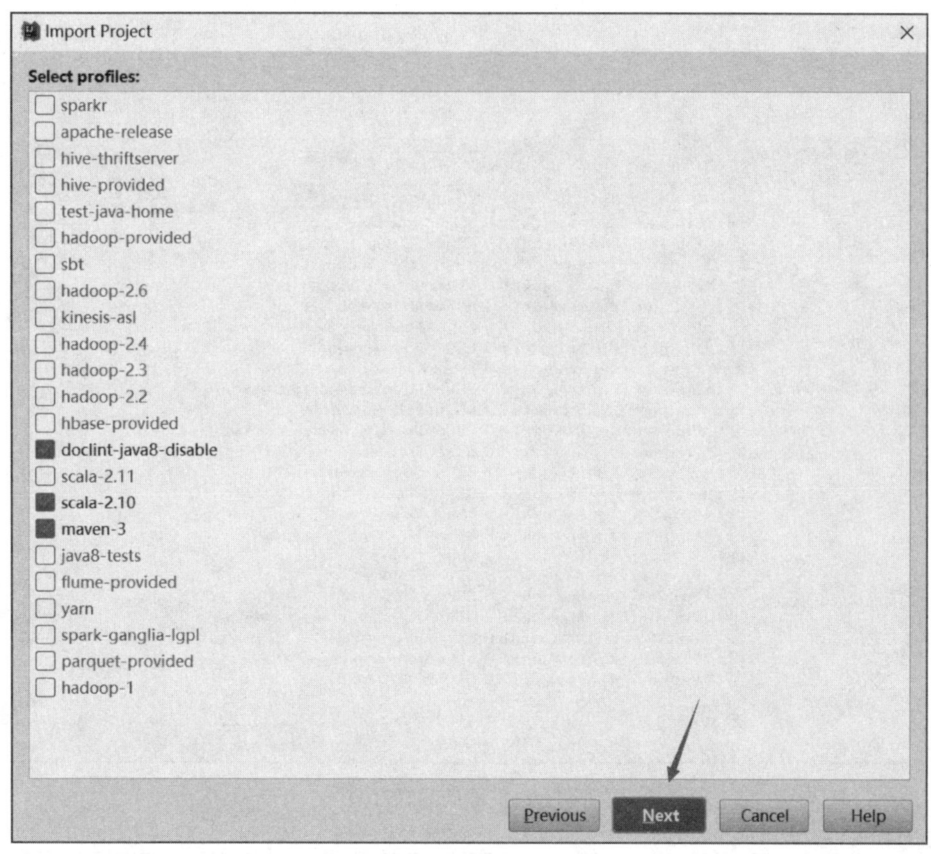

图 11-33 源码导入（1）

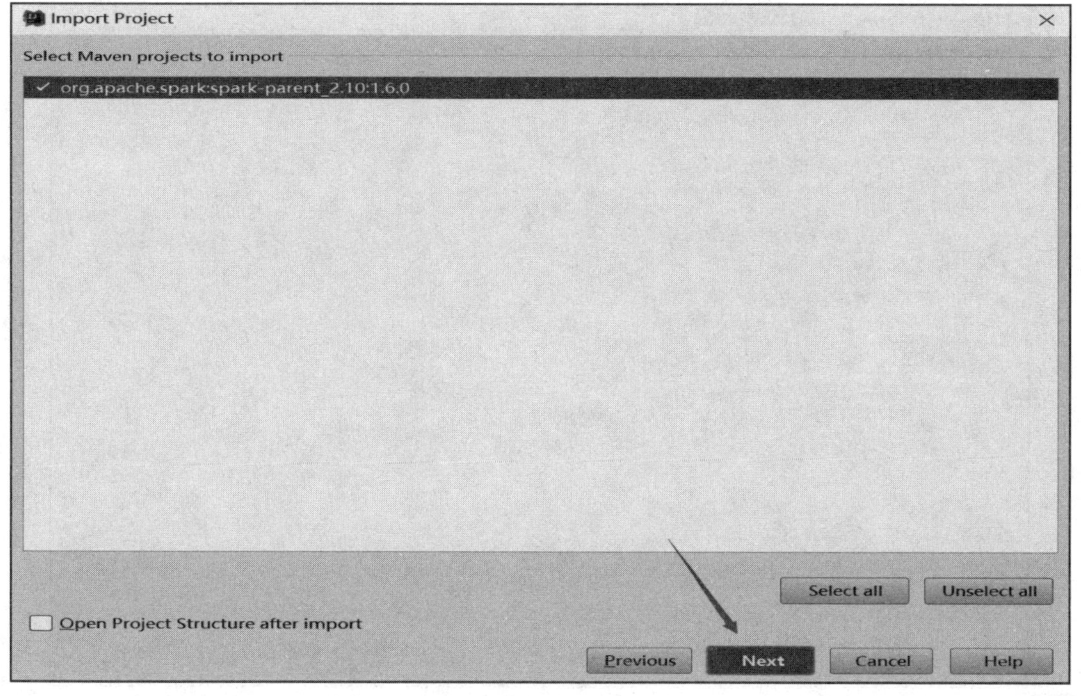

图 11-34 源码导入（2）

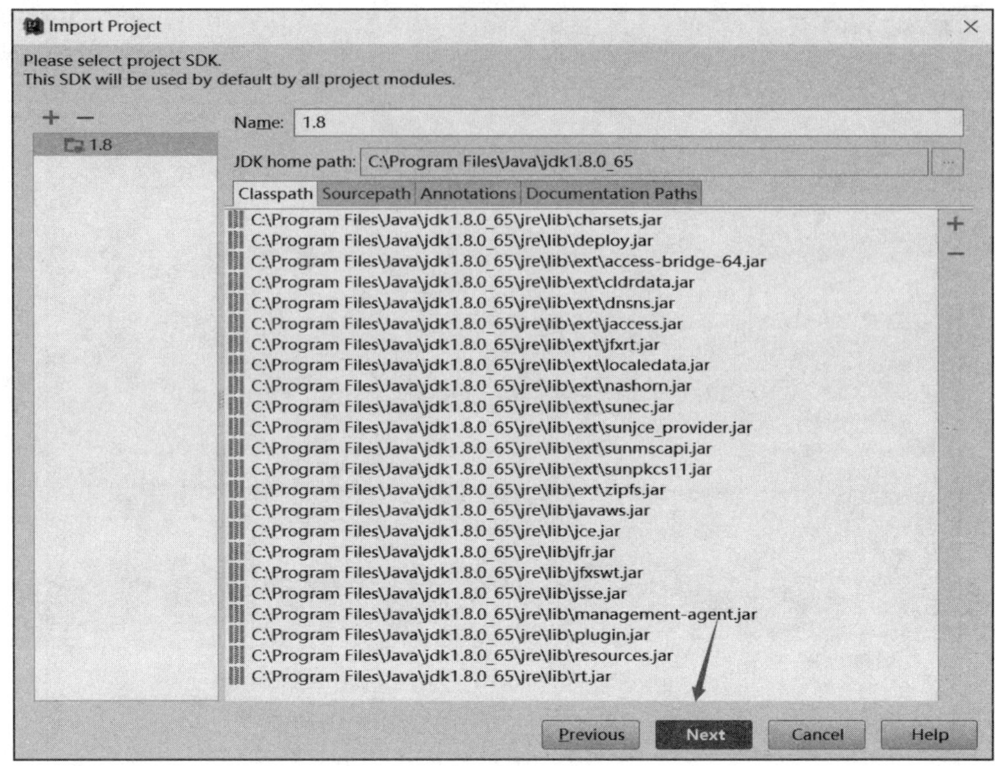

图 11-35　源码导入（3）

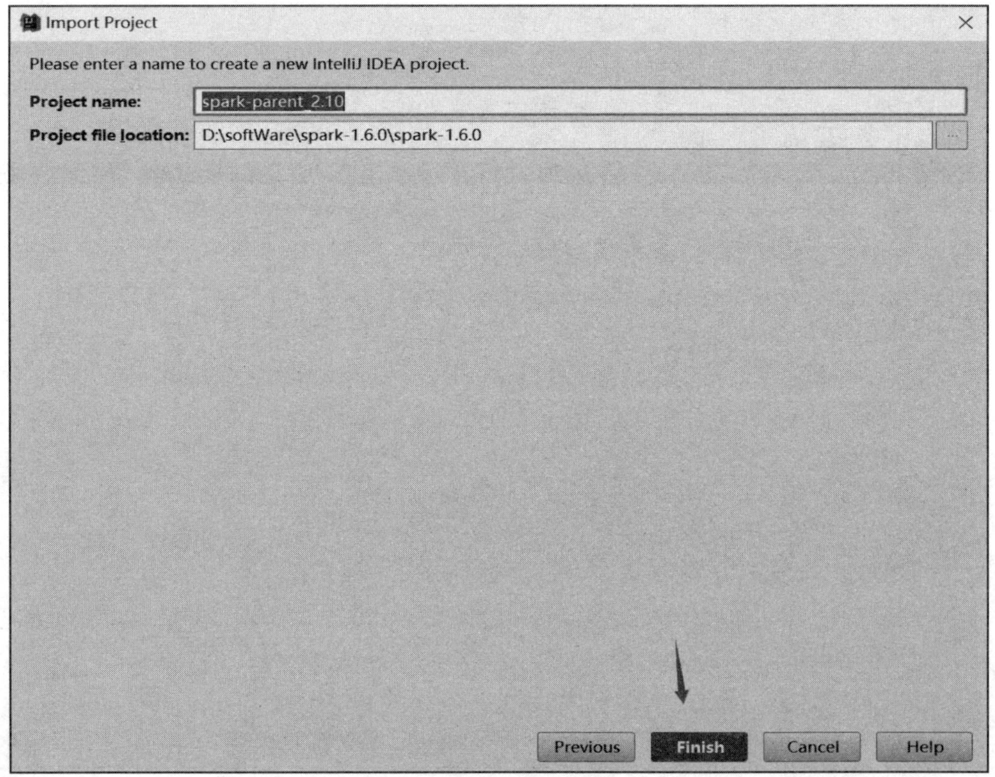

图 11-36　设置导入项目的名字

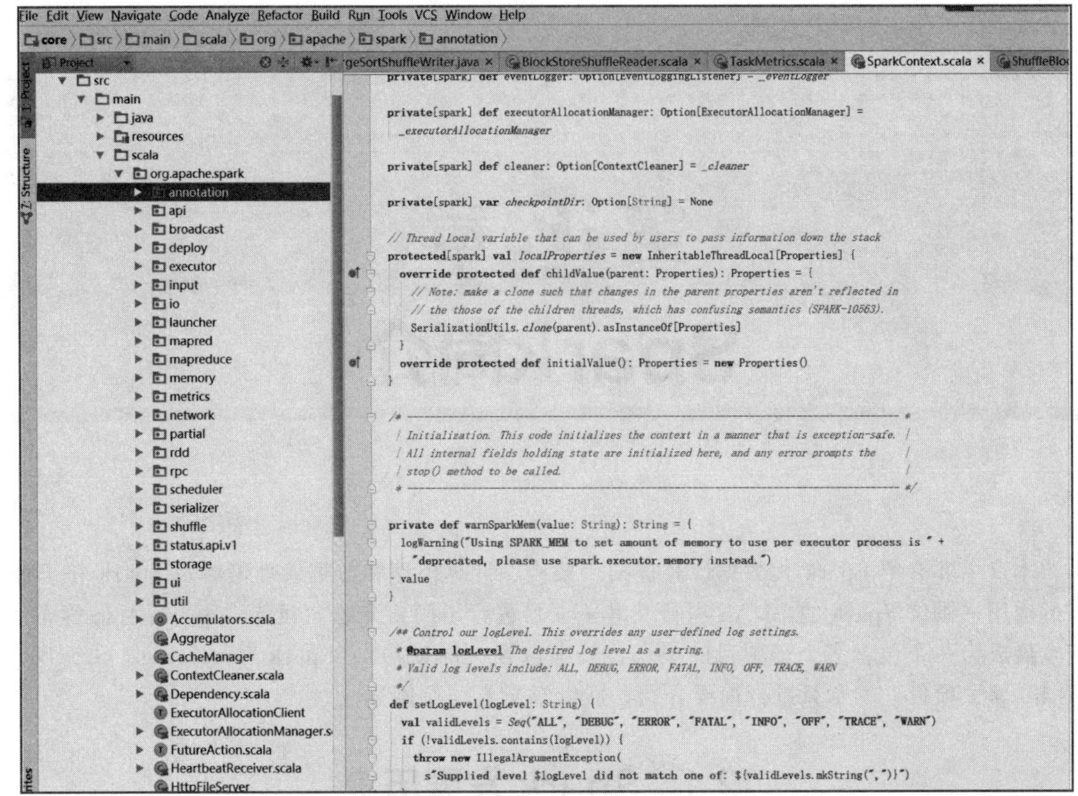

图 11-37　导入源码完成

注意：Ubuntu 系统下源码环境搭建步骤与 Windows 下的源码环境搭建步骤相同。

11.5　小结

本章内容在介绍 idea 软件安装的基础上，详细描述了如果通过 idea 软件进行 Spark 程序开发的各个步骤及过程中容易出现的问题和解决方法，最后通过一个 WordCount 实例让大家加深对 idea 软件使用的理解。

第12章 Spark简介

本章主要介绍Spark大数据计算框架、架构、计算模型和数据管理策略及Spark在工业界的应用。围绕Spark的BDAS项目及其子项目进行了简要介绍。目前，Spark生态系统已经发展成为一个包含多个子项目的集合，其中包含SparkSQL、Spark Streaming、GraphX、MLlib等子项目，本章只进行简要介绍，后续章节再详细阐述。

12.1 Spark发展历史

对于一个具有相当技术门槛与复杂度的平台，Spark从诞生到正式版本的成熟，经历的时间如此之短，让人感到惊诧。

① 2009年：Spark诞生于AMPLab。

② 2010年：开源。

③ 2013年6月：Apache孵化器项目。

④ 2014年2月：Apache顶级项目。

⑤ 2014年2月：大数据公司Cloudera宣称加大Spark框架的投入来取代MapReduce。

⑥ 2014年4月：大数据公司MapR投入Spark阵营，Apache Mahout放弃MapReduce，将使用Spark作为计算引擎。

⑦ 2014年5月：Pivotal Hadoop集成Spark全栈。

⑧ 2014年5月30日：Spark 1.0.0发布。

⑨ 2014年6月：Spark 2014峰会在旧金山召开。

⑩ 2014年7月：Hive on Spark项目启动。

目前AMPLab和Databricks负责整个项目的开发维护，很多公司，如Yahoo!、Intel等参与到Spark的开发中，同时很多开源爱好者积极参与Spark的更新与维护。

AMPLab开发以Spark为核心的BDAS时提出的目标是：one stack to rule them all，也就是说在一套软件栈内完成各种大数据分析任务。

相对于MapReduce上的批量计算、迭代型计算以及基于Hive的SQL查询，Spark可以带来上百倍的性能提升。目前Spark的生态系统日趋完善，Spark SQL的发布、Hive on

Spark 项目的启动以及大量大数据公司对 Spark 全栈的支持，让 Spark 的数据分析范式更加丰富。

在 2013 年来，Spark 进入了一个高速发展期，代码库提交与社区活跃度都有显著增长。以活跃度论，Spark 在所有 Apache 基金会开源项目中，位列前三。相较于其他大数据平台或框架而言，Spark 的代码库最为活跃，如图 12-1 所示。

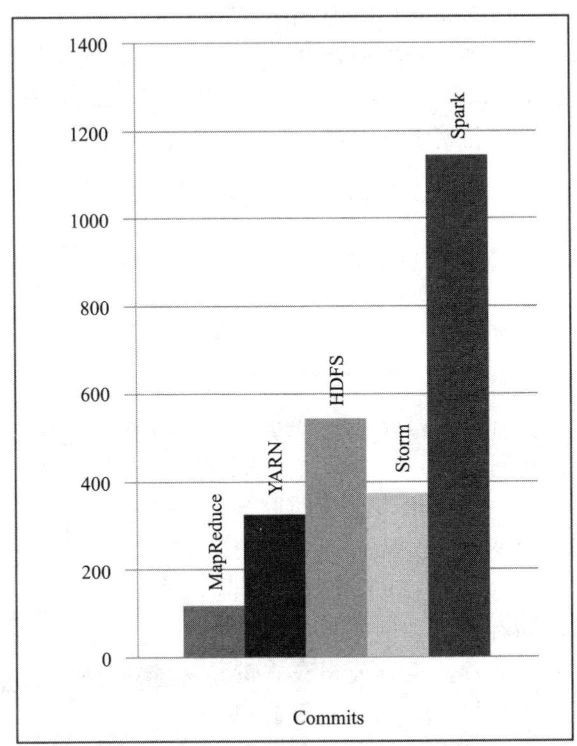

图 12-1　各开源软件代码提交次数比较

图 12-2 则显示了自从 Spark 将其代码部署到 Github 之后的提交数据，一共有 14782 次提交，14 个分支，36 次发布，820 位代码贡献者。

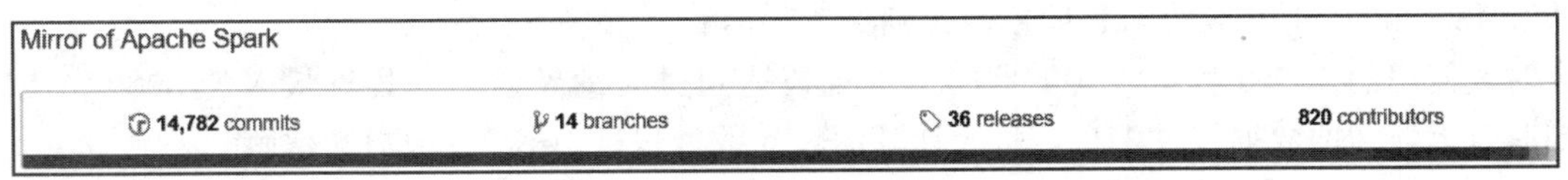

图 12-2　Spark 开发者贡献程度

12.2　Spark 在国内外的使用

随着企业数据量的增长，对大数据的处理和分析已经成为企业的迫切需求。Spark 作为 Hadoop 的替代者，引起学术界和工业界的普遍兴趣，大量应用在工业界落地，许多科研院校开始了对 Spark 的研究。

在学术界，Spark 得到各院校的关注。Spark 源自学术界，最初是由加州大学伯克利分校的 AMPLab 设计开发。国内的中国科学院、中国人民大学、南京大学、华东师范大学等也开始对 Spark 展开相关研究。涉及 Benchmark、SQL、并行算法、性能优化、高可用性等多个方面。

在工业界，Spark 已经在互联网领域得到广泛应用。互联网用户群体庞大，需要存储大数据并进行数据分析，Spark 能够支持多范式的数据分析，解决了大数据分析中迫在眉睫的问题。例如，国外 Cloudera、MapR 等大数据厂商全面支持 Spark，微策略等老牌 BI 厂商也和 Databricks 达成合作关系，Yahoo！使用 Spark 进行日志分析并积极回馈社区，Amazon 在云端使用 Spark 进行分析。国内同样得到很多公司的青睐，淘宝构建 Spark on Yarn 进行用户交易数据分析，使用 GraphX 进行图谱分析。网易用 Spark 和 Shark 对海量数据进行报表和查询。腾讯使用 Spark 进行精准广告推荐。

下面将选取代表性的 Spark 应用案例进行分析，以便于读者了解 Spark 在工业界的应用状况。

根据 2015 Spark 中国峰会的信息了解，以下几大公司对 Spark 的运用已经到了国内的前列。

（1）腾讯公司

深圳市腾讯计算机系统有限公司成立于 1998 年 11 月，由马化腾、张志东、许晨晔、陈一丹、曾李青五位创始人共同创立。是中国最大的互联网综合服务提供商之一，也是中国服务用户最多的互联网企业之一。腾讯多元化的服务包括：社交和通信服务 QQ 及微信/WeChat、社交网络平台 QQ 空间、腾讯游戏旗下 QQ 游戏平台、门户网站腾讯网、腾讯新闻客户端和网络视频服务腾讯视频等。2004 年 6 月 16 日，腾讯公司在香港联交所主板公开上市（股票代号 00700），董事会主席兼首席执行官是马化腾。2015 年，腾讯公司实现总收入 1028.63 亿元，同比增长 30%；腾讯权益持有人应占盈利 288.06 亿元，同比增长 21%。

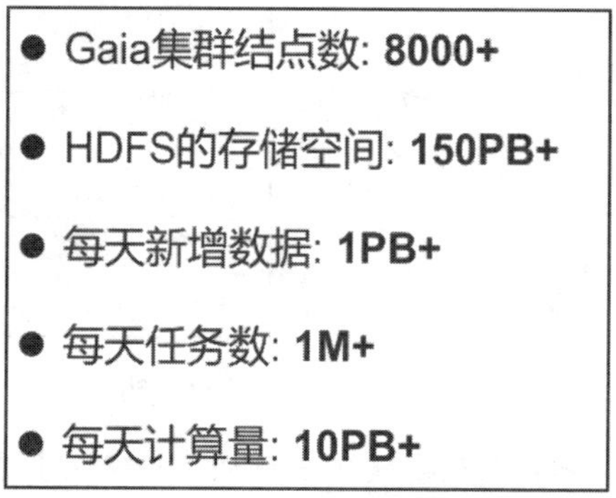

- Gaia集群结点数: **8000+**
- HDFS的存储空间: **150PB+**
- 每天新增数据: **1PB+**
- 每天任务数: **1M+**
- 每天计算量: **10PB+**

图 12-3 腾讯公司 Spark 规模

腾讯公司算是国内较早使用 spark 的公司之一，从 2013 年的 0.6 版本的 spark 开始，到 2015 年峰会时候使用的 1.2.0 版本，其集群的节点规模达到目前世界上单一集群最多的 8000 节点，同时每天新增的数据超过 1PB，每天计算的任务数量超过 10K。如图 12-3 所示。

（2）阿里巴巴公司

阿里巴巴网络技术有限公司（简称：阿里巴巴集团）是以曾担任英语教师的马云为首的 18 人，于 1999 年在中国杭州创立，他们相信互联网能够创造公平的竞争环境，让小企业通过创新与科技扩展业务，并在参与国内或全球市场竞争时处于更有利的位置。阿里巴巴集团经营多项业务，另外也从关联公司的业务和服务中取得经营商业生态系统上的支援。业务和关联公司的业务包括：淘宝网、天猫、聚划算、全球速卖通、阿里巴巴国际交易市场、1688、阿里妈妈、阿里云、蚂蚁金服、菜鸟网络等。

2014 年 9 月 19 日，阿里巴巴集团在纽约证券交易所正式挂牌上市，股票代码“BABA”，创始人和董事局主席为马云。2015 年全年，阿里巴巴总共营业收入为 943.84 亿元人民币，净利润 688.44 亿元人民币。2016 年 4 月 6 日，阿里巴巴正式宣布已经成为全球最大的零售交易平台。

数据挖掘算法有时候需要迭代，每次迭代时间非常长，这是淘宝选择一个更高性能计算框架 Spark 的原因。Spark 编程范式更加简洁也是一大原因。另外，GraphX 提供图计算的能力也是很重要的。

淘宝于 2012 年初开始在 2012 年初开始建设基于 spark0.5（on mesos）版本的 spark on yarn 的淘宝数据挖掘平台，当时只是用于实验，到了 2012 年中，淘宝建立 standalone 模式的 spark0.6，集群规模为 10 台，2013 年 8 月，0.8 版 spark on yarn 200 台机器的集群，到现在，阿里云梯的规模超过 5000 台。

（3）Yahoo!

在 Spark 技术的研究与应用方面，Yahoo! 始终处于领先地位，它将 Spark 应用于公司的各种产品之中。移动 App、网站、广告服务、图片服务等服务的后端实时处理框架均采用了 Spark＋Shark 的架构。在 2013 年，Yahoo! 拥有 72656600 个页面，有上百万的商品类别，上千个商品和用户特征，超过 800 万用户，每天需要处理海量数据。

通过图 12-4 可以看到 Yahoo! 使用 Spark 进行数据分析的整体框架。

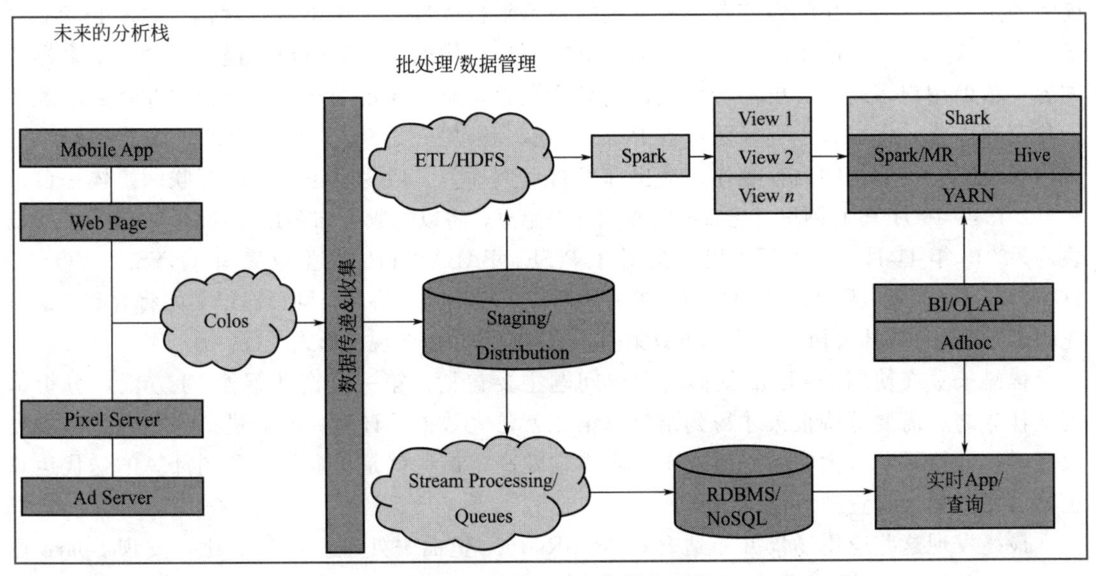

图 12-4　Yahoo! 大数据分析栈

整个数据分析栈构建在 YARN 之上，这是为了让 Hadoop 和 Spark 的任务共存。主要包含两个主要模块。

① 离线处理模块：使用 MapReduce 和 Spark＋Shark 混合架构。由于 MapReduce 适合进行 ETL 处理，还保留 Hadoop 进行数据清洗和转换。数据在 ETL 之后加载进 HDFS/HCat/Hive 数据仓库存储，之后可以通过 Spark、Shark 进行 OLAP 数据分析。

② 实时处理模块：使用 Spark Streaming＋Spark＋Shark 架构进行处理。实时流数据源源不断经过 Spark Steaming 初步处理和分析之后，将数据追加进关系数据库或者 NoSQL 数据库。之后，结合历史数据，使用 Spark 进行实时数据分析。

之所以选择 Spark，Yahoo! 基于以下几点进行考虑。

① 进行交互式 SQL 分析的应用需求。

② RAM 和 SSD 价格不断下降，数据分析实时性的需求越来越多，大数据急需一个内

存计算框架进行处理。

③ 程序员熟悉 Scala 开发，接受 Spark 学习曲线不陡峭。

④ Spark 的社区活跃度高，开源系统的 Bug 能够更快地解决。

⑤ 传统 Hadoop 生态系统的分析组件在进行复杂数据分析和保证实时性方面表现得力不从心。Spark 的全栈支持多范式数据分析能够应对多种多样的数据分析需求。

⑥ 可以无缝将 Spark 集成进现有的 Hadoop 处理架构。

Yahoo! 的 Spark 集群在 2013 年已经达到 9.2TB 持久存储、192GBRAM、112 节点（每节点为 SATA1×500GB（7200 转的硬盘））、400GB SSD（1×400GB SATA 300MB/s）的集群规模。

(4) 优酷土豆

优酷土豆股份有限公司（2015 年 8 月 6 日更名为合一集团）是中国网络视频行业的领军企业，专注于视频领域，旗下拥有中国主要的两个大型视频网站，即优酷和土豆。

2012 年 8 月 23 日由优酷（2010 年于纽约证券交易所上市）和土豆（2011 年于纳斯达克上市）以 100%换股方式合并而成。古永锵出任集团董事长兼首席执行官。优酷土豆拥有庞大的用户群、多元化的内容资源及强大的技术平台优势。优酷土豆月度用户规模已突破 4 亿，意味着已有 1/3 的中国人成为优酷土豆的用户。优酷土豆为用户群提供最全、最多样的内容，帮助用户多终端、更便捷地观赏高品质视频，充分满足用户日益增长的互动需求及多元化视频体验。优酷、土豆现已覆盖 PC、电视、移动三大终端，兼具影视、综艺和资讯三大内容形态，贯通视频内容制作、播出和发行三大环节，成为真正意义的互联网媒体平台。

2015 年 10 月 16 日阿里巴巴宣布将收购优酷土豆，将以每股美国存托股票 26.60 美元的价格收购。2015 年 11 月 6 日，阿里巴巴集团（NYSE：BABA）和优酷土豆集团（NYSE：YOKU）宣布，双方已经就收购优酷土豆股份签署并购协议，根据这一协议，阿里巴巴集团将收购中国领先的多屏娱乐和媒体公司——优酷土豆集团。这项交易将以全现金形式进行。

优酷土豆在使用 Hadoop 集群的突出问题主要包括：第一是商业智能 BI 方面，分析师提交任务之后需要等待很久才得到结果；第二就是大数据量计算，比如进行一些模拟广告投放之时，计算量非常大的同时对效率要求也比较高，最后就是机器学习和图计算的迭代运算也是需要耗费大量资源且速度很慢。

最终发现这些应用场景并不适合在 MapReduce 里面去处理。通过对比，发现 Spark 性能比 MapReduce 提升很多。交互查询响应快，性能比 Hadoop 提高若干倍；模拟广告投放计算效率高、延迟小（同 hadoop 比延迟至少降低一个数量级）；机器学习、图计算等迭代计算，大大减少了网络传输、数据落地等，极大地提高了计算性能。目前 Spark 已经广泛使用在优酷土豆的视频推荐（图计算）、广告业务等。

12.3 Spark 生态系统简介

Hadoop 将 Spark 作为自己生态圈的一部分，但 Spark 完全可以脱离 Hadoop 平台，不单依赖于 HDFS、Yarn，例如它可以使用 Standalone、Mesos 进行集群资源管理，它的包容性使得 Spark 拥有众多的源码贡献者和使用者，其生态系统也日益繁荣。Spark 官方组件如图 12-5 所示。

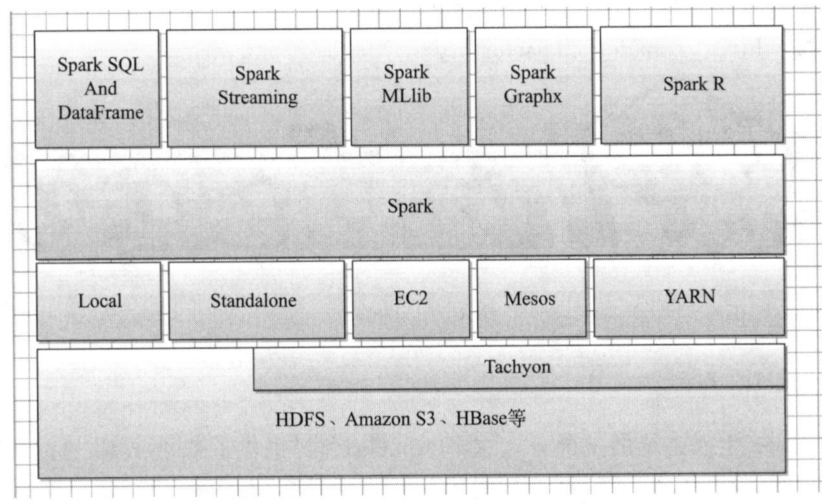

图 12-5 Spark 生态系统

12.3.1 Hadoop 生态系统

作为目前主流的大数据解决方案的重要组成部分，Hadoop 的生态系统是不得不说的一个部分，如图 12-6 所示。

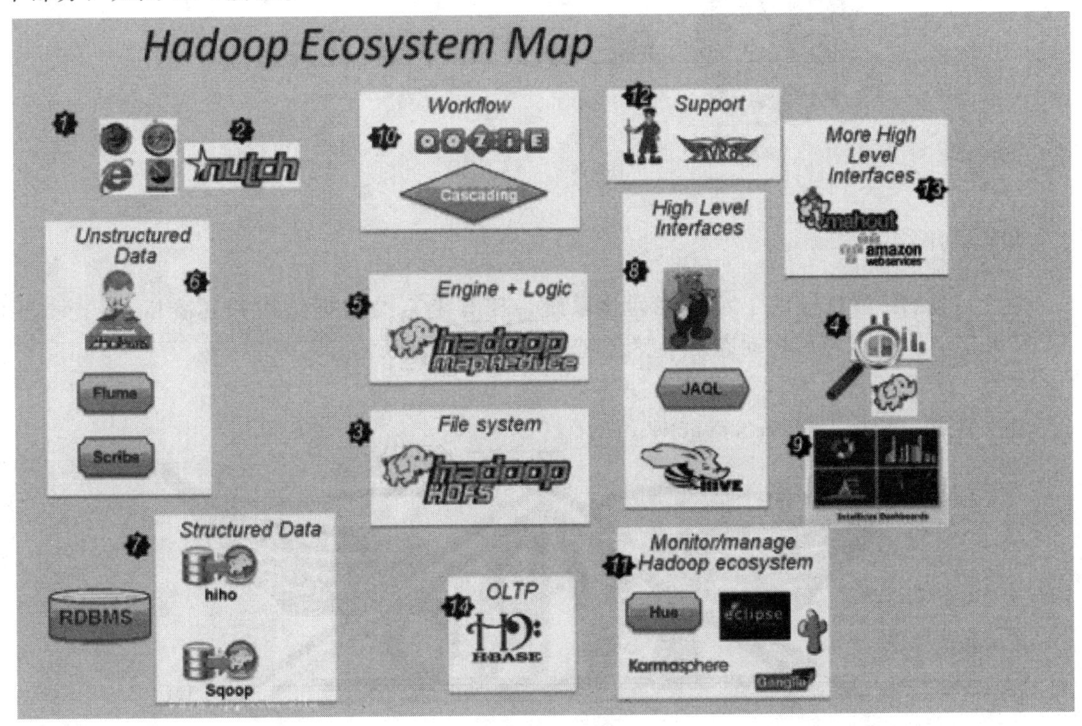

图 12-6 Hadoop 生态系统

(1) Hadoop

Apache 的 Hadoop 项目已几乎与大数据划上了等号。它不断壮大起来，已成为一个完整的生态系统，众多开源工具面向高度扩展的分布式计算。如图 12-7 所示。

支持的操作系统：Windows、Linux 和 OS X。
相关链接：http：//hadoop. apache. org。

图 12-7 Hadoop

(2) Ambari

作为 Hadoop 生态系统的一部分，这个 Apache 项目提供了基于 Web 的直观界面，可用于配置、管理和监控 Hadoop 集群。有些开发人员想把 Ambari 的功能整合到自己的应用程序当中，Ambari 也为他们提供了充分利用 REST（代表性状态传输协议）的 API。如图 12-8所示。

支持的操作系统：Windows、Linux 和 OS X。
相关链接：http：//ambari. apache. org。

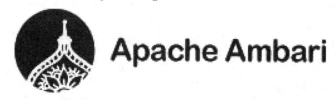

Apache Ambari

图 12-8 Ambari

(3) Avro

这个 Apache 项目提供了数据序列化系统，拥有丰富的数据结构和紧凑格式。模式用 JSON 来定义，它很容易与动态语言整合起来。如图 12-9 所示。

支持的操作系统：与操作系统无关。
相关链接：http：//avro. apache. org。

图 12-9 Avro

(4) Cascading

Cascading 是一款基于 Hadoop 的应用程序开发平台。提供商业支持和培训服务。

支持的操作系统：与操作系统无关。
相关链接：http：//www. cascading. org/projects/cascading/。

(5) Chukwa

Chukwa 基于 Hadoop，可以收集来自大型分布式系统的数据，用于监控。它还含有用于分析和显示数据的工具。如图 12-10 所示。

支持的操作系统：Linux 和 OS X。

相关链接：http：//chukwa. apache. org。

图 12-10 Chukwa

图 12-11 Flume

(6) Flume

Flume 可以从其他应用程序收集日志数据，然后将这些数据送入到 Hadoop。官方网站声称，"它功能强大、具有容错性，还拥有可以调整优化的可靠性机制和许多故障切换及恢复机制。"如图 12-11 所示。

支持的操作系统：Linux 和 OS X。

相关链接：https：//cwiki. apache. org/confluence/display/FLUME/Home。

(7) HBase

HBase 是为有数十亿行和数百万列的超大表设计的，这是一种分布式数据库，可以对大数据进行随机性地实时读取/写入访问。它有点类似谷歌的 Bigtable，不过基于 Hadoop 和 Hadoop 分布式文件系统（HDFS）而建。如图 12-12 所示。

支持的操作系统：与操作系统无关。

相关链接：http：//hbase. apache. org。

(8) Hadoop 分布式文件系统（HDFS）

HDFS 是面向 Hadoop 的文件系统，不过它也可以用作一种独立的分布式文件系统。它基于 Java，具有容错性、高度扩展性和高度配置性。如图 12-13 所示。

支持的操作系统：Windows、Linux 和 OS X。

相关链接：https：//hadoop. apache. org/docs/stable/hadoop-project-dist/hadoop-hdfs/HdfsUserGuide. html。

图 12-12　Hbase

图 12-13　Hadoop

(9) Hive

Apache Hive 是面向 Hadoop 生态系统的数据仓库。它让用户可以使用 HiveQL 查询和管理大数据，这是一种类似 SQL 的语言。如图 12-14 所示。

支持的操作系统：与操作系统无关。

相关链接：http：//hive. apache. org。

图 12-14　Hive

(10) Hivemall

Hivemall 结合了面向 Hive 的多种机器学习算法。它包括诸多高度扩展性算法，可用于数据分类、递归、推荐、k 最近邻、异常检测和特征。

支持的操作系统：与操作系统无关。

相关链接：https：//github. com/myui/hivemall。

(11) Mahout

据官方网站声称，Mahout 项目的目的是“为迅速构建可扩展、高性能的机器学习应用程序打造一个环境。”它包括用于在 Hadoop MapReduce 上进行数据挖掘的众多算法，还包括一些面向 Scala 和 Spark 环境的新颖算法。如图 12-15 所示。

支持的操作系统：与操作系统无关。

相关链接：http：//mahout. apache. org。

图 12-15 Mahout

(12) MapReduce

作为 Hadoop（图 12-16）一个不可或缺的部分，MapReduce 这种编程模型为处理大型分布式数据集提供了一种方法。它最初是由谷歌开发的，但现在也被本文介绍的另外几个大数据工具所使用，包括 CouchDB、MongoDB 和 Riak。

支持的操作系统：与操作系统无关。

相关链接：http：//hadoop. apache. org/docs/current/hadoop-mapreduce-client/hadoop-mapreduce-client-core/MapReduceTutorial. html。

图 12-16 Hadoop

(13) Oozie

这种工作流程调度工具是为了管理 Hadoop 任务而专门设计的。它能够按照时间或按照数据可用情况触发任务，并与 MapReduce、Pig、Hive、Sqoop 及其他许多相关工具整合起来。如图 12-17 所示。

支持的操作系统：Linux 和 OS X。

相关链接：http：//oozie. apache. org。

图 12-17 Oozie

(14) Pig

Apache Pig 是一种面向分布式大数据分析的平台。它依赖一种名为 Pig Latin 的编程语言，拥有简化的并行编程、优化和可扩展性等优点。如图 12-18 所示。

支持的操作系统：与操作系统无关。

相关链接：http：//pig. apache. org。

图 12-18 Pig

(15) Sqoop

企业经常需要在关系数据库与 Hadoop 之间传输数据，而 Sqoop 就是能完成这项任务的一款工具。它可以将数据导入到 Hive 或 HBase，并从 Hadoop 导出到关系数据库管理系统（RDBMS）。如图 12-19 所示。

支持的操作系统：与操作系统无关。

相关链接：http：//sqoop. apache. org。

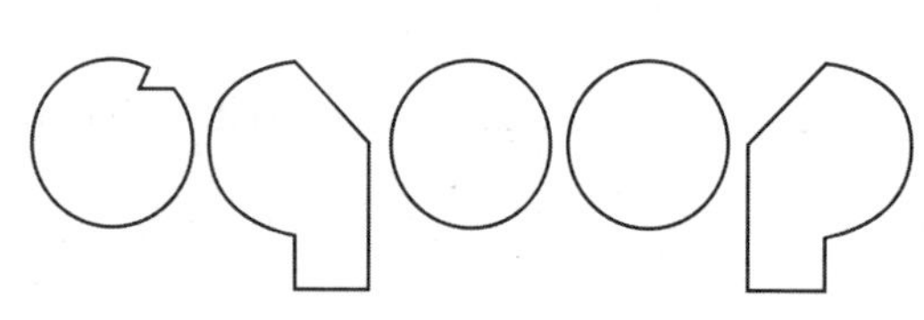

图 12-19 Sqoop

图 12-20 Tez

(16) Tez

Tez 建立在 Apache Hadoop YARN 的基础上，这是“一种应用程序框架，允许为任务构建一种复杂的有向无环图，以便处理数据。”它让 Hive 和 Pig 可以简化复杂的任务，而这些任务原本需要多个步骤才能完成。如图 12-20 所示。

支持的操作系统：Windows、Linux 和 OS X。

相关链接：http：//tez. apache. org。

(17) Zookeeper

这种大数据管理工具自称是“一项集中式服务，可用于维护配置信息、命名、提供分布式同步以及提供群组服务。”它让 Hadoop 集群里面的节点可以彼此协调。如图 12-21 所示。

支持的操作系统：Linux、Windows（只适合开发环境）和 OSX（只适合开发环境）。

图 12-21 Zookeeper

相关链接：http://zookeeper. apache. org

12.3.2 BDAS生态系统

目前，Spark已经发展成为包含众多子项目的大数据计算平台。伯克利将Spark的整个生态系统称为伯克利数据分析栈（BDAS）。其核心框架是Spark，同时BDAS涵盖支持结构化数据SQL查询与分析的查询引擎Spark SQL和Shark，提供机器学习功能的系统MLbase及底层的分布式机器学习库MLlib、并行图计算框架GraphX、流计算框架Spark Streaming、采样近似计算查询引擎BlinkDB、内存分布式文件系统Tachyon、资源管理框架Mesos等子项目。这些子项目在Spark上层提供了更高层、更丰富的计算范式。如图12-22所示。

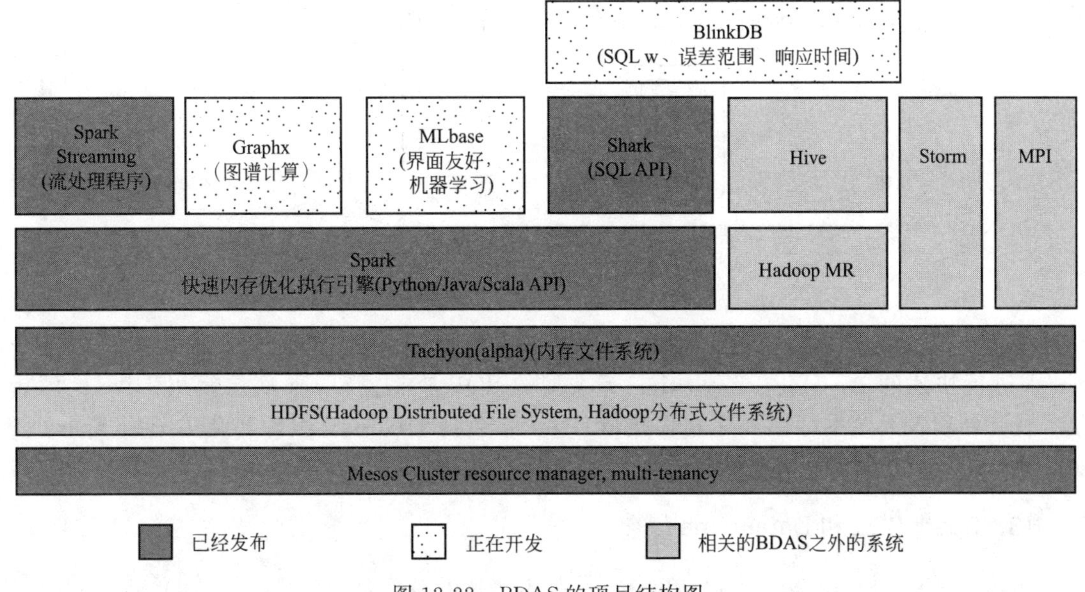

图12-22 BDAS的项目结构图

下面对BDAS的各个子项目进行更详细的介绍。

(1) Spark

Spark是整个BDAS的核心组件，是一个大数据分布式编程框架，不仅实现了MapReduce的算子map函数和reduce函数及计算模型，还提供更为丰富的算子，如filter、join、groupByKey等。Spark将分布式数据抽象为弹性分布式数据集（RDD），实现了应用任务调度、RPC、序列化和压缩，并为运行在其上的上层组件提供API。其底层采用Scala这种函数式语言书写而成，并且所提供的API深度借鉴Scala函数式的编程思想，提供与Scala类似的编程接口。图12-23为Spark的处理流程（主要对象为RDD）。

Spark将数据在分布式环境下分区，然后将作业转化为有向无环图（DAG），并分阶段进行DAG的调度和任务的分布式并行处理。

(2) Shark

Shark是构建在Spark和Hive基础之上的数据仓库。目前，Shark已经完成学术使命，终止开发，但其架构和原理仍具有借鉴意义。它提供了能够查询Hive中所存储数据的一套

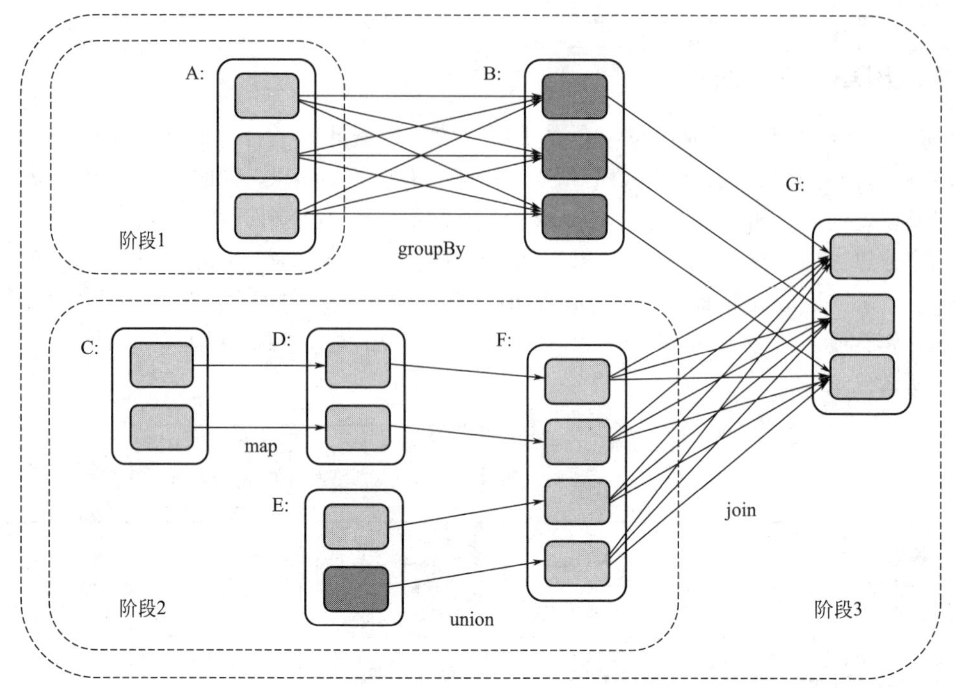

图 12-23　Spark 的任务处理流程图

SQL 接口，兼容现有的 HiveQL 语法。这样，熟悉 Hive QL 或者 SQL 的用户可以基于 Shark 进行快速的 Ad-Hoc、Reporting 等类型的 SQL 查询。Shark 底层复用 Hive 的解析器、优化器以及元数据存储和序列化接口。Shark 会将 Hive QL 编译转化为一组 Spark 任务，进行分布式运算。

（3）Spark SQL 和 DataFrame

2015 年 3 月，Spark 发布了最新的 1.3.0 版本，其中最重要的变化，便是 DataFrame 这个 API 的推出。DataFrame 是一个分布式数据集，在概念上类似于传统数据库的表结构，数据被组织成命名的列，DataFrame 的数据源可以是结构化的数据文件，也可以是 Hive 中的表或外部数据库，也还可以是现有的 RDD。DataFrame 让 Spark 具备了处理大规模结构化数据的能力，在比原有的 RDD 转化方式易用的前提下，计算性能更还快了两倍。这一个小小的 API，隐含着 Spark 希望大一统「大数据江湖」的野心和决心。DataFrame 像是一条联结所有主流数据源并自动转化为可并行处理格式的水渠，通过它 Spark 能取悦大数据生态链上的所有玩家，无论是善用 R 的数据科学家，惯用 SQL 的商业分析师，还是在意效率和实时性的统计工程师。

Spark SQL 提供在大数据上的 SQL 查询功能，类似于 Shark 在整个生态系统的角色，它们可以统称为 SQL on Spark。之前，Shark 的查询编译和优化器依赖于 Hive，使得 Shark 不得不维护一套 Hive 分支，而 Spark SQL 使用 Catalyst 做查询解析和优化器，并在底层使用 Spark 作为执行引擎实现 SQL 的 Operator。用户可以在 Spark 上直接书写 SQL，相当于为 Spark 扩充了一套 SQL 算子，这无疑更加丰富了 Spark 的算子和功能，同时 Spark SQL 不断兼容不同的持久化存储（如 HDFS、Hive 等），为其发展奠定广阔的空间。

（4）Spark Streaming

Spark Streaming 用于进行实时流数据的处理，它具有高扩展、高吞吐率及容错机制，数据来源可以是 Kafka，Flume，Twitter，ZeroMQ，Kinesis 或 TCP，其操作依赖于 discretized stream（DStream），Dstream 可以看作是多个有序的 RDD 组成，因此它也只通过 map，reduce，join window 等操作便可完成实时数据处理，另外一个非常重要的点便是，Spark Streaming 可以与 Spark MLlib、Graphx 等结合起来使用，功能十分强大，似乎无所不能。如图 12-24 所示。

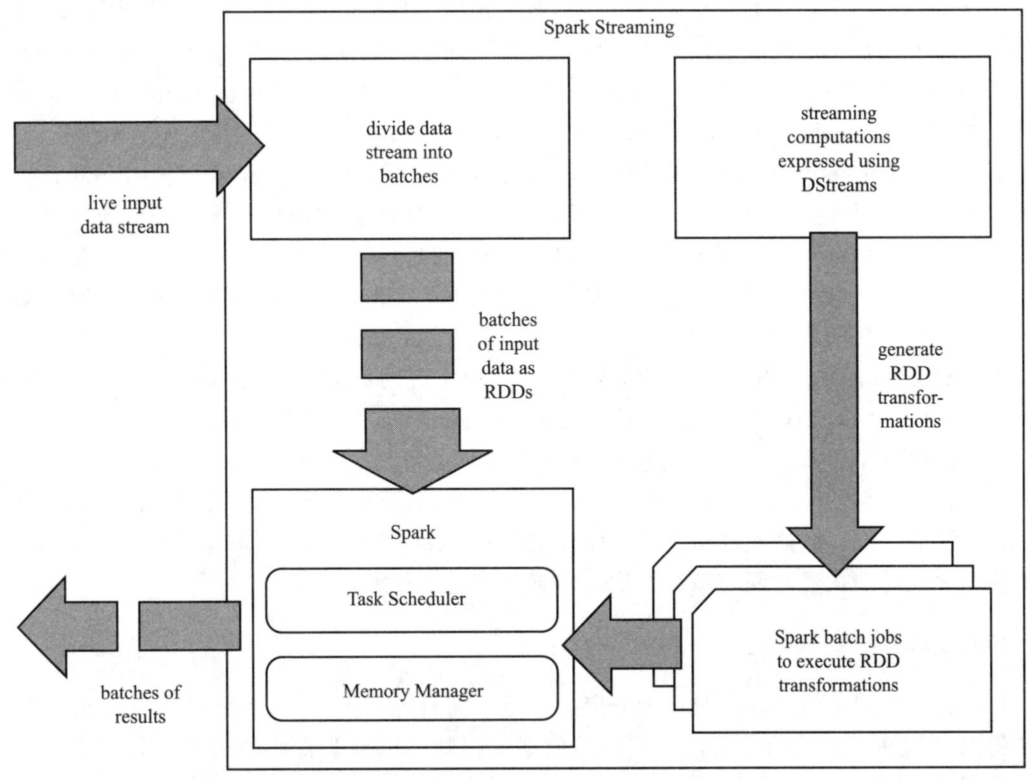

图 12-24　SparkStreaming 流程

（5）GraphX

Spark GraphX 是一个分布式图处理框架，Spark GraphX 基于 Spark 平台提供对图计算和图挖掘的简洁易用而丰富多彩的接口，极大地方便了大家对分布式图处理的需求。

大家都知道，社交网络中人与人之间有很多关系链，例如 Twitter、Facebook、微博、微信，这些都是大数据产生的地方，都需要图计算，现在的图处理基本都是分布式的图处理，而并非单机处理，Spark GraphX 由于底层是基于 Spark 来处理的，因此天然就是一个分布式的图处理系统。

图的分布式或者并行处理其实是把这张图拆分成很多的子图，然后分别对这些子图进行计算，计算的时候可以分别迭代进行分阶段的计算，即对图进行并行计算。

（6）SparkML

Spark 集成了 MLLib 库，其分布式数据结构也是基于 RDD 的，与其它组件能够互通，极大地降低了机器学习的门槛，特别是分布式环境下的机器学习。目前 SparkMLlib 支持下列几种机器学习算法。

① Classification（分类）与 regression（回归）。目前实现的算法主要有：linear models（SVMs，logistic regression，linear regression）、naive Bayes（朴素贝叶斯）、decision trees（决策树）、ensembles of trees（Random Forests and Gradient-Boosted Trees）（组合模型树）、isotonic regression（保序回归）。

② clustering（聚类）。聚类分析是一种探索性的分析，在分类的过程中，人们不必事先给出一个分类的标准，聚类分析能够从样本数据出发，自动进行分类。聚类分析所使用方法的不同，常常会得到不同的结论。不同研究者对于同一组数据进行聚类分析，所得到的聚类数未必一致。目前实现的算法有：k-means、Gaussian mixture、power iteration clustering（PIC）、latent Dirichlet allocation（LDA）、streaming k-means。

③ collaborative filtering（协同过滤）。协同过滤推荐（Collaborative Filtering recommendation）是在信息过滤和信息系统中正迅速成为一项很受欢迎的技术。与传统的基于内容过滤直接分析内容进行推荐不同，协同过滤分析用户兴趣，在用户群中找到指定用户的相似（兴趣）用户，综合这些相似用户对某一信息的评价，形成系统对该指定用户对此信息的喜好程度预测。目前实现的算法只有 alternating least squares（ALS）。

④ dimensionality reduction（特征降维）。目前实现的算法有：singular value decomposition（奇异值分解，SVD）、principal component analysis（主成分分析，PCA）。

除上述机器学习算法之外，还包括一些统计相关算法、特征提取及数值计算等算法。Spark 从 1.2 版本之后，机器学习库作了比较大的发动，Spark 机器学习分为两个包，分别是 mllib 和 ml，ML 把整体机器学习过程抽象成 Pipeline（流水线），避免机器学习工程师在训练模型之前花费大量时间在特征抽取、转换等准备工作上。

（7）SparkR

R 语言在数据分析领域内应用十分广泛，但以前只能在单机环境上使用，Spark R 的出现使得 R 摆脱单机运行的命运，将大量的数据工程师可以以非常小的成本进行分布式环境下的数据分析。

SparkR 是一个 R 语言包，它提供了轻量级的方式使得可以在 R 语言中使用 Apache Spark。在 Spark 1.4 中，SparkR 实现了分布式的 data frame，支持类似查询、过滤以及聚合的操作（类似于 R 中的 data frames：dplyr），但是这个可以操作大规模的数据集。

（8）Tachyon

Tachyon 是一个分布式内存文件系统，可以理解为内存中的 HDFS。为了提供更高的性能，将数据存储剥离 Java Heap。用户可以基于 Tachyon 实现 RDD 或者文件的跨应用共享，并提供高容错机制，保证数据的可靠性。

（9）Mesos

Mesos 是一个资源管理框架，提供类似于 YARN 的功能。用户可以在其中插件式地运行 Spark、MapReduce、Tez 等计算框架的任务。Mesos 会对资源和任务进行隔离，并实现高效的资源任务调度。

（10）BlinkDB

BlinkDB 是一个用于在海量数据上进行交互式 SQL 的近似查询引擎。它允许用户通过在查询准确性和查询响应时间之间做出权衡，完成近似查询。其数据的精度被控制在允许的误差范围内。为了达到这个目标，BlinkDB的核心思想是：通过一个自适应优化框架，随着时间的推移，从原始数据建立并维护一组多维样本；通过一个动态样本选择策略，选择一个适当大小的示例，然后基于查询的准确性和响应时间满足用户查询需求。

12.3.3 其他

(1) Astro

华为开源的Spark SQL on HBase package。Spark SQL on HBase package项目又名Astro，端到端整合了Spark，Spark SQL和HBase的能力，有助于推动帮助Spark进入NoSQL的广泛客户群，并提供强大的在线查询和分析以及在垂直企业大规模数据处理能力。

(2) Apache Zeppelin

开有源的基于Spark的Web交互式数据分析平台，目前Zeppelin还只是孵化项目，它具如下特点：

① 自动流入SparkContext and SQLContext；

② 运行时加载jar包依赖；

③ 停止job或动态显示job进度。

最主要的功能包括：Data Ingestion、Data Discovery、Data Analytics、Data Visualization & Collaboration。

12.4 小结

本章节内容主要带领读者了解Spark发展历史，目前国内外使用情况，以及其生态圈的内容，便于读者根据自己的需要，进行自主的研究和学习。读者通过本章可以初步认识和理解Spark，更为底层的细节将在后续章节详细阐述。

第13章 Spark RDD解密

RDD（Resilient Distributed Dataset）是弹性分布式数据集，该数据集具有容错的功能，可以并行处理的元素，它是一个高度抽象的数据结构，包含多个分区。Spark 建立在抽象的 RDD 之上，使得它可以用一致的方式处理大数据不同的应用场景，把所有需要处理的数据转化成为 RDD，然后对 RDD 进行一系列的算子运算从而得到结果。本章将通过 RDD 的介绍、RDD 的创建方式、RDD API 的解析和 RDD 的持久化及其实战来解密 Spark RDD 的各方面内容。

13.1 浅谈 RDD

RDD（Resilient Distributed Dataset）是弹性分布式数据集，该数据集具有容错的功能，可以并行处理的元素，它是一个高度抽象的数据结构，包含多个分区。其创建方式主要有两大类：外部的文件系统（如共享文件系统，HDFS，HBase）和任何数据源提供 Hadoop Input Format。

此外，RDD 提供了丰富的操作来对集合中的元素进行操作。其支持两种操作类型：Transformations 和 Actions。Transformations 主要是从一个存在的 RDD 产生一个新的 RDD（这也是 Spark 迭代计算的基础），Actions 的操作主要是在数据集上计算之后返回给 Driver。

（1）RDD 依赖

在对 RDD 进行转换操作的过程中，每个操作都会在已有的 RDD 的基础上产生新的 RDD。由于 RDD 的 Lazy 特性，新的 RDD 会依赖于原有的 RDD，这样 RDD 之间就会形成相应的依赖关系。

RDD 的依赖关系分为两大类，如图 13-1 所示。

① 窄依赖（Narrow dependencies）：是指每个父 RDD 的一个 Partition 最多被子 RDD 的一个 Partition 所使用，例如 map、filter、union 等都会产生窄依赖；

② 宽依赖（Wide dependencies）：是指一个父 RDD 的 Partition 会被多个子 RDD 的 Partition 所使用，例如 groupByKey、reduceByKey、sortByKey 等操作都会产生宽依赖。

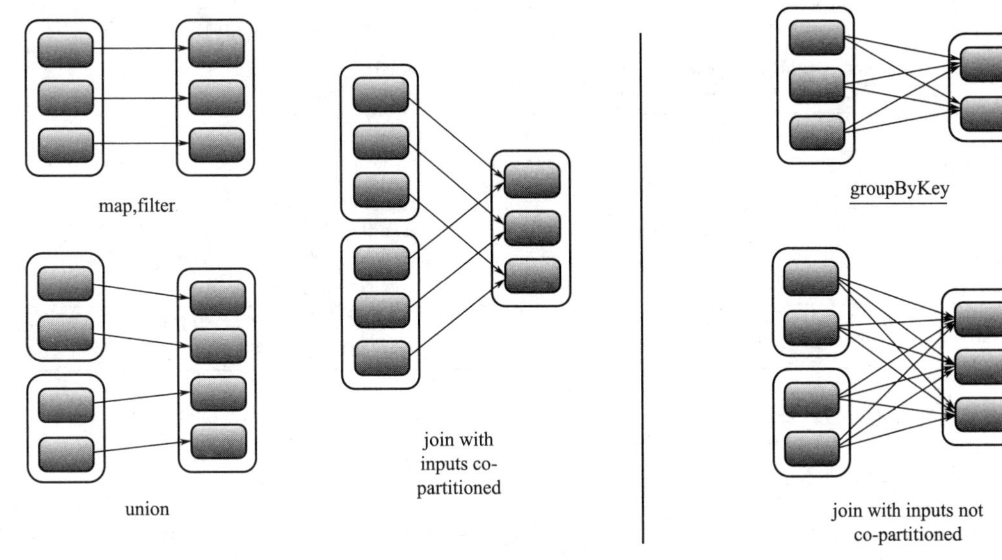

图 13-1 RDD 的依赖关系

实际上，如果父 RDD 的一个 Partition 被一个子 RDD 的 Partition 所使用就是窄依赖，否则的话就是宽依赖。换而言之，如果子 RDD 中的 Partition 对父 RDD 的 Partition 依赖的数量不会随着 RDD 数据规模的改变而改变的话，就是窄依赖，否则的话就是宽依赖。

对于一些特殊的操作，例如 join，如果说 join 操作所使用的每个 Partition 仅仅和已知的 Partition 进行 join，该 join 操作就是窄依赖；其他情况的 join 操作就是宽依赖。

(2) RDD 间的转换关系

RDD 是 Spark 中最重要的部分，是一个不可变、粗粒度的高度抽象的数据集合。在 RDD 抽象数据模型中提供了丰富的转换（Transformation）操作，所有的转换操作都不会实际的执行，由于其 Lazy 特性，只会记录操作的步骤（轨迹），以下列出了 RDD 常用的转换操作，如表 13-1 所示。

表 13-1 RDD 常用的转换操作

转换操作名称	转换操作内容	备注
map	map(f:T=>U):RDD[T]=>RDD[U]	用 f 作用于原有的 RDD 中的每个元素(类型为 T),输出元素类型为 U 的 RDD 集合
filter	filter(f:T=>Bool):RDD[T]=>RDD[T]	过滤复合条件的 RDD 集合
flatMap	flatMap(f:T=>Seq[U]):RDD[T]=>RDD[U]	将 RDD 中的每个集合转化后合并成新的更大的集合
glom	glom():RDD[Array[T]]	将每个分区中的元素形成一个数组
distinct	distinct(numPartitions:Int)(implicitord:Ordering[T]=null):RDD[T]	对每个分区中的元素进行去重
cartesain	cartesian[U:ClassTag](other:RDD[U]):RDD[(T,U)]	对 RDD 中每个数据集中的元素作笛卡尔集
union	union(other:RDD[T]):RDD[T]	将两个 RDD 中的元素合并到一个 RDD 中
mapValues	mapValues[U](f:V=>U):RDD[(K,U)]	仅对集合元素是 K-V 类型中的 value 进行操作

续表

转换操作名称	转换操作内容	备注
subtract	subtract(other:RDD[T]):RDD[T]	去掉两个 RDD 中的相同的元素
sample	sample(withReplacement:Boolean,fraction:Double,seed:Long=Utils. random. nextLong):RDD[T]	对 RDD 中的元素进行采样,之一指定采样的百分比,返回 RDD
takeSample	takeSample(withReplacement:Boolean,num:Int,seed:Long=Utils. random. nextLong):Array[T]	和 Spamle 类似,只是直接返回结果,而非 RDD
groupBy	groupBy[K](f:T=>K,p:Partitioner)(implicitkt:ClassTag[K],ord:Ordering[K]=null):RDD[(K,Iterable[T])]	首先根据传入的 f 产生的 key,形成元素为 K-V 形式的 RDD(PairRDDFunctions),然后调用对 key 值相同的元素进行分组
reduceByKey	reduceByKey(func:(V,V)=>V):RDD[(K,V)]	和 combineByKey 类似,但其返回的 RDD 内部元素类型和原有类型保持一致

说明：Seq [U]，Seq [V] 和 Seq [W] 分别表示类型 U，V 和 W 的序列元素。

(3) RDD 操作类型

RDD 支持两种操作：转换（Transformation）和动作（Action）。其中转换是从已经存在的数据集中创建一个新的数据集，动作是指在数据集上进行计算后返回结果到 Driver。例如，map 是转换操作，是通过函数 f 作用于原有集合的所有元素，得到新的数据集。而 reduce 是一个动作操作，是通过相同的函数来把 RDD 的所有元素聚合起来。

对于转换（Transformation）操作都具有 Lazy 特性，即 Spark 不会立刻进行实际的操作，只会记录执行的流程（或者轨迹），只有出发 Action 操作的时候才会真正执行。这样的设计也是 Spark 高效的一个方面，到真正执行的时候 Spark 根据 DAG 图进行相应的算子合并。

默认的情况下，RDD 的每个动作在执行的时候，都会将之前的数据重新计算一遍，但是为了保证计算的高效性以及计算结果的可重用性。在实际的计算过程中，可以根据实际的情况，在特定的计算环节上进行 persist（或 cache）方法将计算的中间结果持久化到内存或者磁盘上。如果进行了持久化操作，那么在进行 Action 操作的时候，就会从内存或者磁盘将已经计算好的数据取出直接用于后续计算，这样节省了计算步骤和时间，提高整体的计算效率。

(4) RDD 底层实现原理

在 Spark 中 RDD 是一个基本的抽象，它是一个可以被并行计算的不可变的，分片的元素集合。RDD 提供了一系列的基本操作，例如，map，filter，persist 等。

Spark 内部有许多 RDD 的实现，像[[org. apache. spark. rdd. PairRDDFunctions]]只有内部元素为键值对的 RDD 上包含相应的操作；[[org. apache. spark. rdd. DoubleRDDFunctions]]只在内部元素是 Doubles 的 RDD 上提供相应的操作；[[org. apache. spark. rdd. SequenceFileRDDFunctions]]可以被保存为序列化文件。除此之外，也可以自定义 RDD，但是自定义 RDD 必须实现 RDD 接口。

在内部，RDD 都有 5 个重要的特征：一个分区列表、一个计算每个分片的函数、和其他 RDD 一系列的依赖、一个键值 RDDS 分区、一组计算分区的优先位置。

RDD 是一个高度抽象的接口，是整个计算过程中最和基本的操作对象。Spark 中默认的提供了很多种实现来满足各种场景的需要，同时也提供了自定义的方式实现 RDD 的接口。在 RDD 抽象接口包含一些基本的操作。如表 13-2 所示。

表 13-2　RDD 抽象接口包含的一些基本的操作

操作	含义
partitions()	返回一组 Partition 对象
preferredLocations(p)	根据数据存放的位置，返回分区 p 在哪些节点访问更快
dependencies()	返回一组依赖
iterator(p,parentlters)	按照父分区的迭代器，逐个计算分区 p 的元素
partitioner()	返回 RDD 是否 hash/range 分区的元数据信息

(5) RDD 的并行度

RDD 在创建的时候会被划分成多个分区（Partition），在 Spark 进行任务调度的时候会为每个分区创建一个任务。默认情况下，该任务会在集群的一个节点上进行计算（即 Partition 的数量决定了 Stage 中 Task 的数量）。RDD 的并行度的设置要根据实际的情况，如果并行度过低，会导致集群中资源不能够充分利用，形成资源空闲，不能充分发挥集权的优势。如果并行度过高，需要为其创建的任务数也会随之增加，同时会增加读写的次数。

在运行 Spark 程序的时候，RDD 会尝试基于集群的状况来设置分区的数目，所谓的集权状况一般是指集群中低层存储系统的并行度。例如 HDFS 上存储的文件区块。此外，也可以通过 parallelize 的重载方法来手动设置分片的数目。RDD 的并行度也具有继承性，如果没有对 RDD 的分区进行设置，在默认的情况下，子 RDD 会根据父 RDD 的 Partition 决定。例如，当进行 map 操作时，子 RDD 的 Partition 数量会与父 Partition 是保持一致，而进行 Union 操作时，由于只是将两个子 RDD 合并，因此子 RDD 的 Partition 数量是父 Partition 数量和。

(6) RDD 的弹性特性

RDD 之所以被称为弹性数据集，其主要体现在如下几个方面。

① 自动的进行内存和磁盘数据存储的切换。RDD 是基于内存的，但是当内存放不下的时候，会将一部分数据放到磁盘，前提是持久化级别设置成 MEMORY _ AND _ DISK。

② 基于 Lineage 的高效容错。当计算步骤很多的时候，如果是其中某个环节出错，可能从指定的位置恢复已经计算的数据，不需要重新计算。当然可恢复的前提是在相应的位置进行了计算结果数据的持久化。

③ Task 如果失败会自动进行特定次数的重试。

④ Stage 如果失败会自动进行特定次数的重试，而且只会只计算失败的分片。

⑤ checkpoint（持久化）和 persist（检查点）。当就算过程中，有的计算相对复杂，计算链条相对较长或者其结果被经常访问的时候，可以将其结果进行缓存，以便后续直接访问，以此来节省计算时间，提高整体运行速度。

⑥ 数据调度，DAG Task 和资源管理无关。Spark 集群中任务调度和资源调度是分开的。

⑦ 数据分片的高度。在实际的计算过程中，当数据的分片过小时会造成当个节点上的数据很大，容易造成 OOM；而当数据分片过多时，又会导致文件的频繁读取同时计算计算资源在单位时间内资源利用率不高的问题。所以可以根据不同的情况设置不同的分片数量，

提高或者降低并行度。

13.2 创建 RDD 的几种常用方式

（1）使用程序中的集合创建 RDD

RDD 的数据来源可以是程序中的集合，在 Spark 中可以通过 parallelize 和 makeRDD 将集合转化成 RDD，SparkContext 中的 parallelize 方法可以指定分区数的个数，如果没有指定，默认情况下是程序分配到的 CPUCORES 的个数。makeRDD 方法除了实现 parallelize 方法以外，还可以指定 RDD 分区中的优先位置，源码如下所示。

```
def makeRDD[T: ClassTag](
seq: Seq[T],
numSlices: Int = defaultParallelism): RDD[T] = withScope {
parallelize(seq, numSlices)
}
def makeRDD[T: ClassTag](seq: Seq[(T, Seq[String])]): RDD[T] = withScope {
assertNotStopped()
val indexToPrefs = seq.zipWithIndex.map(t => (t._2, t._1._2)).toMap
new ParallelCollectionRDD[T](this, seq.map(_._1), seq.size, indexToPrefs)
}
```

在程序中使用集合来创建 RDD，一般只是用来程序测试的。程序代码如下所示。

```
scala> val numbers = 1 to 20
numbers: scala.collection.immutable.Range.Inclusive = Range(1, 2, 3, 4, 5, 6, 7, 8, 9, 10, 11, 12, 13,
14, 15, 16, 17, 18, 19, 20)
scala> val rdd = sc.parallelize(numbers)
rdd: org.apache.spark.rdd.RDD[Int] =ParallelCollectionRDD[0] at parallelize at <console>:29
scala>  val rdd = sc.makeRDD(numbers)
rdd: org.apache.spark.rdd.RDD[Int] = ParallelCollectionRDD[1] at makeRDD at <console>:29
```

（2）使用本地文件系统创建 RDD

RDD 的数据来源也可以是本地的文件系统，这对于程序中需要进行相对较大的数据量测试是很有必要的。在 Spark 中可以通过 textFile 方法来读取本地文件系统创建 RDD（当然该方法不仅限于本地文件系统，其他情况将在下节内容中说明），其中的路径可以是文件也可以是目录，此外，textFile 提供一个重载的方法，使用该方法创建 RDD 时，可以指定分片数。Spark 为每个文件的每个文件块（块在 HDFS 默认 64MB）创建一个分区。在指定分片的时候，分片的数目不能少于文件块的数目。

使用本地系统创建 RDD，一般用于测试大量的数据文件。程序代码如下所示。

```
scala>  val rdd = sc.textFile("file:///data/spark/data")
16/02/17 11:04:22 INFO storage.MemoryStore: Block broadcast_12 stored as values in memory
(estimated size 229.3 KB, free 3.1 MB)
16/02/17 11:04:22 INFO storage.MemoryStore: Block broadcast_12_piece0 stored as bytes in
memory (estimated size 19.7 KB, free 3.2 MB)
16/02/17 11:04:22 INFO storage.BlockManagerInfo: Added broadcast_12_piece0 in memory on
192.168.1.124:42406 (size: 19.7 KB, free: 517.2 MB)
```

```
16/02/17 11:04:22 INFO spark.SparkContext: Created broadcast 12 from textFile at <console>:27
rdd: org.apache.spark.rdd.RDD[String] = MapPartitionsRDD[28] at textFile at <console>:27
scala> rdd.partitions.size
16/02/17 11:04:41 INFO mapred.FileInputFormat: Total input paths to process : 2
res13: Int = 2
```

(3) 使用 HDFS 创建 RDD

RDD 的数据来源可以是 HDFS，而且从 HDFS 上读取数据来创建 RDD 的方式也是目前 Spark 生产系统中最常用的创建方式。由于 Spark 的思想来源于 Hadoop 的 MapReduce，对于 Hadoop 中的 HDFS 也提供了很好的兼容。Spark 中从 HDFS 创建 RDD 的方法和从本地文件的方法相似，不同的是文件是来源于 HDFS 文件系统。其实除了 HDFS 文件系统外，Spark 兼容 hadoop 支持的所有文件 URI 方式。在执行 textFile 方法时。其内部会调用了 hadoopFile 方法，hadoopFile 方法中会实例化 HadoopRDD，创建 HadoopRDD 的同时会指定其输入的格式、键类型、值类型以及分区值。

使用 HDFS 创建 RDD，是生产环境下最常使用的方式。程序代码如下所示。

```
scala>  val rdd = sc.textFile("hdfs://master:9000/spark/data")
16/02/17 11:36:28 INFO storage.MemoryStore: Block broadcast_1 stored as values in
memory (estimated size 229.3 KB, free 478.2 KB)
16/02/17 11:36:28 INFO storage.MemoryStore: Block broadcast_1_piece0 stored as bytes
in memory (estimated size 19.7 KB, free 498.0 KB)
16/02/17 11:36:28 INFO storage.BlockManagerInfo: Added broadcast_1_piece0 in
memory on 192.168.1.124:59193 (size: 19.7 KB, free: 517.4 MB)
16/02/17 11:36:28 INFO spark.SparkContext: Created broadcast 1 from textFile at
<console>:27
rdd: org.apache.spark.rdd.RDD[String] = MapPartitionsRDD[3] at textFile at <console>:27
```

(4) 使用数据流创建 RDD

数据流也是 RDD 最常见的数据来源。Spark 可以接收实时的数据流，这些数据流的来源包括 kafka、flume、Twitter、ZeroMQ、Kinesis 等。Spark Streaming 接收实时的输入数据流，然后将这些数据按照时间切分，供 Spark 引擎处理，Spark 引擎通过一系列的计算生成相应的数据，最后把数据推送到文件系统、数据库或者 dashboards 中，如图 13-2 所示。

图 13-2　Spark Streaming 输入和输出

在使用数据流创建 RDD 需要依赖于 Spark Streaming 的相关类，在写程序之前需要导入相关的类。程序代码如下所示。

```
/**
 * 创建Spark的配置对象SparkConf,设置Spark程序的运行时的配置信息
 * 例如：通过setMaster来设置程序要链接的Spark集群的Master的URL
 * local[2]:代表Spark本地运行,2表示需要2个cores
 */
val conf = new SparkConf().setMaster("local[2]").setAppName("RDD from Stream")
//根据conf创建StreamingContext， Seconds(1)表示每隔一秒钟对数据流切割作为一个批次
val ssc = new StreamingContext(conf, Seconds(1))
//通过ssc上下文创建DStream，这个DStream将去监听TCP源（主机为localhost，端口为9999）获取流式数据
val lines = ssc.socketTextStream("localhost", 9999)
// 将每一行的数据按照空格切分
val words = lines.flatMap(_.split(" "))
```

(5) 使用其他方式创建 RDD

Spark 对 RDD 的数据来源提供了广泛的支持，除了上几节中提到的创建方式外，RDD 也可从其他的数据库上创建 RDD，例如 HBase、Hive、MySQL 等。

以下通过 HBase 来读取其中的表数据然后创建 RDD。程序代码如下所示。

```
//实例化Spark的配置文件，并根据其创建SparkContext对象
val conf = new SparkConf().setMaster("local[2]").setAppName("RDD from HBase")
val sc = new SparkContext(conf)
//初始化HBase配置
val hconf = HBaseConfiguration.create()
val tableName = "DT_STUDENT"
//设置需要获取数据的表名，hconf可以设置相应的过滤条件
hconf.set(TableInputFormat.INPUT_TABLE,tableName)
//创建HBase访问对象
val admin = new HBaseAdmin(hconf)
//从HBase获得的数据并转化成RDD
val hBaseRDD = sc.newAPIHadoopRDD(hconf,classOf[TableInputFormat],
classOf[org.apache.hadoop.hbase.io.ImmutableBytesWritable],
classOf[org.apache.hadoop.hbase.client.Result])
hBaseRDD.count()
sc.stop() //关闭Spark上下文
admin.close() //关闭连接
```

13.3 Spark RDD API 解析及其实战

RDD 的接口中提供的丰富的转换接口和操作接口，这些接口都是在实际开发中经常用到的。在转换操作中，例如 map、flatMap、filter 等转换操作实现了 monad 模式，可以自定义内部的处理函数。操作接口中，例如 reduce、join、reduceByKey 等会触发 Job 进行任务的执行。本节将对开发中经常用的接口进行详细的讲解。

(1) map[U:ClassTag](f:T=>U):RDD[U]

通过函数 f 将元素类型为 T 的集合转化成元素类型为 U 的集合。map 操作不会改变 RDD

的分区数目，返回 MapPartitionsRDD 类型的 RDD。如图 13-3 所示。其中每个方块代表一个分区。

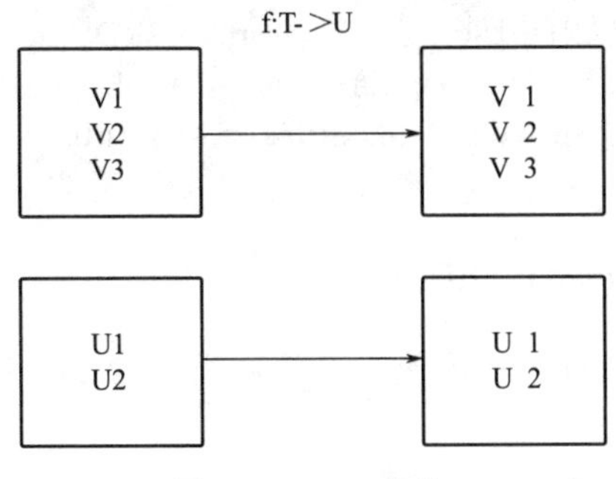

图 13-3　map 操作

以下代码通过集合创建 RDD，并通过 map 操作对边前后的 partitions 数量。程序代码如下所示。

```
scala> val rdd = sc.parallelize(1 to 100)
rdd: org.apache.spark.rdd.RDD[Int] = ParallelCollectionRDD[4] at parallelize at <console>:29
scala> rdd.partitions.size
res3: Int = 8 //操作前RDD的分区数
scala> val mapRDD = rdd.map(_ * 2)
mapRdd: org.apache.spark.rdd.RDD[Int] = MapPartitionsRDD[5] at map at <console>:31
scala> mapRdd.partitions.size
res4: Int = 8 //操作前RDD的分区数
```

从上述结果（res3 和 res4）中可以看出，map 的操作并没有改变 partitions 的数量。

(2) flatMap[U: ClassTag](f:T=>TraversableOnce[U]): RDD[U]

类似于 map，但是在进行元素遍历处理时返回的是多个元素，最终返回 MapPartitionsRDD 类型的 RDD。如图 13-4 所示。

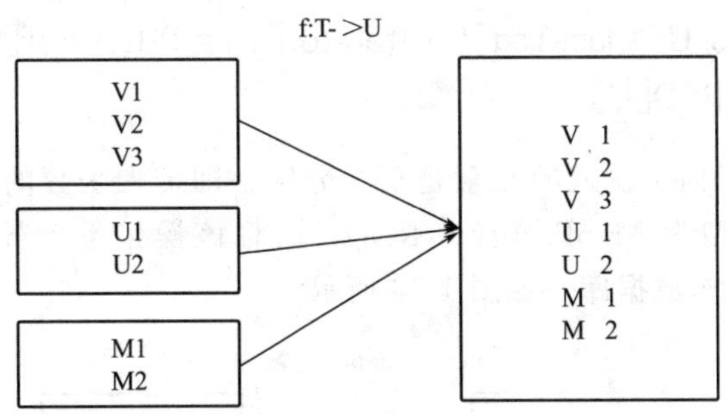

图 13-4　flatMap 操作

以下代码通过本地文件创建 RDD，并通过 flatMap 将每行按照空格分隔成多个单词。程序代码如下所示。

```
scala> val rdd = sc.textFile("file:///data/spark/data")
16/02/23 10:37:50 INFO storage.MemoryStore: Block broadcast_0 stored as values in memory
(estimated size 229.2 KB, free 229.2 KB)
16/02/23 10:37:50 INFO storage.MemoryStore: Block broadcast_0_piece0 stored as bytes
in memory (estimated size 19.7 KB, free 249.0 KB)
```

```
16/02/23 10:37:50 INFO storage.BlockManagerInfo: Added broadcast_0_piece0 in memory on 192.168.1.124:46442 (size: 19.7 KB, free: 517.4 MB)
16/02/23 10:37:50 INFO spark.SparkContext: Created broadcast 0 from textFile at <console>:27
rdd: org.apache.spark.rdd.RDD[String] = MapPartitionsRDD[9] at textFile at <console>:27
scala> val flatMapRDD = rdd.flatMap(line=>line.split(" "))
flatMapRDD: org.apache.spark.rdd.RDD[String] = MapPartitionsRDD[10] at flatMap at <console>:29
```

(3) filter(f:T=>Boolean): RDD[T]

传入的函数要求返回值是 Boolean 类型，返回类型为 MapPartitionsRDD 的 RDD。如图 13-5 所示。

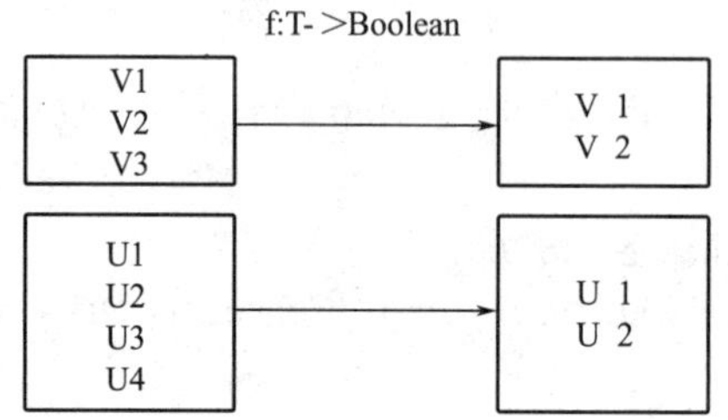

图 13-5 filter 操作

以下代码通过集合创建 RDD，并通过 filter 过滤出元素为偶数的集合。程序代码如下所示。

```
scala> val rdd = sc.parallelize(1 to 10)
rdd: org.apache.spark.rdd.RDD[Int] = ParallelCollectionRDD[11] at parallelize at <console>:27
scala> val filterRDD = rdd.filter(_ % 2 == 0)
filterRDD: org.apache.spark.rdd.RDD[Int] = MapPartitionsRDD[12] at filter at <console>:29
scala> filterRDD.collect
res8: Array[Int] = Array(2,4,6,8,10)
```

(4) mapPartitions[U: ClassTag](f: Iterator[T]=>Iterator[U],preservesPartitioning:Boolean=false): RDD[U]

和 map 功能类似，但是输入的元素是整个分区，即传入函数的操作对象是每个分区的 Iterator 集合，返回类型为 MapPartitionsRDD，而且该操作不会导致 Partitions 数量的变化。该方法一般用于操作数据库。如图 13-6 所示。

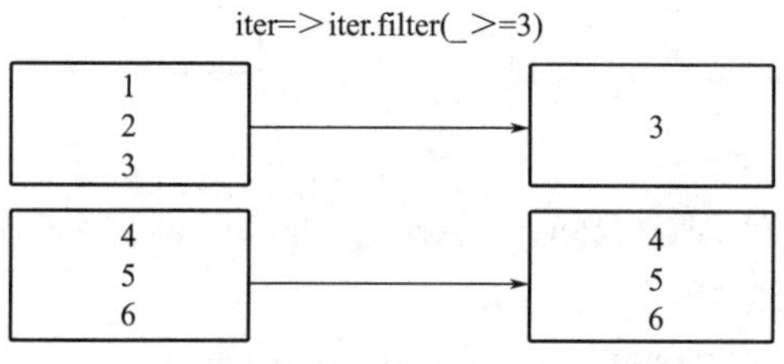

图 13-6 mapPartitions 操作

以下代码通过集合创建 RDD，并通过 mapPartitions 对每个分区中的元进行过滤。程序代码如下所示。

```
scala> val rdd = sc.parallelize(1 to 10)
rdd: org.apache.spark.rdd.RDD[Int] = ParallelCollectionRDD[0] at parallelize at <console>:27
scala> val mapPartitionsRDD = rdd.mapPartitions(iter=>iter.filter(_>3))
mapPartitionsRDD: org.apache.spark.rdd.RDD[Int] = MapPartitionsRDD[1] at mapPartitions at <console>:29
scala> mapPartitionsRDD.collect
res0: Array[Int] = Array(4, 5, 6, 7, 8, 9, 10)
```

(5) glom(): RDD[Array[T]]

将每个分区转化成数组，返回类型为 MapPartitionsRDD 的 RDD。如图 13-7 所示。

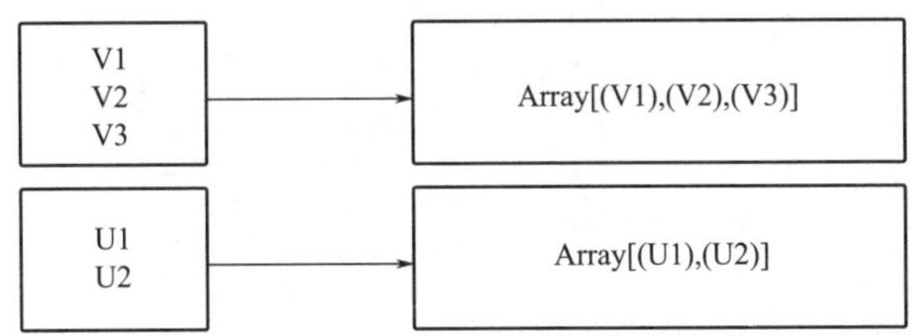

图 13-7 glom 操作

以下代码通过集合创建 RDD，并通过 glom 对每个分区中的元素转化成数组。程序代码如下所示。

```
scala> val rdd = sc.parallelize(1 to 10)
rdd: org.apache.spark.rdd.RDD[Int] = ParallelCollectionRDD[0] at parallelize at <console>:27
scala> val glomRDD = rdd.glom()
glomRDD: org.apache.spark.rdd.RDD[Array[Int]] = MapPartitionsRDD[2] at glom at<console>:29
```

从上述代码中可以看成 glomRDD 返回的是元素是 Array [Int]。

(6) distinct (numPartitions: Int) (implicitord: Ordering [T] = null): RDD[T]

该操作主要是将 RDD 中每个 partitioner 内部重复的元素去除掉，返回 MapPartitionsRDD 类型 RDD，同时该方法可以改变分区的 Partitions 数量。如图 13-8 所示。

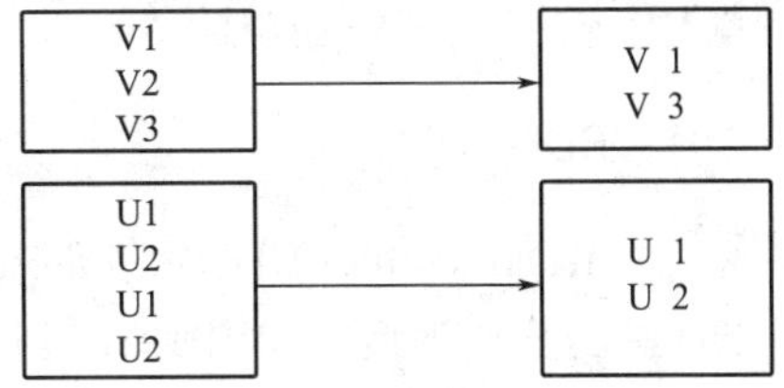

图 13-8 distinct 操作

以下代码通过集合创建 RDD，并通过 distinct 将每个分区中的重复元素过滤掉。程序代码如下所示。

```
scala> val rdd = sc.parallelize(Array(1,2,2,4,5,5,7,6,8))
rdd: org.apache.spark.rdd.RDD[Int] = ParallelCollectionRDD[3] at parallelize at <console>:27
scala> val distinctRDD = rdd.distinct(2)
distinctRDD: org.apache.spark.rdd.RDD[Int] = MapPartitionsRDD[6] at distinct at <console>:29
scala> distinctRDD.collect
res0: Array[Int] = Array(4, 6, 8, 2, 1, 7, 5)
```

(7) cartesian[U: ClassTag] (other: RDD[U]): RDD[(T,U)]

该操作是在两个 RDD 之间，最终会将两个 RDD 中分区元素的笛卡尔积以内部元素类型为 Tuple 形式的 RDD（CartesianRDD）返回。如图 13-9 所示。

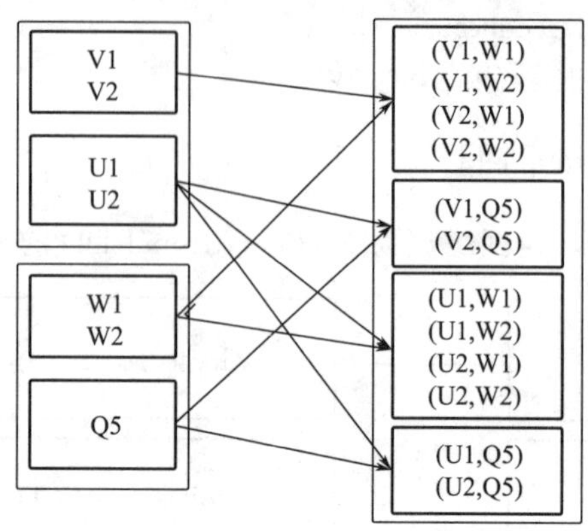

图 13-9　cartesian 操作

以下代码通过集合创建两个 RDD，并通过 cartesian 将两个 RDD 内元素的笛卡尔积返回。程序代码如下所示。

```
scala> val rdd1 = sc.parallelize(Array("A1","A2","A3","A4","A5"))
rdd1: org.apache.spark.rdd.RDD[String] = ParallelCollectionRDD[7] at parallelize at <console>:27
scala> val rdd2 = sc.parallelize(Array("B1","B2","B3","B4","B5"))
rdd2: org.apache.spark.rdd.RDD[String] = ParallelCollectionRDD[8] at parallelize at <console>:27
scala> val cartesianRDD = rdd1.cartesian(rdd2)
16/02/23 14:03:18 INFO spark.ContextCleaner: Cleaned accumulator 1
cartesianRDD: org.apache.spark.rdd.RDD[(String, String)] = CartesianRDD[9] at cartesian at
<console>:31
scala> cartesianRDD.collect
res0: Array[(String, String)] = Array((A1,B1), (A1,B2), (A1,B3), (A1,B4), (A1,B5), (A2,B1),
(A2,B2), (A2,B3), (A2,B4), (A2,B5), (A3,B1), (A3,B2), (A3,B3), (A3,B4), (A3,B5), (A4,B1),
(A4,B2), (A4,B3), (A4,B4), (A4,B5), (A5,B1), (A5,B2), (A5,B3), (A5,B4), (A5,B5))
```

(8) union(other: RDD[T]): RDD[T]

该操作将两个 RDD 合并成一个 RDD（UnionRDD）。union 和＋＋运算相同，主要用于不同的数据来源，但要保证 RDD 中的数据类型。如图 13-10 所示。

以下代码通过集合创建两个 RDD，并通过 union 将两个 RDD 合并成一个 RDD。程序代码如下所示。

```
scala> val rdd1 = sc.parallelize(Array("A1","A2","A3","A4","A5"))
rdd1: org.apache.spark.rdd.RDD[String] = ParallelCollectionRDD[7] at parallelize at<console>:27
scala> val rdd2 = sc.parallelize(Array("B1","B2","B3","B4","B5"))
rdd2: org.apache.spark.rdd.RDD[String] = ParallelCollectionRDD[8] at parallelize at<console>:27
scala> val unionRDD = rdd1.union(rdd2)
unionRDD: org.apache.spark.rdd.RDD[String] = UnionRDD[10] at union at <console>:31
scala> unionRDD.collect
res0: Array[String] = Array(A1, A2, A3, A4, A5, B1, B2, B3, B4, B5)
```

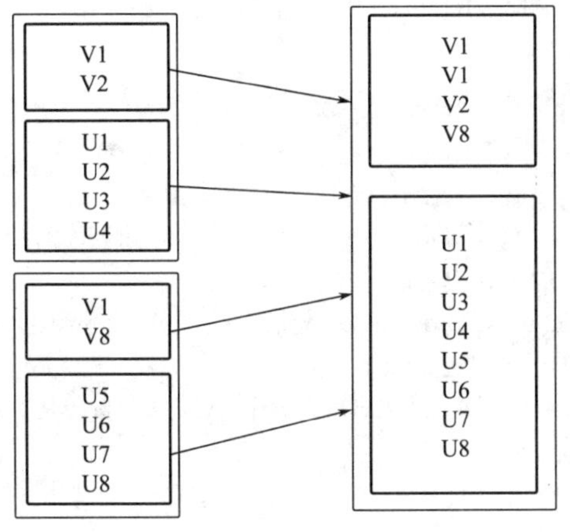

图 13-10　union 操作

(9) mapValues[U](f:V=>U): RDD[(K,U)]

该方法是位于 RDD 的具体实现（PairRDDFunctions）中，其只对内部元素是 K-V 形式的 RDD 中的 Value 进行操作，不会对 Key 操作，返回类型为 MapPartitionsRDD 的 RDD。如图 13-11 所示。

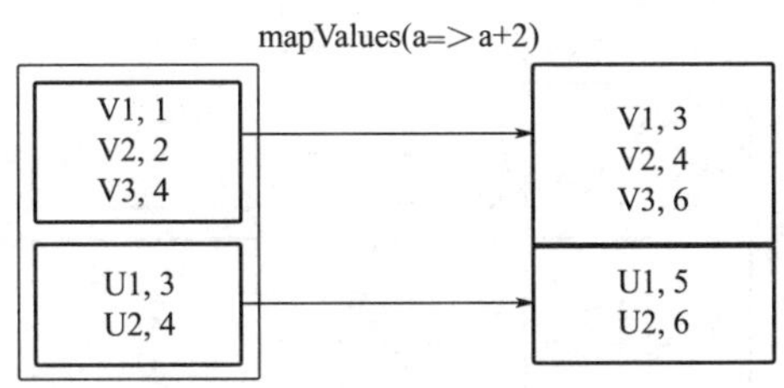

图 13-11　mapValues 操作

以下代码通过集合创建 RDD，并通过 mapValues 对其中的 value 进行乘 2 操作。程序代码如下所示。

```
val rdd = sc.parallelize(Array(("A",1),("B",2),("C",3)))
val mapValuesRDD = rdd.mapValues(a=>a*2)
scala> mapValuesRDD.collect
res0: Array[(String, Int)] = Array((A,2), (B,4), (C,6))
```

(10) subtract (other:RDD[T]): RDD[T]

该方法是在两个 RDD 间进行的，其主要获取两个 RDD 之间的差集。如图 13-12 所示。

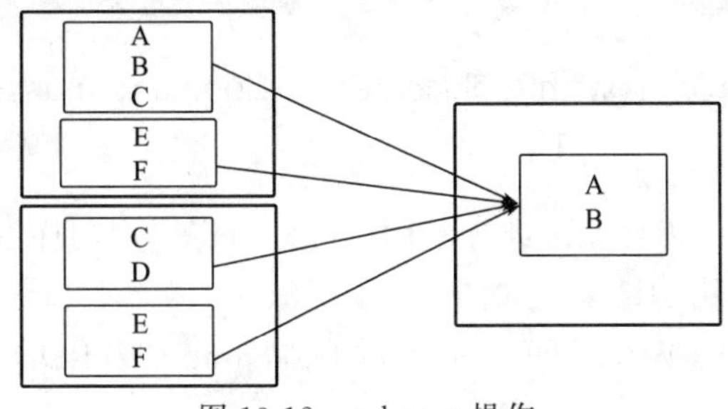

图 13-12　subtract 操作

以下代码通过集合创建两个 RDD，并通过 subtract 将第一个集合中包含第二个集合的元素移除掉。程序代码如下所示。

```
scala> val rdd1 = sc.parallelize(Array("A","B","C","D"))
rdd1: org.apache.spark.rdd.RDD[String] = ParallelCollectionRDD[13] at parallelize at <console>:27
scala> val rdd2 = sc.parallelize(Array("C","D","E","F"))
rdd2: org.apache.spark.rdd.RDD[String] = ParallelCollectionRDD[14] at parallelize at <console>:27
scala> val subtractRDD = rdd1.subtract(rdd2)
subtractRDD: org.apache.spark.rdd.RDD[String] = MapPartitionsRDD[18] at subtract at <console>:31
scala> val subtractRDD = rdd2.subtract(rdd1)
subtractRDD: org.apache.spark.rdd.RDD[String] = MapPartitionsRDD[22] at subtract at <console>:31
scala> subtractRDD.collect
res3: Array[String] = Array(A, B)
```

（11） sample（withReplacement: Boolean, fraction: Double, seed: Long = Utils.random.nextLong）:RDD[T]

该方法是对集合中的元素进行取样，返回 RDD 中元素集合的子集，可以指定其取出元素放回的百分比以及随机的种子元素，返回的 RDD 类型为 PartitionwiseSampledRDD。如图 13-13 所示。

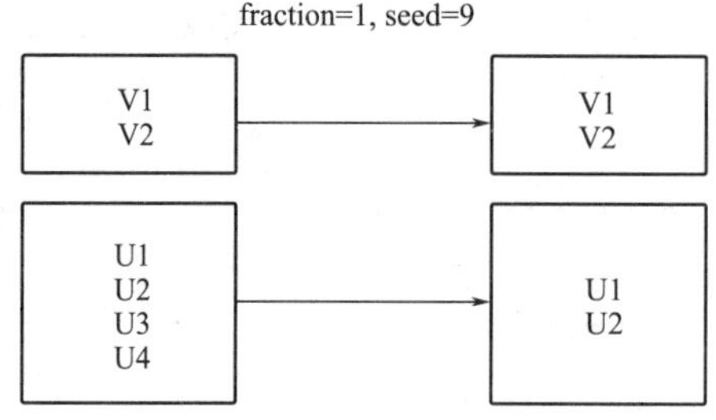

图 13-13 sample 操作

以下代码通过集合创建两个 RDD，并通过 sample 从 RDD 集合中随机取出元素。程序代码如下所示。

```
scala> val rdd = sc.parallelize(Array("A","B","C","D"))
rdd: org.apache.spark.rdd.RDD[String] = ParallelCollectionRDD[0] at parallelize at<console>:27
scala> val sampleRDD = rdd.sample(true,0.5,3)
sampleRDD: org.apache.spark.rdd.RDD[String] = PartitionwiseSampledRDD[1] at sample at <console>:29
scala> sampleRDD.collect
res0: Array[String] = Array(C, D)
```

（12） takeSample（withReplacement: Boolean, num: Int, seed: Long = Utils.random.nextLong）:Array[T]

该方法也是对 RDD 中集合元素进行采用，可以指定采用的样本个数，同时其最终的产生的是 Array，包含其采用的结果。如图 13-14 所示。

以下代码通过集合创建两个 RDD，并通过 takeSample 从 RDD 集合中随机取出 3 元素。程序代码如下所示。

```
1. scala> val rdd = sc.parallelize(Array("A","B","C","D"))
2. rdd: org.apache.spark.rdd.RDD[String] = ParallelCollectionRDD[0] at parallelize at <console>:27
3. scala>val takeSampleRDD = rdd.takeSample(true,3,3)
4. takeSampleRDD: Array[String] = Array(A, A, D)
```

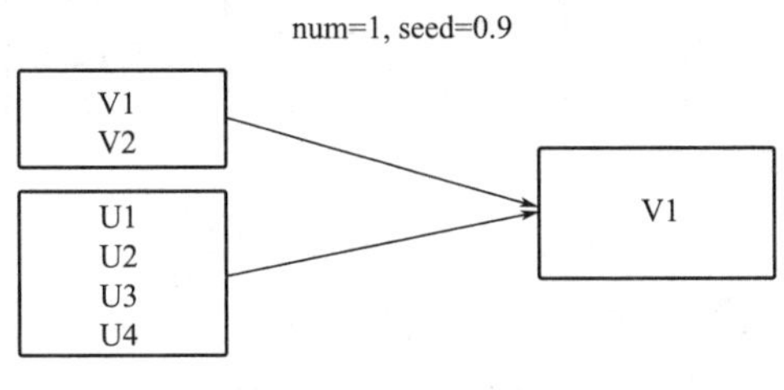

图 13-14 takeSample 操作

(13) groupBy[K](f:T=>K,p:Partitioner)(implicitkt:ClassTag[K],ord:Ordering[K]=null):RDD[(K,Iterable[T])]

该方法首先根据传入的 f 产生的 key，形成元素为 K-V 形式的 RDD（PairRDDFunctions），然后调用 groupByKey 对 key 值相同的元素进行分组。如图 13-15 所示。

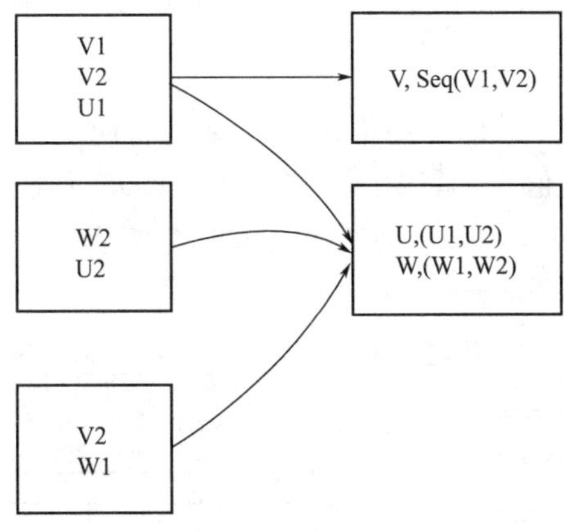

图 13-15 groupBy 操作

以下程序通过集合创建 RDD，并将集合中的没给元素的首字母作为每个元素的 key，然后根据该 key 进行分组。程序代码如下所示。

```
1. scala> val rdd = sc.parallelize(Array("V1","V2","U1","W2","U2","V2","W1"))
2. rdd: org.apache.spark.rdd.RDD[String] = ParallelCollectionRDD[0] at parallelize at <console>:27
3. scala> val groupByRDD = rdd.groupBy(_.substring(0,1))
4. groupByRDD: org.apache.spark.rdd.RDD[(String, Iterable[String])] = ShuffledRDD[2] at
   groupByat <console>:29
5. res0: Array[(String, Iterable[String])] = Array((U,CompactBuffer(U1, U2)),
   (V,CompactBuffer(V1, V2, V2)), (W,CompactBuffer(W2, W1)))
```

(14) partitionBy(partitioner: Partitioner): RDD[(K, V)]

该方法是位于 RDD 的具体实现（PairRDDFunctions）中，其只适用于内部元素是 K-V 形式的 RDD，主要是将 RDD 进行重新分区，如果分区结果与之前的一直则返回自身，否则会产生 ShuffledRDD 类型的 RDD。如图 13-16 所示。

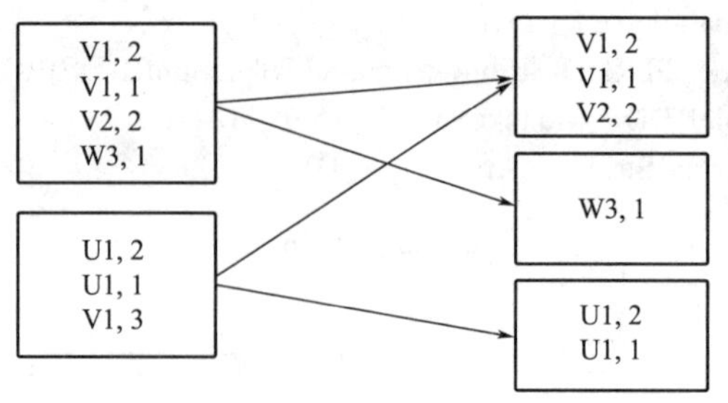

图 13-16 partitionBy 操作

以下通过集合创建 RDD，通过 partitionBy 对 RDD 进行分区。程序代码如下所示。

```
scala> val rdd = sc.parallelize(Array(("V1",2),("V1",1),("V2",2),("W3",1),("U1",2),("U1",1),("U1",3)))
rdd: org.apache.spark.rdd.RDD[(String, Int)] = ParallelCollectionRDD[4] at parallelize at <console>:30
scala> import org.apache.spark._
import org.apache.spark._
scala> val partitionByRDD = rdd.partitionBy(new HashPartitioner(3))
partitionByRDD: org.apache.spark.rdd.RDD[(String, Int)] = ShuffledRDD[6] at
partitionBy at <console>:32
```

(15) cogroup[W](other:RDD[(K,W)],partitioner:Partitioner):RDD[(K,(Iterable[V],Iterable[W]))]

该方法只适用于元素类型为 K-V 的 RDD，主要是将两个 RDD 中的元素 Key 值相同的元素行合并形成新的 K-V 键值对，其中 value 是每个 RDD 元素集合的迭代器构成的 Tuple 类型的元素，返回 MapPartitionsRDD 类型的 RDD。如图 13-17 所示。

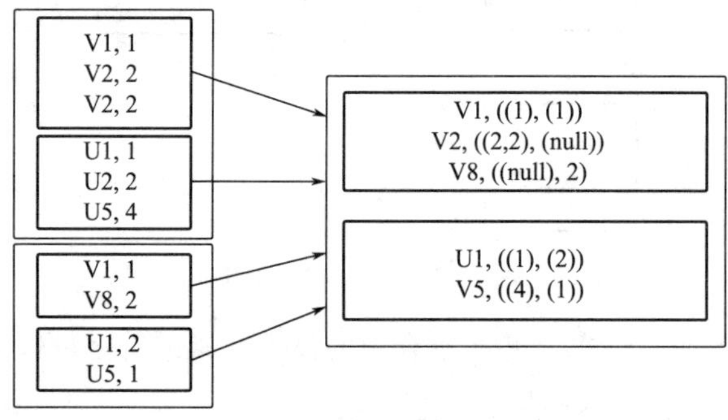

图 13-17 cogroup 操作

以下通过集合创建两个 RDD，通过 cogroup 将两个 RDD 构建成新的 RDD 集合。程序代码如下所示。

```
scala> val rdd1 = sc.parallelize(Array(("V1",1),("V2",2),("V2",2),("U1",2),("U2",1),("U5",4)))
rdd1: org.apache.spark.rdd.RDD[(String, Int)] = ParallelCollectionRDD[7] at parallelize at <
console >:30
scala> val rdd2 = sc.parallelize(Array(("V1",1),("V8",2),("U1",2),("U5",1)))
rdd2: org.apache.spark.rdd.RDD[(String, Int)] = ParallelCollectionRDD[8] at parallelize at
<console>:30
scala> val cogroupRDD = rdd1.cogroup(rdd2)
cogroupRDD: org.apache.spark.rdd.RDD[(String, (Iterable[Int], Iterable[Int]))] = MapPartitionsRDD
[10] at cogroup at <console>:34
```

(16) combineByKey[C](createCombiner:V=>C,mergeValue:(C,V)=>C,mergeCombiners:(C,C)=>C):RDD[(K,C)]

该方法只适用于元素类型为 K-V 形式的 RDD，它将每个分区中的元素按照 Key 合并。最后返回 ShuffledRDD 类型的 RDD。如图 13-18 所示。

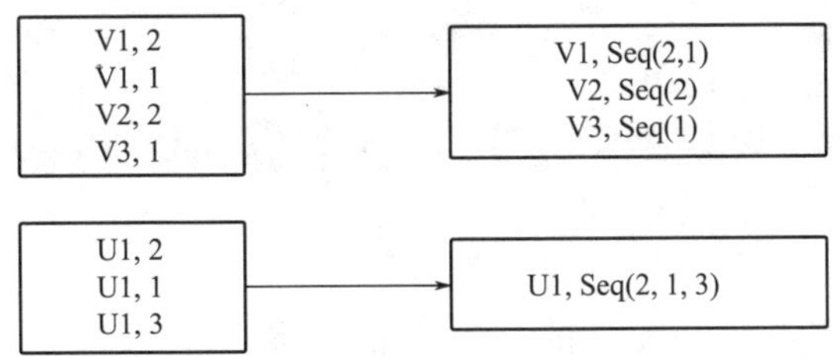

图 13-18 combineByKey 操作

以下通过集合创建 RDD，并通过将 Key 相同的元素组合成为 Key，以其中的元素为集合的键值对作为 RDD 内部元素的新类型。程序代码如下所示。

```
scala> val rdd =sc.parallelize(Array(("V1",1),("V1",2),("V2",2),("U3",1),("U1",1),("U2",2),("V2",2),("U3",1)))
rdd:org.apache.spark.rdd.RDD[(String,Int)] = ParallelCollectionRDD[0] at parallelize at <console>:27
scala> val combineByKeyRDD = rdd.combineByKey((v : Int) =>List(v), (c : List[Int], v : Int) => v::c,  (c1 : List[Int],c2:List[Int])=> c1:::c2 )
combineByKeyRDD: org.apache.spark.rdd.RDD[(String, List[Int])] = ShuffledRDD[2] at combineByKey at <console>:29
scala> combineByKeyRDD.collect
res0: Array[(String, List[Int])] = Array((V1,List(1,2)),(U1,List(1)),(V2,List(2,2)),(U2,List(2)),(U3,List(1,1)))
```

(17) reduceByKey(func:(V,V)=>V):RDD[(K,V)]

该方法和 combineByKey 类似，但其返回的 RDD 内部元素类型和原有类型保持一致。如图 13-19 所示。

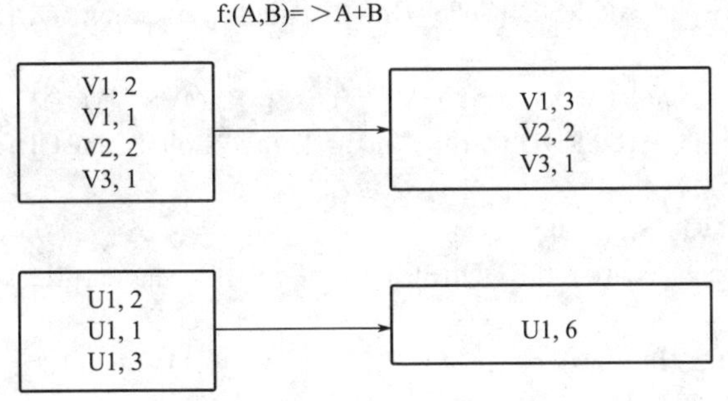

图 13-19 reduceByKey 操作

以下通过集合创建 RDD，并通过将 Key 相同的 Value 相加。程序代码如下所示。

```
scala> val rdd = sc.parallelize(Array(("V1",1),("V1",2),("V2",2),("U3",1),("U1",1),("U2",2),("V2",2),("U3",1)))
rdd: org.apache.spark.rdd.RDD[(String, Int)] = ParallelCollectionRDD[0] at parallelize at <console>:27
scala> val reduceByKey = rdd.reduceByKey(_+_)
reduceByKey: org.apache.spark.rdd.RDD[(String, Int)] = ShuffledRDD[1] at reduceByKey at <console>:29
scala> val reduceByKeyRDD = rdd.reduceByKey(_+_)
reduceByKeyRDD: org.apache.spark.rdd.RDD[(String, Int)] = ShuffledRDD[2] at reduceByKey at <console>:29
scala> reduceByKeyRDD.collect
res0: Array[(String, Int)] = Array((V1,3), (U1,1), (V2,4), (U2,2), (U3,2))
```

(18) join[W](other:RDD[(K,W)]):RDD[(K,(V,W))]

该方法只适用于元素类型为 K-V 的 RDD，他是将两个 RDD 中 Key 相同的元素先合并成以 Key 为 Key，以每个 RDD 中该 Key 的元素的集合为集合（内部是调用 cogroup 方法），追后将集合中的元素取出形成一个大的集合。如图 13-20 所示。

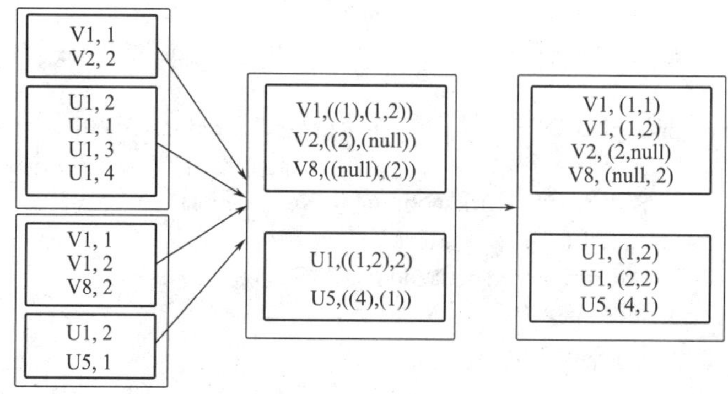

图 13-20　join 操作

以下通过集合创建两个 RDD，并通过 join 将两个 RDD 内部元素进行合并。程序代码如下所示。

```
scala> val rdd1 = sc.parallelize(Array(("V1",1),("V1",2),("V2",2),("U1",1),("U1",2),("U1",3),("U5",4)))
rdd1: org.apache.spark.rdd.RDD[(String, Int)] = ParallelCollectionRDD[3] at parallelize at <console>:27
scala> val rdd2 = sc.parallelize(Array(("V1",1),("V1",2),("V8",2),("U1",2),("U5",1)))
rdd2: org.apache.spark.rdd.RDD[(String, Int)] = ParallelCollectionRDD[4] at parallelize at <console>:27
scala> val joinRDD = rdd1.join(rdd2)
joinRDD: org.apache.spark.rdd.RDD[(String, (Int, Int))] = MapPartitionsRDD[7] at join at <console>:31
res0: Array[(String, (Int, Int))] = Array((U5,(4,1)), (V1,(1,1)), (V1,(1,2)), (V1,(2,1)), (V1,(2,2)), (U1,(1,2)), (U1,(3,2)), (U1,(2,2)))
```

PairRDDFunctions 中还提供了如 leftOuterJoin、rightOuterJoin 等方法，它们和 join 的关系如同数据库中的 join 和 leftjoin，rightjoin 类似，此处不再详述。

13.4 RDD的持久化解析及其实战

数据的持久化是Spark中提升性能的重要环节，同时也是容错的重要环节。RDD的持久化是将每个节点上的计算结果保存在内存（或者磁盘）中。如果后续的操作用到该持久化的结果时就不需要重新计算，这样极大地节省了计算时间和资源，挺高整体计算的性能。

在Spark中对数据进行持久的方式有两种：persist和cache。本质上cache只是persist存储级别为StorageLevel. MEMORY _ ONLY的特殊情况。这两种方式仅仅是对相关的RDD进行标记，并不会真正的执行操作，只有当真正的动作操作被触发的时候才会执行。同样的，这也是由于其Lazy特性导致的。其中cache具有容错性，即如果相关节点上的持久化数据丢失，一方面可以通过其副本（缓存时可以指定副本数）找到，另一方面即便丢失，也只有相关的缓存数据丢失，该部分会被自动计算，并不需要对整体进行计算。

RDD的持久化机制可以让开发人员根据不同的场景灵活的指定缓存的级别。在Spark中默认提供以下几种缓存的级别，如表13-3所示。

表13-3 Spark默认的缓存级别

操作	含　义
NONE	不进行持久化。
DISK_ONLY	数据反序列化存储，并只存放在磁盘。
DISK_ONLY_2	和DISK_ONLY相似，但存储两份副本。
MEMORY_ONLY	数据进行反序列化存储，并只存放在内存。默认采用的方式。
MEMORY_ONLY_2	和MEMORY_ONLY相似，但存储两份副本。
MEMORY_ONLY_SER	数据序列化存储，这种方式节省空间但会增加CPU的性能消耗。
MEMORY_ONLY_SER_2	和MEMORY_ONLY_SER相似，但存储两份副本。
MEMORY_AND_DISK	数据反序列化存储，内存不够会存储在磁盘。
MEMORY_AND_DISK_2	和MEMORY_AND_DISK相似，但存储两份副本。
MEMORY_AND_DISK_SER	数据序列化存储，内存不够会存储在磁盘，节省空间但增加CUP消耗。
MEMORY_AND_DISK_SER_2	和MEMORY_AND_DISK_SER相似，但存储两份副本。
OFF_HEAP	数据序列化存储在Tachyon中。Tachyon的这种方式在jvm进行gc时不会对该区域进行垃圾回收。

RDD在持久化时，有时需要对结果进行多副本备份（如MEMORY _ ONLY _ 2），其目的主要是为了防止某些计算特别复杂和耗时的操作产生的数据丢失。如果采用多副本的方式，其中一份数据丢失可以较快从其他机器的内存中读取数据，节省计算时候。这样通过空间换时间的方式实现局部容错，不会影响整体计算的时间。

在Spark整个计算链条中，并不是每步都需要进行持久化操作。相反如果每个步骤都进行持久化操作，会消耗大量的内存（或者磁盘）。因此RDD的持久化需要在特殊的情况下才进行，一般情况下会在如下几个环节进行持久化操作。

① 计算特别耗时的步骤。这个很容易理解，主要是为了避免在重新计算是浪费大量的计算时间。

② 计算链条特别长的情况。计算链条特别长，其实也包含了耗时的情况，以及中间计

算的复杂性。

③ Checkpoint 所在的 RDD。Checkpoint 是沿着 DAG 图从后往前追溯过程中触发的，如果该 RDD 没有进行持久化操作，那么 Checkpoint 执行时会触发新的 Job，导致重新计算。

④ Shuffle 之后。该操作是集群内部需要进行数据（计算结果）的传输，为了防止在传输过程中数据的丢失，需要进行持久化操作。

⑤ Shuffle 之前。该步骤操作是 Spark 框架中默认完成的。

以下针对 RDD 是否持久化来对比执行过程中所需要的计算时间。

⑥ 未进行 persist 操作时的执行时间。

```
scala> var rdd = sc.textFile("hdfs://master:9000/spark/data").flatMap{ line => line.split(" ")}.map
{word => (word,1)}.reduceByKey(_+_);
……
16/02/28 14:28:47 INFO scheduler.DAGScheduler: Job 0 finished: collect at <console>:30, took
4.936123 s
```

⑦ 进行 persist 操作时的执行时间。

```
scala> rdd.persist
res1: org.apache.spark.rdd.RDD[(String, Int)] = ShuffledRDD[4] at reduceByKey at <console>:27
scala> rdd.collect
……
16/02/28 14:44:56 INFO scheduler.DAGScheduler: Job 1 finished: collect at <console>:30, took
0.193746 s
```

通过以上对比可以看出，RDD 在执行持久化操作后，再次执行会很大程度上提高运算速度（实际上是从持久化的数据中读取），在实际生产环境中在适当的位置进行数据持久化，会在整个计算链条中极大地提升性能。

13.5 小结

本章节内容主要描述了以下内容：RDD 之间的依赖、常见转换关系、操作类型、底层实现原理，并行度和弹性特性，创建 RDD 的几种常用方式，以及 RDDAPI 和持久化的解析和相关案例代码实战。通过本章节内容的学习，可以大致了解 SparkRDD 的本质，为后续章节的深入学习打下基础。

第14章 Spark程序之分组TopN开发实战解析

在当前这个数据技术（Data Technology，DT）时代下，随着大量移动端接入互联网和物联网的兴起，数据产出量巨大并且以惊人的速度在增长，以传统的单台机器模式已经无法适应数据量的海量存储和快速计算。Hadoop在这个大背景下应运而生，它提供了HDFS分布式存储和分布式的计算框架MapReduce（后来改为Yarn框架），短短的几年时间里，它已无处不在，事实上它已经成了大数据技术的代名词。然而在人们越来越多地使用Hadoop提供的MapReduce框架处理大数据的时候，却发现它存在许多天生的缺陷，如效率低、编程模型不够灵活、只适合做离线计算等。Spark的出现无疑让诸多大数据计算的从业者和爱好者眼前一亮，它基于内存，并且提供了更加丰富的算子可以更高效和灵活地处理大数据。

在日常生活中经常遇到要统计分类中取前几名的数据，比如在电商网站经常看到在每个销售频道页面较显眼的位置列举前十个最畅销的商品，以吸引浏览者的眼球，促进交易的迅速达成。类似的案例还有很多，本章将动手实战这类统计算法的Spark实现——分组TopN，涉及的主要内容主要包括分别用Scala和Java编写分组TopN程序，并在本地模式运行，以及彻底解密Scala之分组TopN运行的原理。

环境要求：Win7、JDK1.7、Maven3、Scala 2.10、IntelliJ IDEA14。

14.1 分组TopN动手实战

本节将在IntelliJ IDEA新建Maven工程并且动手实战Java和Scala两种语言的分组TopN实现，以及在Local模式运行程序并验证结果。

14.1.1 Java之分组TopN开发实战

A. 准备素材文件TopNGroup.txt

为了简化分组TopN的分析，这里提供了少量的测试数据，数据信息包含以下两个字段，字段名及其描述如下。

① Name：名称。

② Score：得分。

每一列分别对应前面的各个字段：Name，Score。将 TopNGroup. txt 存放在 D 盘根目录下，也可以由自己指定位置。内容如图 14-1 所示。

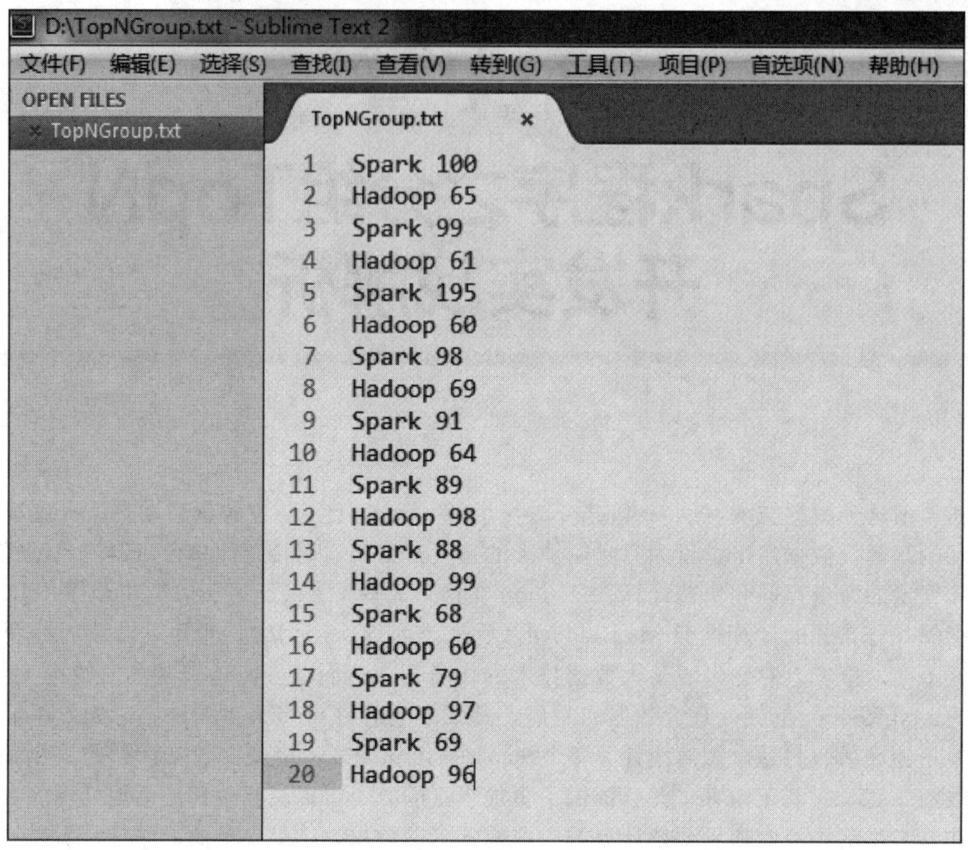

图 14-1 TopNGroup. txt 内容

B. 在 IntelliJ IDEA 新建 Java 版的 TopNGroup 工程，步骤与图中序号匹配，具体步骤如下所示。

(1) 创建 Project

启动 IDEA，点击 Create New Project 创建一个新 Project，如图 14-2 所示。

(2) 选择 Maven 工程

使用 Maven 来管理项目的构建，在本例中主要是在 pom. xml 导入 Spark 的 spark-core 核心包，如图 14-3 中序号 2 所示。

(3) 勾选 Create from archetype

maven archetype 是一个原型构建框架，可以把一些重复性的配置代码放到 archetype 里，不用每次都从头去搭建项目，因此 archetype 可以快速勾勒出项目骨架。要使用 archetype，则需要勾选 “Create from archetype”，如图 14-3 中序号 3 所示。

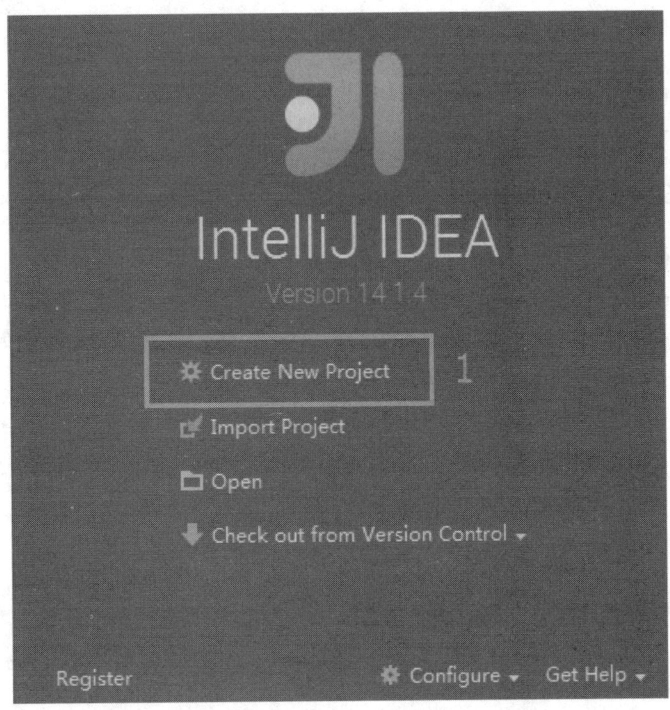

图 14-2　IntelliJ IDEA 创建新 Project 界面

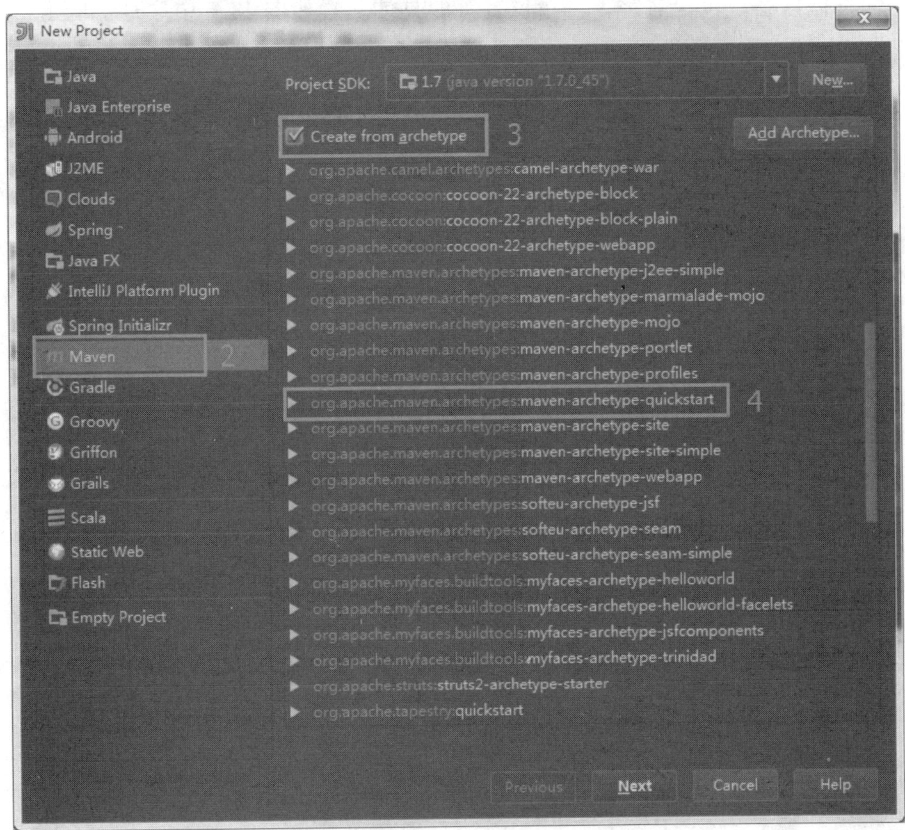

图 14-3　IntelliJ IDEA 的创建 Project 界面

(4) 选择创建项目的模板

本例中仅需要一个拥有 main 方法的启动类，所以选择 archetype 为 org. apache. maven. archetypes：maven-archetype-quickstart，如图 14-3 中序号 4 所示。

(5) 设置工程的坐标信息

每个 artifact 都是由 groupId：artifactId：version 组成的唯一标识符，可以理解为坐标，每个 Maven 项目都有一个唯一的坐标信息。其中标示符的三个参数含义如下。

① groupId：是项目创建团体或组织的唯一标志符，通常是域名倒写，如 groupId 为 org. apache. maven. plugins 就是为所有 Maven 插件预留的。

② artifactId：是项目 artifact 唯一的基地址名。

③ version：artifact 的版本，通常能看见为类似 0. 0. 1-SNAPSHOT，其中 SNAPSHOT 表示项目开发中，为开发版本。

工程的坐标信息如图 14-4 中序号 5 所示。

(6) 设置工程相关 Maven 的信息

设置 Maven 插件版本，配置信息以及本地仓库的位置，如图 14-5 中序号 6 所示。

(7) 设置工程名及存储位置

工程的具体信息，包含名字，目录位置等的具体配置如图 14-6 中序号 7 所示。

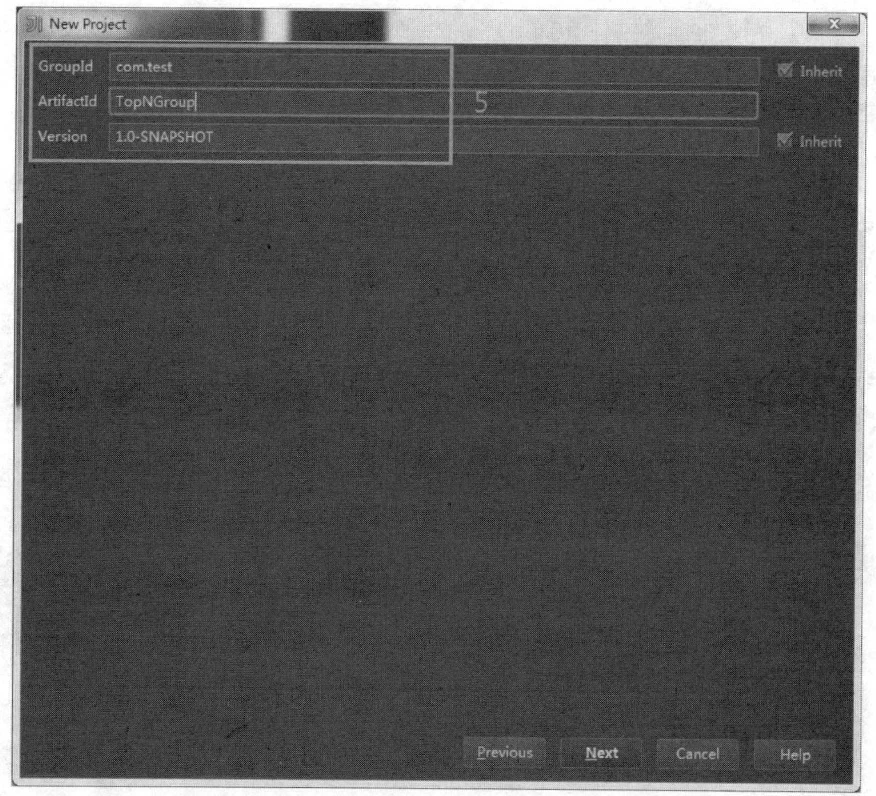

图 14-4　IntelliJ IDEA 的创建 Project 的配置坐标界面

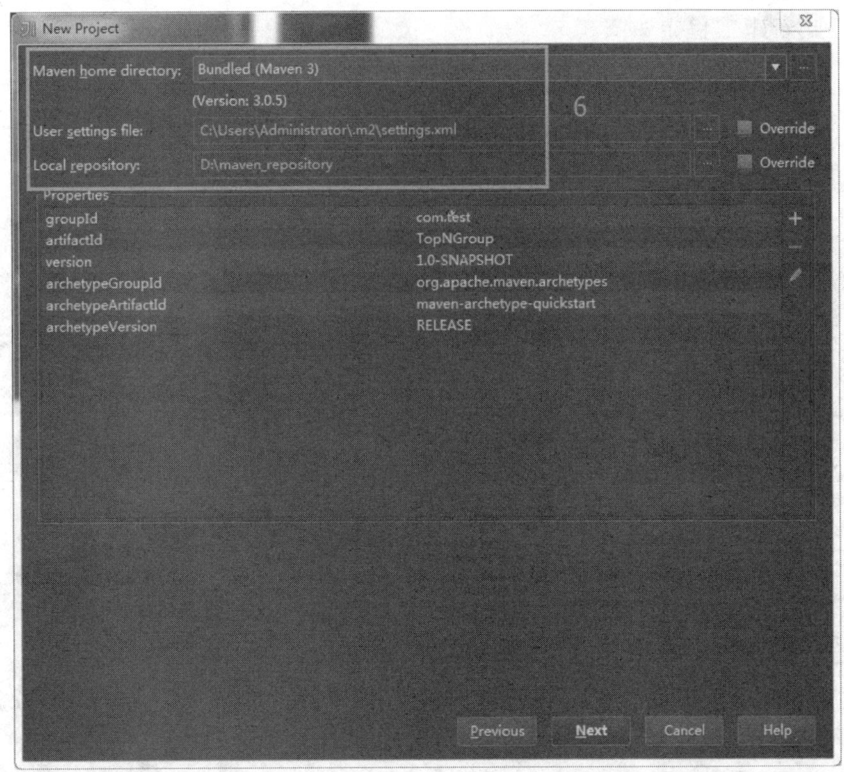

图 14-5　IntelliJ IDEA 的创建 Project 的相关 Maven 信息界面

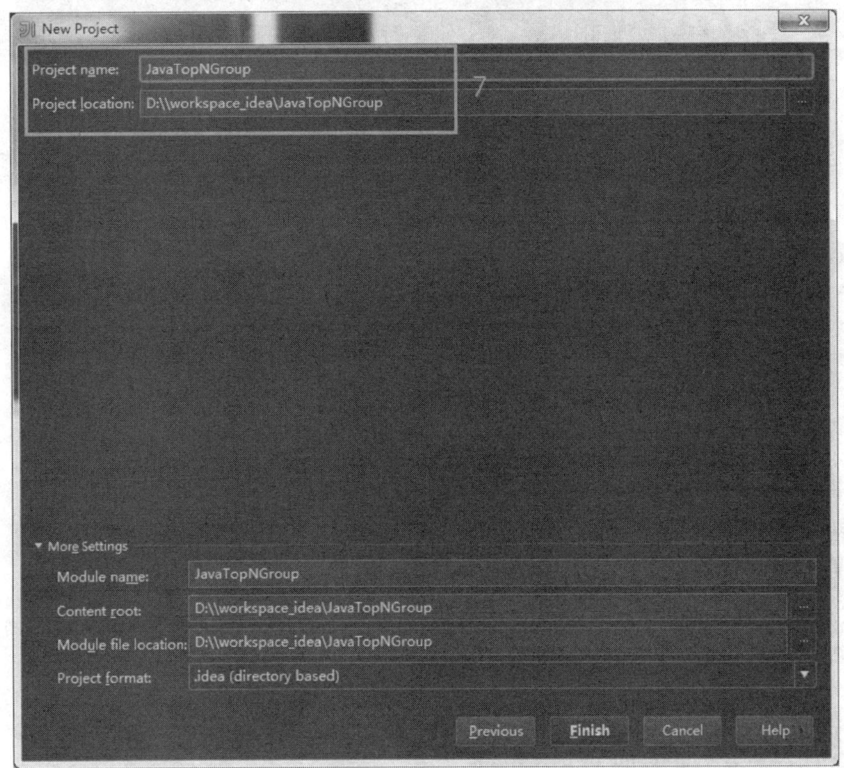

图 14-6　IntelliJ IDEA 的创建 Project 配置工程名及工程的存储位置界面

(8) 至此新建工程完毕

C. 添加工程的依赖包

① 在 pom. xml 添加 Spark 核心包依赖 spark-core，如图 14-7 中序号 1 所示。

② 重新导入 Maven 工程的依赖，并等待下载 jar 包结束，如图 14-7 中序号 2 所示。

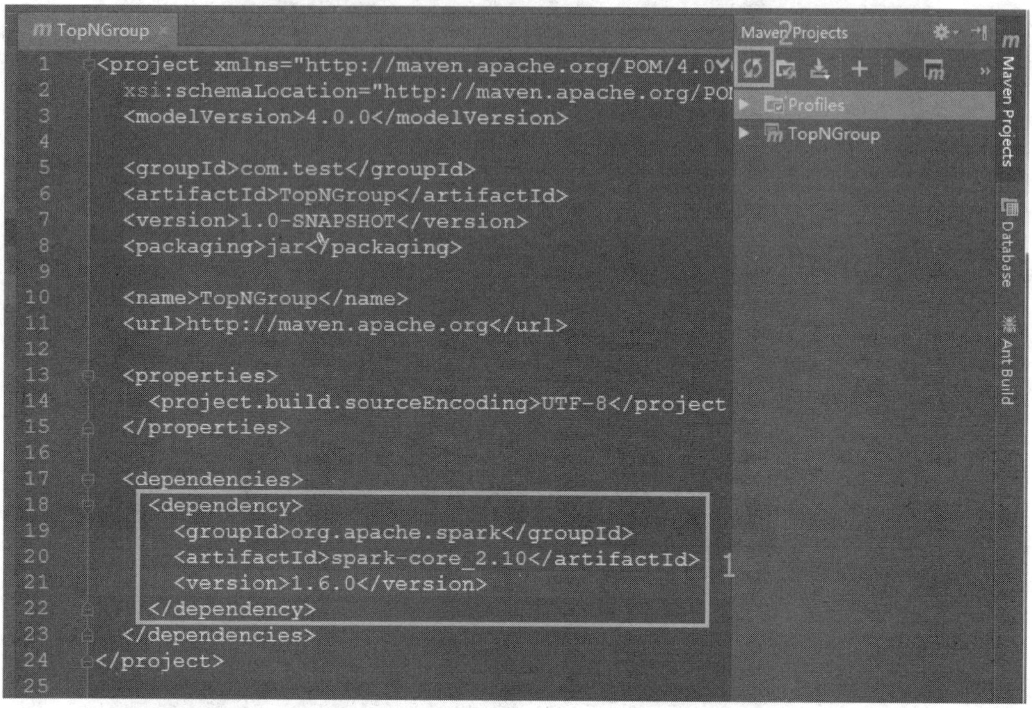

图 14-7　pom. xml 添加 Spark 核心包依赖 spark-core

D. 编写分组 TopN 的 Java 代码 TopNGroup

TopNGroup. java 代码如下。

```
package com.test;
import org.apache.spark.SparkConf;
import org.apache.spark.api.java.JavaPairRDD;
import org.apache.spark.api.java.JavaRDD;
import org.apache.spark.api.java.JavaSparkContext;
import org.apache.spark.api.java.function.PairFunction;
import org.apache.spark.api.java.function.VoidFunction;
import scala.Tuple2;
import java.util.Arrays;
import java.util.Iterator;
/**
 * 使用Java开发本地运行的Spark TopNGroup程序
 * Created by DT大数据梦工厂 on 2016/2/16.
 */
public class TopNGroup {
  public static void main(String[] args) {
      /**
```

```
     *第一步：创建Spark的配置对象SparkConf，设置Spark程序的运行时的配置信息，
     *例如通过setMaster来设置程序要链接的Spark集群的Master的URL，如果设置
     *为local，这代表Spark程序在本地运行，特别适合于机器配置条件非常差（例如
     *只有1G的内存）的初学者
     */
    SparkConf sparkConf = new SparkConf().setAppName("JavaTopNGroup").setMaster("local");
    /**
     *第二步：创建SparkContext对象
     * SparkContext是Spark程序所有功能的唯一入口，无论是采用Scala、Java、Python、R等必须有一个SparkContext（不用语言具体的类会不一样）
     * SparkContext核心作用：初始化Spark应用程序运行所需要的核心组件，包括DAGScheduler，TaskScheduler，SchedulerBackend
     *同时还会负责Spark程序往Master注册程序等
     * SparkContext是整个Spark应用程序中最为至关重要的一个对象
     */
    JavaSparkContext ctx = new JavaSparkContext(sparkConf);//其底层实际上就是Scala的SparkContext
    /**
     *第三步：根据具体的数据来源（HDFS、HBase、Local Fs、DB、S3等）JavaRDD
     * JavaRDD的创建基本有三种方式：根据外部的数据来源（例如HDFS）、根据Scala集合、有其他的RDD操作
     * 数据会被RDD划分成为一系列的Partitions，分配到每个Partition的数据属于一个Task的处理范畴
     */
    JavaRDD<String> lines = ctx.textFile("file://D://TopNGroup.txt");
    // 把每行数据变成符合要求的Key-Value的方式
    JavaPairRDD<String, Integer> pairs = lines.mapToPair(new PairFunction<String, String, Integer>() {
        @Override
        public Tuple2<String, Integer> call(String line) {
            String[] splitedLine = line.split("");
            return new Tuple2<String, Integer>(splitedLine[0], Integer.valueOf(splitedLine[1]));
        }
    });
    JavaPairRDD<String, Iterable<Integer>> groupedPairs = pairs.groupByKey();
    final JavaPairRDD<String, Iterable<Integer>> top5 = groupedPairs.mapToPair(new PairFunction<Tuple2<String, Iterable<Integer>>, String, Iterable<Integer>>() {
        @Override
        public Tuple2<String, Iterable<Integer>> call(Tuple2<String, Iterable<Integer>> groupedData) throws Exception {
            Integer[] top5 = new Integer[5];//保存Top5的数据
            String groupedKey = groupedData._1();//获取分组的组名key
            Iterator<Integer> groupValue = groupedData._2().iterator();//获取每组的内容集合
            while (groupValue.hasNext()) {//查看是否有下一个元素，如果有则继续进行循环
                Integer value = groupValue.next();//获取当前循环的元素本身内容value
                for (int i = 0; i < 5; i++) {
                    if (top5[i] == null) {
                        top5[i] = value;
```

```
                        break;
                    } else if (value > top5[i]) {
                        for (int j = 4; j > i; j--) {
                            top5[j] = top5[j - 1];
                        }
                        top5[i] = value;
                        break;
                    }
                }
            }
            return new Tuple2<String, Iterable<Integer>>(groupedKey, Arrays.asList(top5));
        }
    });
    // 打印分组后的TopN
    top5.foreach(new VoidFunction<Tuple2<String, Iterable<Integer>>>() {
        @Override
        public void call(Tuple2<String, Iterable<Integer>> topped) throws Exception {
            System.out.println("Group Key : " + topped._1());
            Iterator<Integer> toppedValue = topped._2().iterator();
            while (toppedValue.hasNext()) {
                    Integer value = toppedValue.next();
                System.out.println(value);
            }
            System.out.println("******************************");
        }
    });
    //    关闭JavaSparkContext
    ctx.stop();
  }
}
```

E. 运行程序

右键 TopNGroup. java，选择运行该类的 main 方法，如图 14-8 所示。

F. 查看结果

运行之后的结果显示在 IDEA 的控制台中，可以看出该结果有两个分组，每个分组显示前五的 Score 得分，正确地实现了分组 TopN 的功能，如图 14-9 所示。

14.1.2 Scala 之分组 TopN 开发实战

(1) 在 IntelliJ IDEA 新建 Scala 版的 TopNGroup 的工程

步骤与前面的新建 Java 版的 TopNGroup 的工程类似，这里只罗列粗略步骤。

① 在 File 菜单下单击 New 下的 Project...，创建一个新的 Project，如图 14-10 中序号 1 所示。

② 选择 Maven 工程，如图 14-11 中序号 2 所示。

③ 勾选 Create from archetype，如图 14-11 中序号 3 所示。

④ 点击 Add Archetype...，列表没有现成的 archetype 可以使用，所以需要新增 archetype，如图 14-11 中序号 4 所示。

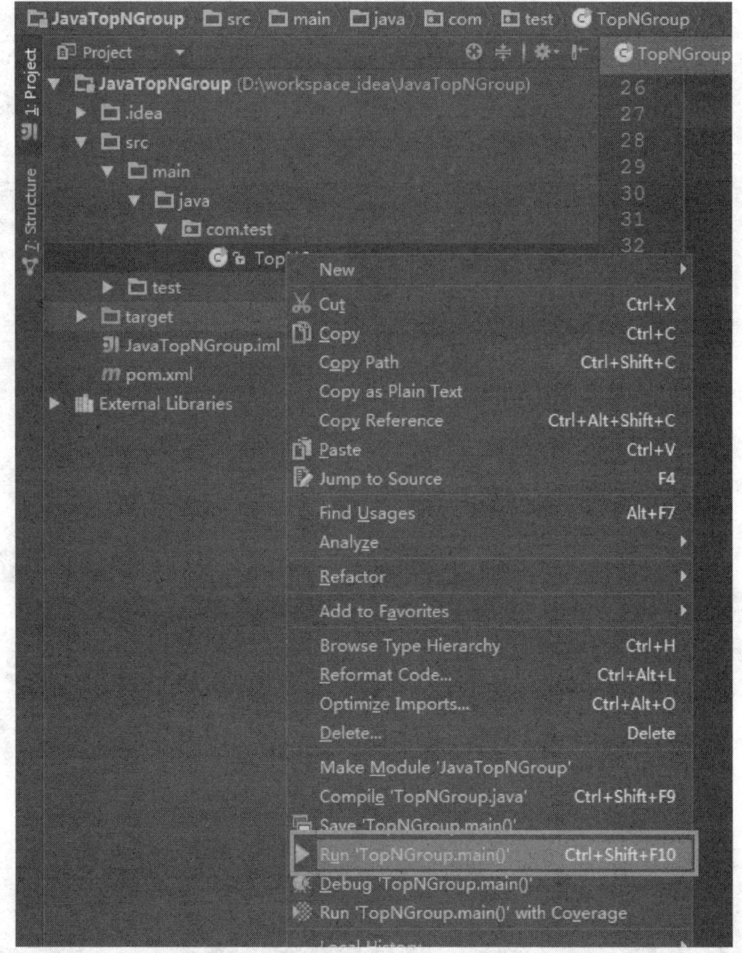

图 14-8 右键执行

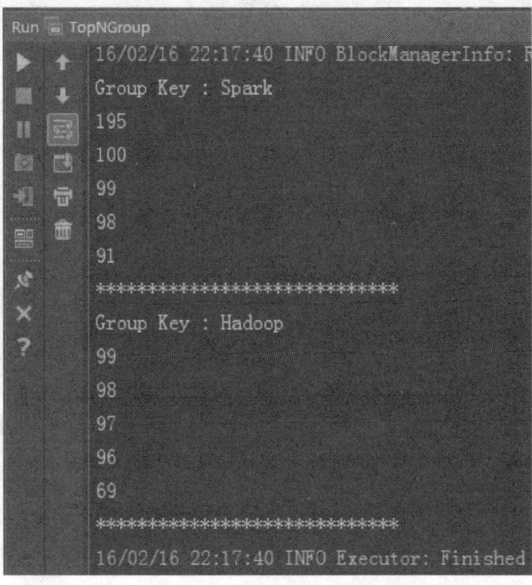

图 14-9 TopNGroup 执行结果

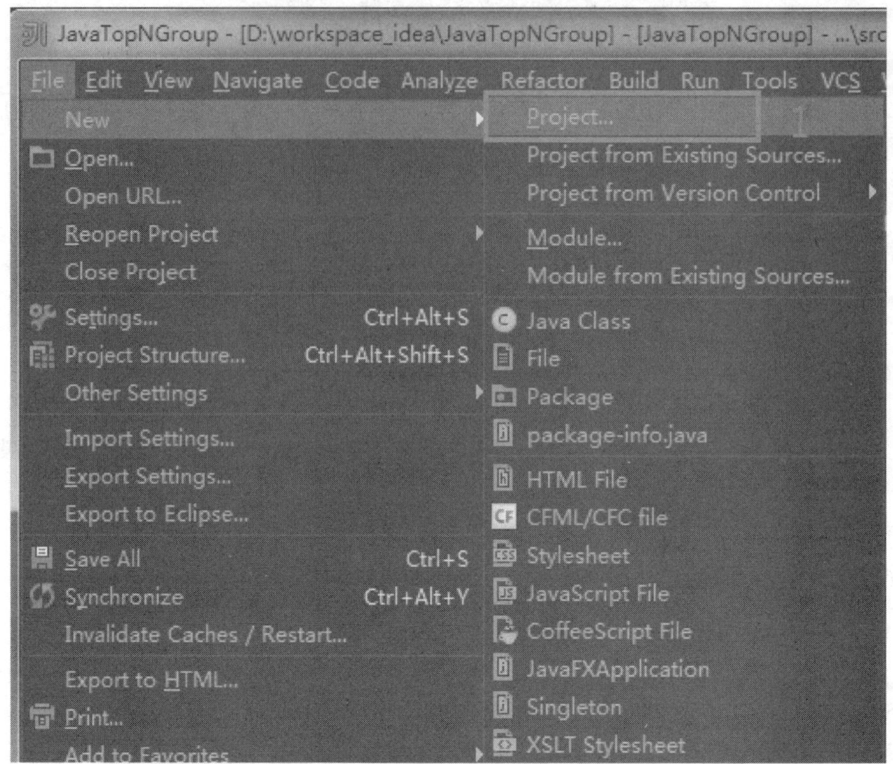

图 14-10　新建工程

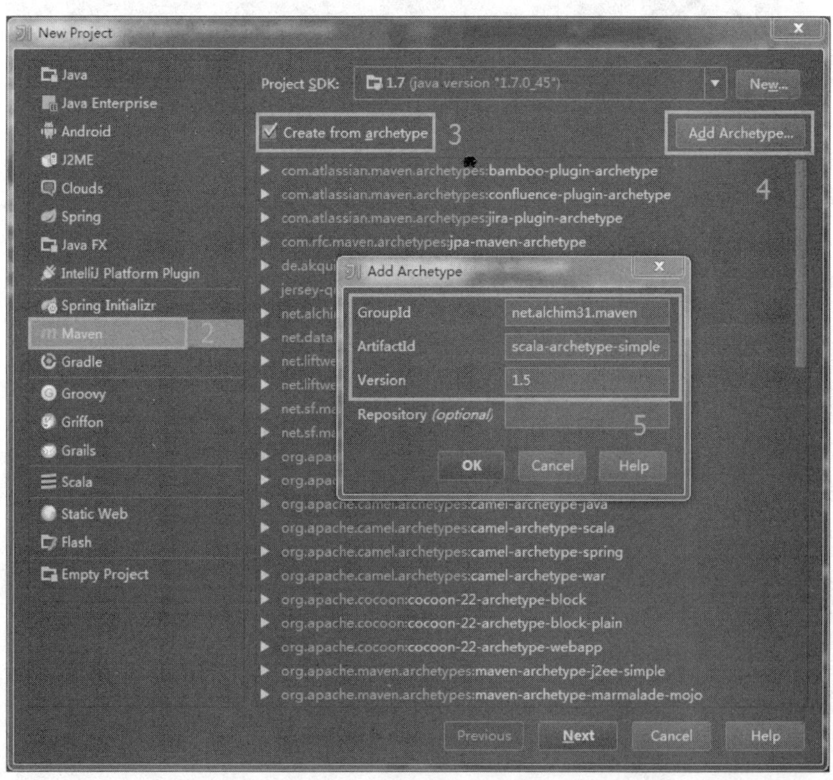

图 14-11　IntelliJ IDEA 的创建 Project 界面

⑤ 设置相关的 archetype 坐标信息，archetype 的坐标信息如图 14-11 中序号 5 所示。

⑥ 设置工程的坐标信息，如图 14-12 中序号 6 所示。

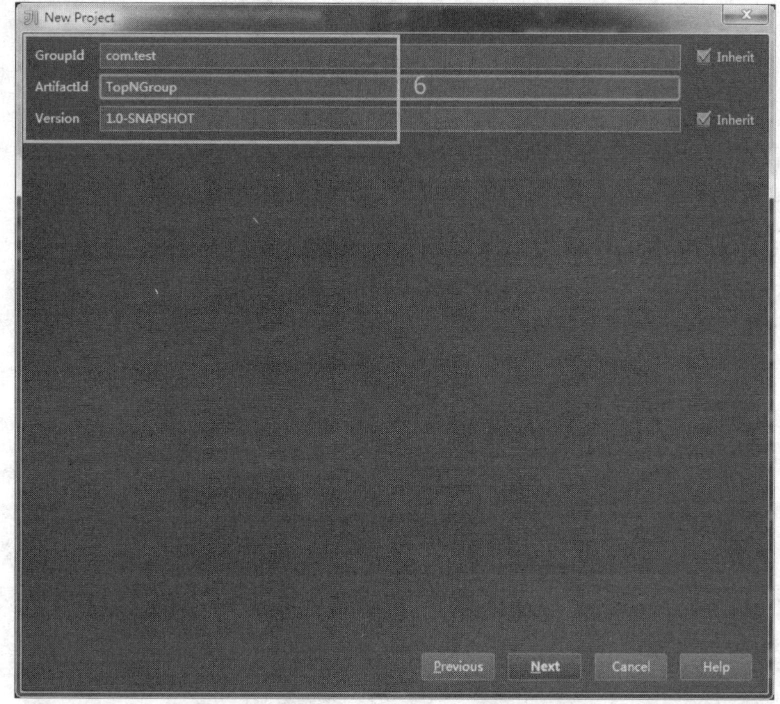

图 14-12　IntelliJ IDEA 的创建 Project 的配置坐标界面

⑦ 设置工程相关 Maven 的信息，如图 14-13 中序号 7 所示。

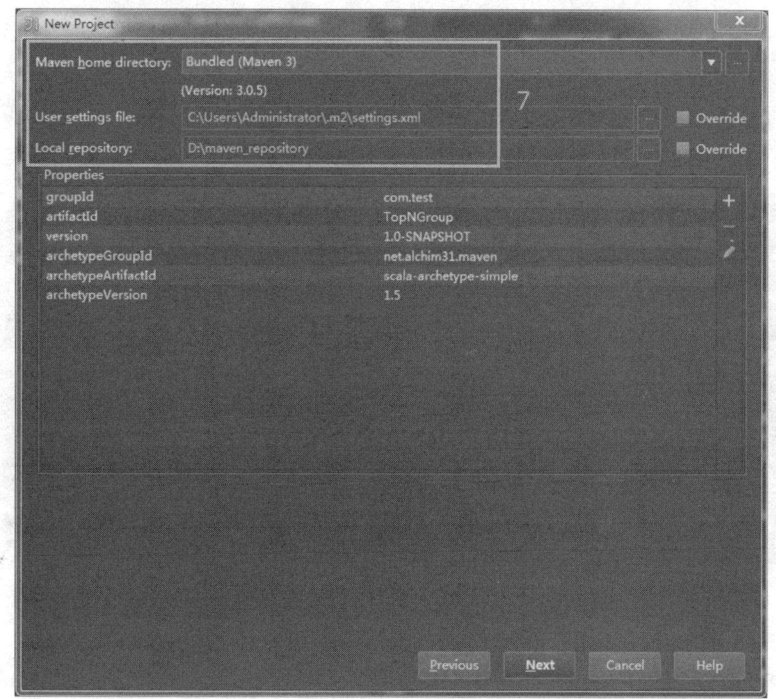

图 14-13　IntelliJ IDEA 的创建 Project 的相关 Maven 信息界面

⑧ 设置工程名及存储位置，如图 14-14 中序号 8 所示。

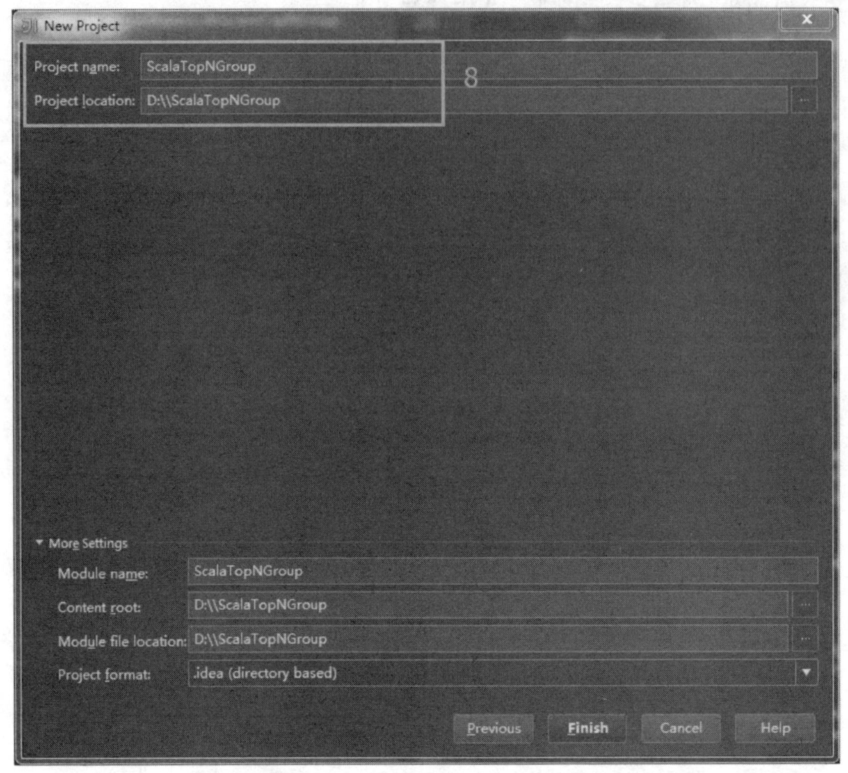

图 14-14　IntelliJ IDEA 的创建 Project 配置工程名及工程的存储位置界面

⑨ 至此新建工程完毕。

(2) 添加工程的依赖包

① 在 pom. xml 添加 Spark 核心包依赖 spark-core，如图 14-15 中序号 1 所示。

② 重新导入 Maven 工程的依赖，并等待下载 jar 包结束，如图 14-15 中序号 2 所示。

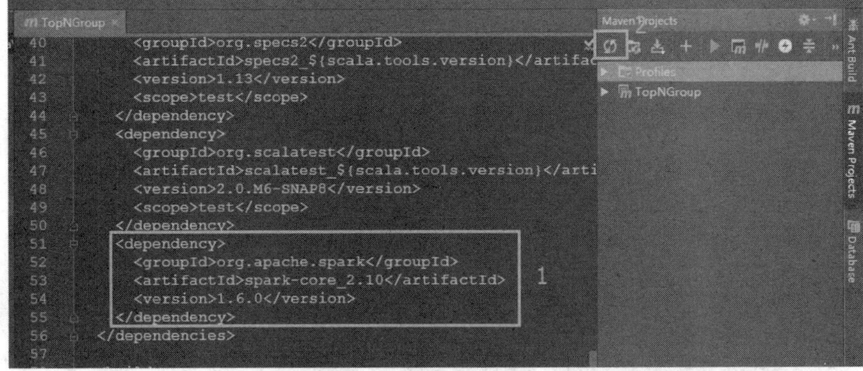

图 14-15　pom. xml 添加 Spark 核心包依赖 spark-core

(3) 编写分组 TopN 的 Scala 代码 TopNGroup

TopNGroup. scala 代码如下所示。

```
package com.test
import org.apache.spark.{SparkContext, SparkConf}
/**
 *使用Scala开发本地运行的Spark TopNGroup程序
 * Created by DT大数据梦工厂 on 2016/2/16.
 */
object TopNGroup {
  def main(args: Array[String]) {
    val conf = new SparkConf().setAppName("ScalaTopNGroup").setMaster("local")
    val sc = new SparkContext(conf)
    val lines = sc.textFile("file://D://TopNGroup.txt")
    val top5= lines.map { line =>
      val splitedLine = line.split("")
      (splitedLine(0), splitedLine(1).toInt)
    }.groupByKey().map { groupedData =>
      val groupedKey = groupedData._1
      val top5: List[Int] = groupedData._2.toList.sortWith(_ > _).take(5)
      Tuple2(groupedKey, top5)
    }
    top5.foreach{topped =>
      Systemout.println("Group Key : " + topped._1)
      val toppedValue: Iterator[Int] = topped._2.iterator
      while (toppedValue.hasNext) {
        val value: Integer = toppedValue.next
        System.out.println(value)
      }
      System.out.println("******************************")
    }
    sc.stop()
  }
}
```

(4) 执行计算，右键 TopNGroup. scala，选择运行该 Object 的 main 方法

(5) 查看执行结果，结果与前面一致，如图 14-16 所示

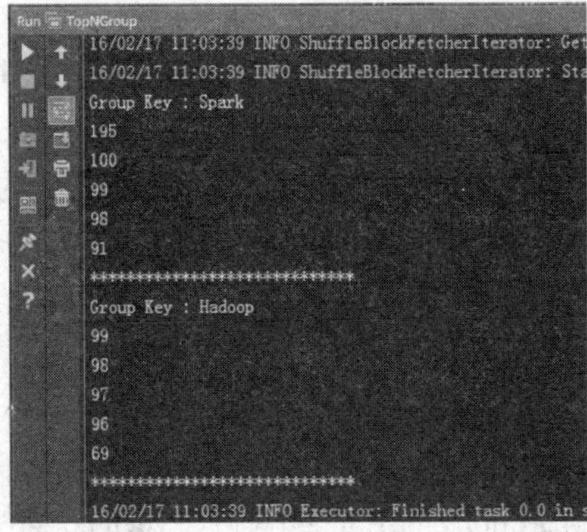

图 14-16 TopNGroup 执行结果

14.2 Scala之分组 TopN 运行原理解密

本节将深入了解分组 TopN 的所使用的 RDD，RDD 有两种：Transformations 与 Actions。

① Transformations 操作是 Lazy 的，也就是说从一个 RDD 转换生成另一个 RDD 的操作不是马上执行，Spark 在遇到 Transformations 操作时只会记录相应的操作，并不会去执行，需要等到有 Actions 操作的时候才会真正启动计算过程进行计算。此类 RDD 有：map、filter、groupBy、join 等。

② Actions 操作会返回结果或把 RDD 数据写到存储系统中。Actions 是触发 Spark 启动计算的动因，此类 RDD 有：foreach、count、collect、save 等。

14.2.1 textFile

(1) 功能描述

从 HDFS（以 hdfs://master: port 开头）、本地文件系统（本章所使用的，以 file: // 开头），或者在 Hadoop 上支持的文件系统（例如 Tachyon，以 tachyon://master: port 开头）读取文本文件。

(2) 定义

```
def textFile(
   path: String,
   minPartitions: Int = defaultMinPartitions): RDD[String] = withScope {
 assertNotStopped()
 hadoopFile(path, classOf[TextInputFormat], classOf[LongWritable], classOf[Text],
   minPartitions).map(pair => pair._2.toString)
}
```

textFile 各参数含义如下。

path 参数用于指定读取文件的路径，有三种指定方式，如下。

① 一个文件路径，这时候只装载指定的文件。例如：textFile("/myPath/directory/test. txt")，加载 test. txt 文件。

② 一个目录路径，这时候只装载指定目录下面的所有文件（不包括子目录下面的文件）。例如：textFile("/myPath/directory/")，加载 directory 目录下的所有文件。

③ 通配符，这时候加载多个文件或者加载多个目录下面的所有文件。例如：textFile("/myPath/directory/ *. txt")，加载 directory 目录下的所有 txt 格式的文件。

minPartitions 代表分片数，也就是任务的并行度。但当没有设置该属性时，默认值为 defaultMinPartitions。

查看 SparkContext 中的 defaultMinPartitions 源码。

```
/**
 * Default min number of partitions for Hadoop RDDs when not given by user
 * Notice that we use math.min so the "defaultMinPartitions" cannot be higher than 2.
```

```
 * The reasons for this are discussed in https://github.com/mesos/spark/pull/718
 */
def defaultMinPartitions: Int = math.min(defaultParallelism, 2)
```

跳转到 defaultParallelism 源码。

```
/** Default level of parallelism to use when not given by user (e.g. parallelize and makeRDD). */
def defaultParallelism: Int = {
  assertNotStopped()
  taskScheduler.defaultParallelism
}
```

跳转到 taskScheduler. defaultParallelism 源码。

```
// Get the default level of parallelism to use in the cluster, as a hint for sizing jobs.
def defaultParallelism(): Int
```

单击方法前的箭头，获取具体子类的实现，当前只有一个实现的子类，即 TaskSchedulerImpl。如图 14-17 所示。

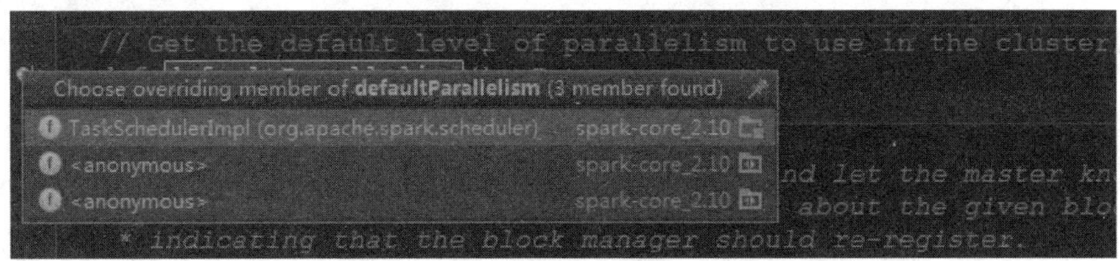

图 14-17　获取 taskScheduler 方法的具体实现

跳转到 TaskSchedulerImpl 源码。

```
override def defaultParallelism(): Int = backend.defaultParallelism()
```

继续跳转到 backend. defaultParallelism，然后跳转到具体子类的实现，继承的全部子类如图 14-18 所示。

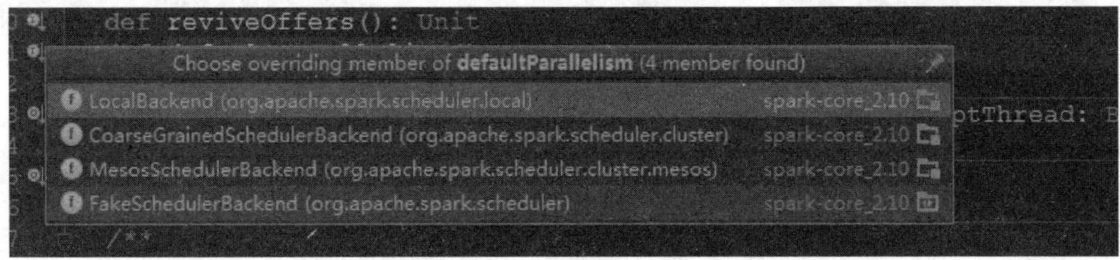

图 14-18　backend 具体实现子类列表

如 LocalBackend（对应的集群模式，本例采用 Local 模式）的源码。

```
override def defaultParallelism(): Int =
scheduler.conf.getInt("spark.default.parallelism", totalCores)
```

可以看到 defaultParallelism 的配置属性为“spark. default. parallelism”，spark. default. parallelism 可以在 $SPARK_HOME/conf/spark-default. conf 文件配置或 SparkConf 属性指定。但当没有设置该属性时，默认值为 totalCores，totalCores 是本机的总内核数。

📖 默认的并行度 defaultMinPartitions 只是作为加载文件时分区的最小参考值，实际的分区数由加载文件时的 Splits 数决定的，即文件的 Blocks 数。

(3) textFile 解密

textFile 方法调用 hadoopFile 时，对应的 InputFormat 设置为 TextInputFormat，即文本文件的输入。其中 hadoopFile 返回 HadoopRDD [K，V] 类型的 RDD，其中 K 代表该分片的文本内容在文本中的偏移量，V 代表该分片的文本内容。因为 textFile 是基于文本内容进行计算的，所以 HadoopRDD 经 map 过滤掉偏移量，最终 textFile 得到 String 类型的文本内容的 RDD。

📖 一个分片对应一个文件中多行内容，不能跨多个文件。

14.2.2 map

(1) 功能描述

将函数 f 作用于 RDD 的每个元素。

(2) 定义

```
/**
 *Return a new RDD by applying a function to all elements of this RDD.
 */
def map[U: ClassTag](f: T => U): RDD[U] = withScope {
  val cleanF = sc.clean(f)
  new MapPartitionsRDD[U, T](this, (context, pid, iter) => iter.map(cleanF))
}
```

map 中 f 参数含义为对每个元素进行的转换操作。

(3) map 解密

① lines 后面的 map 作用：对分片的每一行文本内容按照空格分隔，返回（类名，得分）的 Tuple2 元组。

② groupByKey 后面的 map 作用：先利用 sortWith(_>_)对每一个类名的得分集合进行降序排序，然后再取排序好的集合的前五个元素，得到 top5 的元素，最后返回（类名，Seq[Int](top5))Tuple2 元组。

14.2.3 groupByKey

(1) 功能描述

在一个由（K，V）对组成的数据集上调用，返回一个（K，Seq [V] ）对的数据集。

(2) 定义

```
/**
 * Group the values for each key in the RDD into a single sequence. Hash-partitions the
 * resulting RDD with the existing partitioner/parallelism level. The ordering of elements
 * within each group is not guaranteed, and may even differ each time the resulting RDD is
 * evaluated.
 *
 * Note: This operation may be very expensive. If you are grouping in order to perform an
 * aggregation (such as a sum or average) over each key, using [[PairRDDFunctions.aggregateByKey]]
 * or [[PairRDDFunctions.reduceByKey]] will provide much better performance.
 */
def groupByKey(): RDD[(K, Iterable[V])] = self.withScope {
  groupByKey(defaultPartitioner(self))
}
```

由于 groupByKey 是宽依赖，宽依赖往往意味着 shuffle 操作，将数据写入本地磁盘，同时宽依赖也是 Spark 划分 Stage 的依据。groupByKey 是针对 Key-Value 元素类型的 RDD 的操作，Key-Value 元素类型的 RDD 针对 shuffle 提供了精细的分区器 partitioner，可以通过设置分区器，来指定元素如何分区的策略，以及分区数。

defaultPartitioner 源码如下所示。

```
/**
 * Choose a partitioner to use for a cogroup-like operation between a number of RDDs.
 *
 * If any of the RDDs already has a partitioner, choose that one.
 *
 * Otherwise, we use a default HashPartitioner. For the number of partitions, if
 * spark.default.parallelism is set, then we'll use the value from SparkContext
 * defaultParallelism, otherwise we'll use the max number of upstream partitions.
 *
 * Unless spark.default.parallelism is set, the number of partitions will be the
 * same as the number of partitions in the largest upstream RDD, as thisshould
 * be least likely to cause out-of-memory errors.
 *
 * We use two method parameters (rdd, others) to enforce callers passing at least 1 RDD.
 */
def defaultPartitioner(rdd: RDD[_], others: RDD[_]*): Partitioner = {
  val bySize = (Seq(rdd) ++ others).sortBy(_.partitions.size).reverse
  for (r <- bySize if r.partitioner.isDefined && r.partitioner.get.numPartitions > 0) {
    return r.partitioner.get
  }
  if (rdd.context.conf.contains("spark.default.parallelism")) {
    new HashPartitioner(rdd.context.defaultParallelism)
  } else {
    new HashPartitioner(bySize.head.partitions.size)
  }
}
```

从代码可以看到，groupByKey 默认采用 HashPartitioner 方式进行 shuffle。当没有设置分区数时，会使用“spark. default. parallelism”属性值作为默认的分区数。

（3）groupByKey 解密

groupByKey 对（类名，得分）进行分组，将相同的类名的得分合并在一起，组成 Seq［Int］（得分 1，得分 2.....）的集合，然后采用 HashPartitioner 分区器将数据写入到本地磁盘。

① 功能描述：将函数 f 作用于 RDD 的每个元素，此处将会触发 Job。

② 定义：

```
/**
 * Applies a function f to all elements of this RDD.
 */
defforeach(f: T => Unit): Unit = withScope {
  val cleanF = sc.clean(f)
  sc.runJob(this, (iter: Iterator[T]) => iter.foreach(cleanF))
}
```

foreach 中 f 参数含义为对每个元素进行的转换操作。

foreach 调用了 SparkContext 的 runJob 方法，runJob 源码如下：

```
/**
 * Run a function on a given set of partitions in an RDD and pass the results to the given
 * handler function. This is the main entry point for all actions in Spark.
 */
def runJob[T, U: ClassTag](
    rdd: RDD[T],
    func: (TaskContext, Iterator[T]) => U,
    partitions: Seq[Int],
    resultHandler: (Int, U) => Unit): Unit = {
    if (stopped.get()) {
     throw new IllegalStateException("SparkContext has been shutdown")
    }
    val callSite = getCallSite
    val cleanedFunc = clean(func)
    logInfo("Starting job: " + callSite.shortForm)
    if (conf.getBoolean("spark.logLineage", false)) {
      logInfo("RDD's recursive dependencies:\n" + rdd.toDebugString)
    }
    dagScheduler.runJob(rdd, cleanedFunc, partitions, callSite, resultHandler, localProperties.get)
    progressBar.foreach(_.finishAll())
    rdd.doCheckpoint()
}
```

③ foreach 解密：map 与 foreach 看似差不多，两个都是把函数 f 作用于每个元素，但实质上有很大区别。map 是 Transformations 操作，并不会立即执行。而 foreach 是 Actions 操作，会导致作业的执行，前面提到的算子 textFile、map、groupByKey 在 foreach 触发后才真正地执行。

14.3 小结

本章主要介绍分组 TopN 程序，首先是在 IntelliJ IDEA 动手实战 Java 和 Scala 两种语言实现分组 TopN 算法，采用本地运行模式运行程序并验证了执行结果的正确性。接着从源码的角度深入理解 textFile、map、groupByKey、foreach 算子的功能及本质，对如何编写 Spark 程序及对基础算子运用有一个清楚的认识。

第15章 MasterHA工作原理解密

高可用性（High Availability，HA）是当前很多企业对于应用系统一个很重要的非功能性要求之一，Spark作为大型分布式处理系统当然也不例外。为了保证HA，Spark在架构设计的时候对于架构中的每个重要组件如Master、Worker等都设计了相应的策略。这一章介绍Spark MasterHA的4种策略以及着重解密如何通过ZOOKEEPER这种企业经常使用的策略来保证Spark Master HA。

15.1 Spark需要Master HA的原因

对于Standalone模式的Spark集群而言，由于其采用Master/Slaves的架构，这种架构不可避免地会带来SPOF（单点故障，Single Point of Failure），而对于一些大型的分布式系统，这种情况是不能接受的，要保证集群高度的稳定性和容错性。为此Spark提供了集群的不同的重启策略，保证系统的正常运行，从Spark的源码可以看出，其中相应的策略包含如下几种方式。

① ZOOKEEPER：企业经常使用，实时恢复。

② FILESYSTEM：低延迟恢复。

③ CUSTOM：自定义恢复。

④ NONE：不采用HA，默认就是Master/Slaves。

具体采用哪种恢复策略，可以通过spark-env.sh的RECOVERY _ MODE属性进行设置。

15.2 Spark Master HA的实现

(1) ZOOKEEPER实现解密

通过ZOOKEEPER实现Master的高可用性是企业在生产环境下常用的方式。目前一般会采用3台机器作ZOOKEEPER服务（如图15-1所示）。使用这种方式会创建ZOO-

KEEPER 实例化引擎（ZooKeeperPersistenceEngine），该引擎实现 PersistenceEngine 抽象类，内部提供了一系列的方法，便于在集群恢复的时候从 ZOOKEEPER 中读取元数据进行集群恢复。

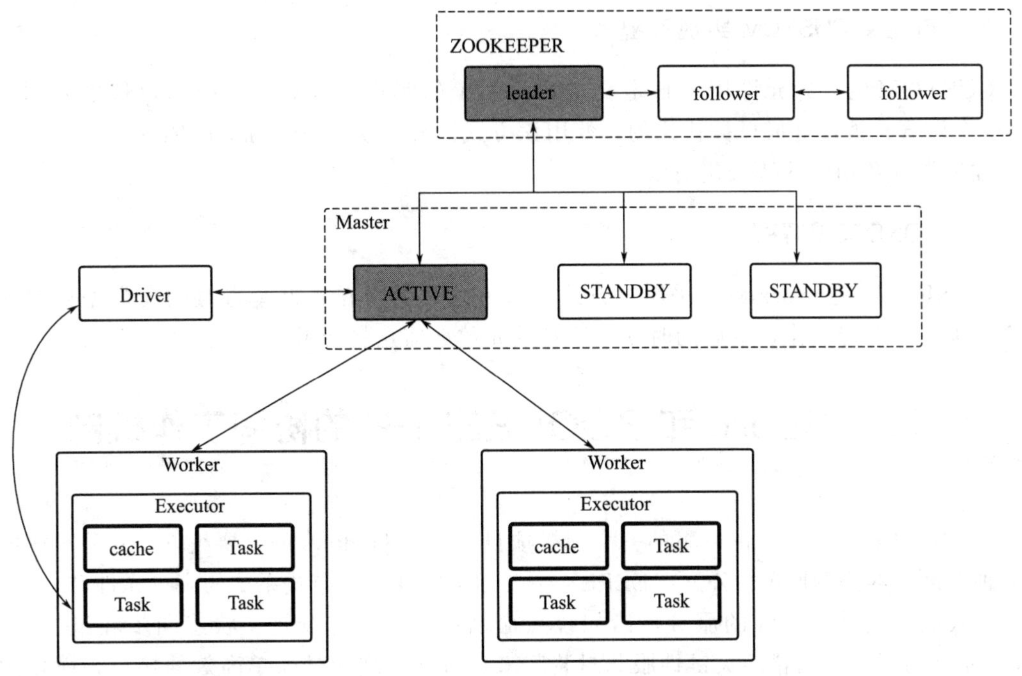

图 15-1 ZOOKEEPER 实现 Spark Master HA 的原理

ZOOKEEPER 会自动化管理 Masters 的切换，整个集群运行时的元数据（包括 Workers、Drivers、Applications、Executors）会被保存在 ZOOKEEPER 中，ZOOKEEPER 集群中的当前状态为 leader 的是目前正在工作的，而 spark 集群中状态为 Active 的 Master 管理着当前的集群。

当前 Active 级别的 Master 出现故障时，ZOOKEEPER 会从状态为 Standby 的 Master 中选择出一台作为 ActiveMaster。在从 Standby 被选为 Active 后，需要从 ZOOKEEPER 中获取相应的元数据信息，对集群进行恢复。由于任务在提交时已经向 Cluster Manager 提交并获取了相应的计算资源，因此在此切换过程中所有已经运行的任务正常运行不受影响。但是由于 Master 没有切换完成，此时集群不能接受新的任务。

使用 ZOOKEEPER 实现 Master 的 HA，不需要手动操作集群，整个切换过程是由 ZOOKEEPER 自动完成的。

(2) FILESYSTEM 实现解密

FILESYSTEM 的方式会创建 FileSystemRecoveryModeFactory，在其内部会根据 spark. deploy. recoveryDirectory 所设置的文件目录来创建目录，将整个集群运行时的元数据保存在该目录下。同时 FileSystemRecoveryModeFactory 中还会创建文件系统的持久化引擎（FileSystemPersistenceEngine），该持久化引擎中提供了 persist（将数据持久化到磁盘文件）、unpersist（将文件从磁盘上移除）、read（从磁盘上读取数据）、以及序列化和反序列化等方法。集群恢复时会通过 readPersistedData 方法，读取持久化在磁盘的元数据进行集群恢复。

在这种策略下。在 Master 出现故障后需要手动重新启动机器，机器启动后，Master 的状态会立即成为 Active，对外提供服务（包括接受应用程序提交的请求、接受新的 Job 运行的请求）。

（3）自定义 CUSTOM 实现解密

CUSTOM 的方式允许用户自定义 MasterHA 的实现，这对于高级用户特别有用。在 Spark 中该方式被设计成可插拔式的，使用该方式同样要实现 StandaloneRecoveryModeFactory 抽象类以及相应的持久化引擎。

（4）NONE 实现解密

NONE 策略是 Spark 默认采用的方式，该方式只会将在集群运行时的元数据保存到临时文件中，不会持久化，重启的时候，所有的元数据信息将会丢失。

15.3 Spark 和 ZOOKEEPER 的协同工作机制

ZOOKEEPER 是 Apache 下的一个顶级项目，也是 Hadoop 的重要组件。它分布式应用的一种高性能的协调服务。ZOOKEEPER 主要用于实现在分布式系统的高可靠性和易用性，为分布式应用提供一致性的服务，诸如统一命名服务，配置管理，状态同步和组服务等。Zookeeper 将复杂易出错的关键性服务封装起来，对外呈现的是简单高效的接口，简化的用户的开发难度，不需要关心分布式系统在开发时的同步和一致性问题。

ZOOKEEPER 的两大主要流程：选举 Leader 和同步数据。

（1）ZOOKEEPER 对 Spark 的管理

ZOOKEEPER 是企业中最常用的解决 Spark 集群（也是其他的分布式系统采用的常用方式）单点故障的方式。

对于 ZOOKEEPER 对 Spark 的管理，如图 15-2 所示。首先在 ZOOKEEPER 集群中会有一个处于 leader 状态的 ZOOKEEPER 服务器处于工作状态，ZOOKEEPER 会从状态为 STANDBY 状态的 Master 中选举出一个 Master，该 Master 会通过 ZOOKEEPER 持久化引擎（ZooKeeperPersistenceEngine）去 ZOOKEEPER 中读取 Spark 集群的元数据信息（包括 Drivers、Applitions、Workers、Excutors）。

Master 对读取到的信息进行判断，如果有相关的信息，那么将相应的信息重新注册到该 Master 中。然后要进行一致性的校验，即所获取的信息是否和集群中的信息是一致的。验证的时候首先会将 Applications 和 Workers 等状态标记为 UNKOWN，然后会向 Applications 和 Workers 发送现在被选举出来的 Master 的信息（MasterChanged 其中包括 Master 的引用 masterRef 和 masterWebUiUrl）。此时接收到信息的 Applications 和 Workers 会通过 masterRef 发送消息（WorkerSchedulerStateResponse 其中包括 workerId、executors、drivers）给 Master。

Master 接收到来自 Applications 和 Workers 后，如果是 Worker，那么 Master 会根据传递过来的 workerId 的内存（workerId）中的 worker，并将其 state 设置成 ALIVE，同时会将合法的 Executor 信息重新添加到 Worker。如果是 Drivers，会判断 driverId 在 Master

的内存中是否存在，若存在会将 driverId 对应的 Driver 的状态设置成 RUNNING，并将该 driver 信息记录到 worker 中。最终会调用 completeRecovery 方法来处理没有响应的 Drivers、Workers 和 Applications。处理完成后，会将被选举出来的 Master 的 state 设置成 RecoveryState. ALIVE。

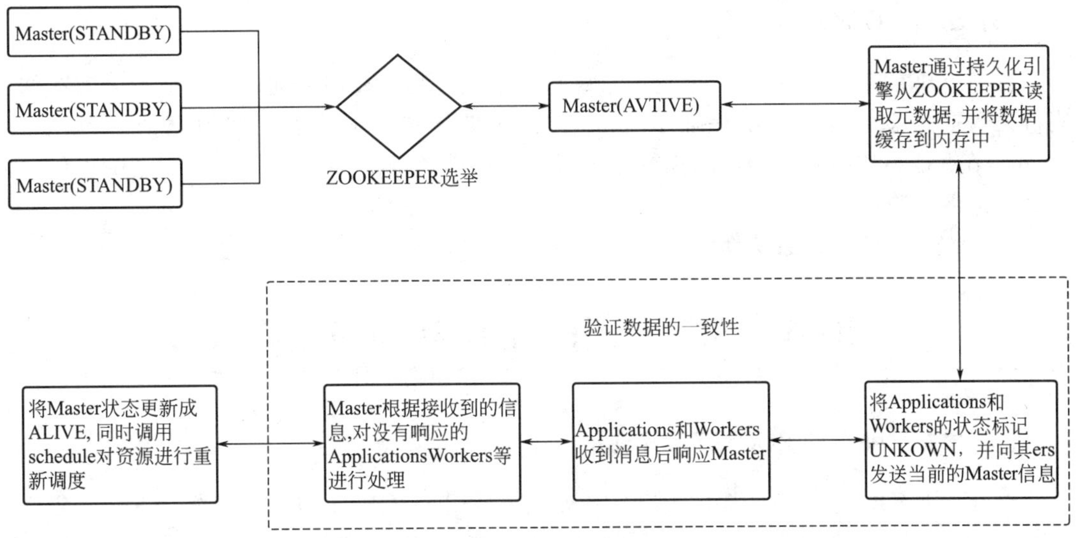

图 15-2　ZOOKEEPER 对 Spark 的管理

最后 Master 会调用 schedule 方法调度当前可以使用的资源给正在等待的 Applications。

(2) 通过 ZOOKEEPER 选举

LeaderElection 机制是 ZOOKEEPER 中重要的功能，默认的实现算法是 FastLeaderElection。

如图 15-3 所示，ZOOKEEPER 在刚启动的时候，每个节点都会读取 myid 中的数据（对应的 id），读取完数据之后，会向其他的服务器发送自己选举的 Leader（一般推荐自

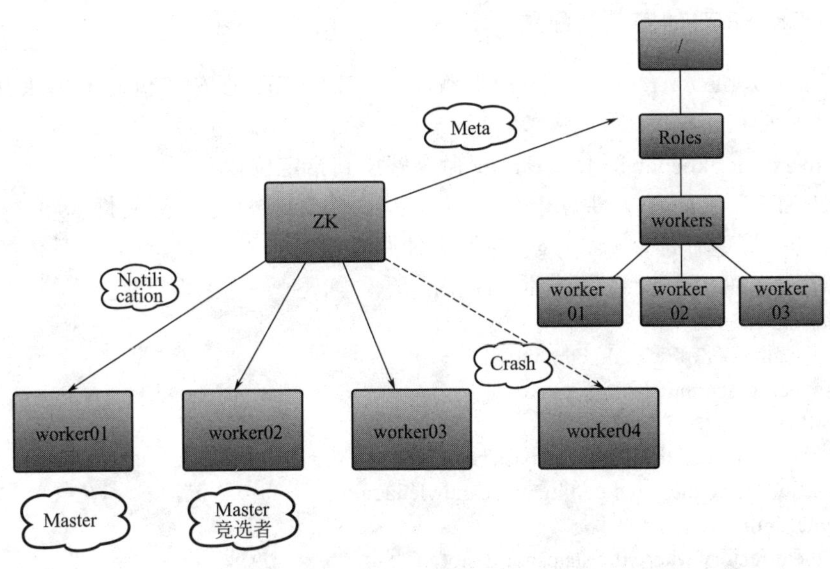

图 15-3　ZOOKPER 选举 Leader 的机制

己），同时接受其他服务器发来的信息（这个时候还不知道谁是 Leader），更新相关的选举信息，该信息会不断地更新，最后以逻辑时钟最大为准，这样每台服务器中得到相同的选举结果。在执行选举过程之后，会产生 Leader，其他的节点根据得到的选举信息，尝试连接 Leader。

（3）通过 ZOOKEEPER 恢复

在 ZOOKEEPER 服务器之间存在心跳机制，该机制主要用来同步各个节点之间存储的集群元数据信息。这种机制保证了 Leader 节点和 Follower 节点之间数据的一致性。当 Leader 节点遇到故障时，会再次执行选举过程，选出新的 Leader，而新的节点中也保存的数据是也最新的。

Spark 中的 Master 会从新的 Leader 中读取元数据进行恢复。

15.4 ZOOKEEPER 实现应用实战

（1）添加 Spark 的 ZOOKEEPER 支持

Spark 对 ZOOKEEPER 提供了支持，如果要使用 ZOOKEEPER 来对管理 Spark 集群中的 Master，需要添加在 SparkEnv 中添加 ZOOKEEPER 支持，将 Master 的恢复策略（Dspark. deploy. recoveryMode）设置成 ZOOKEEPER，同时指定 Master，去掉 SPARK _ MASTER _ IP 的设置。具体配置如下。

```
#主节点的IP地址
#export SPARK_MASTER_IP=master
#配置spark的zookeeper支持信息
export SPARK_DAEMON_JAVA_OPTS="-Dspark.deploy.recoveryMode=ZOOKEEPER
            -Dspark.deploy.zookeeper.url=master:2182,worker1:2182,worker2:2182
            -Dspark.deploy.zookeeper.dir=/data/zookeeper/spark"
```

（2）ZOOKEEPER 的安装及配置

从 http：//zookeeper. apache. org/下载 ZOOKEEPER 的安装包，本书采用的版本是 zookeeper-3. 4. 6. tar. gz。

使用 tar-zxvfzookeeper-3. 4. 6. tar. gz 命令解压到指定位置。

ZOOKEEPER 的配置文件位于 conf 下，默认的有个配置实例文件 zoo _ sample. cfg。将该文件拷贝一份重命名为 zoo. cfg。打开文件并对相关的属性进行配置。具体配置如下。

```
# The number of milliseconds of each tick
tickTime=2000
# The number of ticks that the initial
# synchronization phase can take
initLimit=10
# The number of ticks that can pass between
# sending a request and getting an acknowledgement
syncLimit=5
# the directory where the snapshot is stored.
# do not use /tmp for storage, /tmp here is just
```

```
# example sakes.
dataDir=/data/zookeeper/data
dataLogDir=/data/zookeeper/log
# the port at which the clients will connect
clientPort=2182
```

initLimit：zookeeper 集群中的包含多台 server，其中一台为 leader，集群中其余的 server 为 follower，initLimit 参数配置初始化连接时，follower 和 leader 之间的最长心跳时间。此时该参数设置为 5，说明时间限制为 5 倍 tickTime，即 5 * 2000＝10000ms＝10s。

syncLimit：该参数配置 leader 和 follower 之间发送消息，请求和应答的最大时间长度。此时该参数设置为 2，说明时间限制为 2 倍 tickTime，即 4000ms。

dataDir：数据存放目录。

dataLogDir：日志数据存放路径。

server. X：其中 X 是一个数字，表示这是第几号 server，X 是该 server 所在的 IP 地址。B 配置该 server 和集群中的 leader 交换消息所使用的端口，C 配置选举 leader 时所使用的端口。由于配置的是伪集群模式，所以各个 server 的 B、C 参数必须不同。

配置 ZOOKEEPER 的环境变量，在/etc/profile 在文件的末尾添加如下内容。

```
export ZOOKEEPER_HOME=/usr/app/zookeeper
export PATH=$PATH:$ZOOKEEPER_HOME/bin
```

注意：dataDir 和 dataLogDir 要手动创建目录保存数据，如果采用默认的方式，在 ZOOKEEPER 重启后里面的数据将会丢失。

日志信息的配置的 conf/log4j. properties。

启动测试：

```
[root@YH121 bin]# ./zkServer.sh  status
JMX enabled by default
Using config: /software/zookeeper/bin/.../conf/zoo.cfg
Mode: standalone
```

(3) 使用 ZOOKEPPER 实现 HA

使用 ZOOKEEPER 实现 Spark 集群的 HA，目前采用三台 ZOOKEEPER 服务器，三台 Master。

ZOOKEEPER 集群的安装需要在配置文件中指定所有的服务器地址，配置如下。

```
# The number of milliseconds of each tick
tickTime=2000
# The number of ticks that the initial
# synchronization phase can take
initLimit=10
# The number of ticks that can pass between
# sending a request and getting an acknowledgement
syncLimit=5
# the directory where the snapshot is stored.
# do not use /tmp for storage, /tmp here is just
# example sakes.
dataDir=/data/zookeeper/data
```

```
#日志数据
dataLogDir=/data/zookeeper/log
# the port at which the clients will connect
clientPort=2182
#集群环境
server.0=master:2888:3888
server.1=worker1:2888:3888
server.2=worker2:2888:3888
```

配置完成后需要将该机器的配置拷贝到其他机器上，使用 scp 命令将 ZOOKEEPER 的安装目录、环境变量配置以及工作目录拷贝到其他的节点上。

① 拷贝安装目录 scp-r/usr/app/zookeeper root@worker1：/usr/app/zookeeper。

② 拷贝环境变量配置 scp /etc/profile root@worker1：/etc/profile。

③ 拷贝工作目录 scp-r /data/zookeeper root@worker1：/data/zookeeper。

最后修改 myid 文件。到/data/zookeeper/myid（如果没有可以手动创建）里面修改 server 的 id，在修改 id 的时候要保证整个集群中的 id 不重复。

至此，ZOOKEEPER 的环境已安装完成，接下来将 Spark 集群的恢复策略设置成 ZOOKEEPER 集群，需要在 SparkEnv 中添加如下配置。

```
export SPARK_DAEMON-JAVA_OPTS="Dspark.deploy.recoveryMode=ZOOKEEPER
      -Dspark.deploy.zookeeper.url=Master:2181,Worker1:2181,Worker2:2181
      -Dspark.deploy.zookeeper.dir=/spark"
```

其中 Dspark. deploy. recoveryMode 指定 Spark 的恢复策略是 zookeeper 管理；-Dspark. deploy. zookeeper. url 指定集群中 Master 的机器；-Dspark. deploy. zookeeper. dir 指定 zookeeper的工作目录。同时注释掉 SPARK _ MASTER _ ID。

(4) 启动集群

要启动集群首先要到每台 ZOOKEEPER 上去启动相应的服务，依次到 master、worker1、worker2 上启动 ZOOKEEPER。

```
[root@master / ]# zkServer.sh start
JMX enabled by default
Using config: /usr/app/zookeeper/bin/../conf/zoo.cfg
Starting zookeeper ... STARTED
```

如果此时使用 zkServer. shstatus 查看状态时，会显示如下信息。

```
[root@master / ]# zkServer.sh status
JMX enabled by default
Using config: /usr/app/zookeeper/bin/../conf/zoo.cfg
Error contacting service. It is probably not running.
```

这是由于 ZOOKEEPER 集群中配置的其他节点并未启动造成的（此时查看 ZOOKEEPER 日志会发现该机器在不断尝试连接其他的 ZOOKEEPER 服务器）。此时，到其他的节点将 ZOOKEEPER 服务启动后，查看 ZOOKEEPER 的状态，显示以下信息。

```
[root@master / ]# zkServer.sh status
JMX enabled by default
Using config: /usr/app/zookeeper/bin/../conf/zoo.cfg
Mode: follower
```

其中有一台的 Mode 状态为 leader，其他两台的状态为 follower。

此时去启动 Spark 的集群，同样到其他两台机器上手动启动 Master。启动结束后，可以通过浏览器查看 Master 的状态（访问地址 http://IP:8080/）。

① Active 状态的 Master：

Spark 1.6.0 **Spark Master at spark://master:7077**

URL: spark://master:7077
REST URL: spark://master:6066 *(cluster mode)*
Alive Workers: 2
Cores in use: 12 Total, 0 Used
Memory in use: 4.0 GB Total, 0.0 B Used
Applications: 0 Running, 0 Completed
Drivers: 0 Running, 0 Completed
Status: ALIVE

② STANDBY 状态的 Master：

Spark Master at spark://worker1:7077

URL: spark://worker1:7077
REST URL: spark://worker1:6066 *(cluster mode)*
Alive Workers: 0
Cores in use: 0 Total, 0 Used
Memory in use: 0.0 B Total, 0.0 B Used
Applications: 0 Running, 0 Completed
Drivers: 0 Running, 0 Completed
Status: STANDBY

(5) Master HA 测试

主要从两个方面对于集群进行测试，ZOOKEEPER 集群和 Spark 集群。

ZOOKEEPER 集群测试。测试过程中会采用手动关闭状态为 leader 状态的进程来模拟故障，首先观察下关闭之前各节点的状态，其中 master 和 worker2 两台机器的状态为 follower，worker1 的状态为 leader。关闭掉 worker1 上的 ZOOKEEPER 进程之后，worker2 的状态变为 leader。Master 仍然是 follower，worker1 的状态为 notrunning。

① 关闭前：

```
[root@master sbin]# zkServer.sh status
JMX enabled by default
Using config: /usr/app/zookeeper/bin/../conf/zoo.cfg
Mode: follower
```

```
[root@worker1  sbin]#  zkServer.sh  status
JMX enabled by default
Using config:  /usr/app/zookeeper/bin/../conf/zoo.cfg
Mode: leader
```

```
[root@worker2  sbin]#  zkServer.sh  status
JMX enabled by default
Using config:  /usr/app/zookeeper/bin/../conf/zoo.cfg
Mode: follower
```

② 关闭后：

```
[root@master  sbin]#  zkServer.sh  status
JMX enabled by default
Using config:  /usr/app/zookeeper/bin/../conf/zoo.cfg
Mode: follower
```

```
[root@worker1  sbin]#  zkServer.sh  status
JMX enabled by default
Using config:  /usr/app/zookeeper/bin/../conf/zoo.cfg
Error contacting service.  It is probably not running.
```

```
[root@worker2  sbin]#  zkServer.sh  status
JMX enabled by default
Using config:  /usr/app/zookeeper/bin/../conf/zoo.cfg
Mode:  leader
```

Spark 集群测试。同样的，通过手动关闭状态的 Active 的 Master 来观察各个节点（已启动 Master 进程并在 SparlEnv 中指定的机器）上 Master 的状态。

① 关闭前：

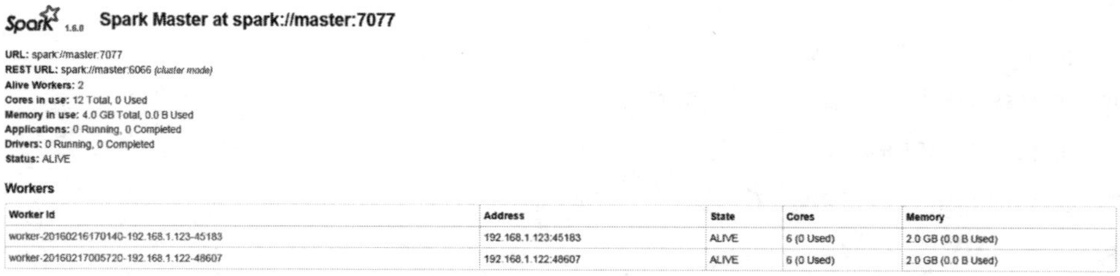

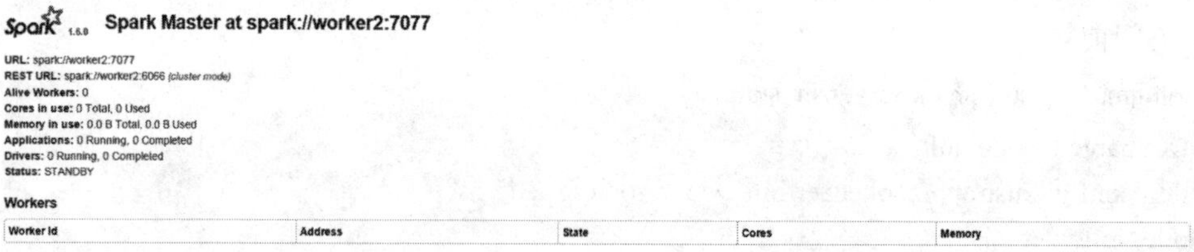

Spark 1.6.0 Spark Master at spark://master:7077

URL: spark://master:7077
REST URL: spark://master:6066 *(cluster mode)*
Alive Workers: 2
Cores in use: 12 Total, 0 Used
Memory in use: 4.0 GB Total, 0.0 B Used
Applications: 0 Running, 0 Completed
Drivers: 0 Running, 0 Completed
Status: ALIVE

Workers

Worker Id	Address	State	Cores	Memory
worker-20160216170140-192.168.1.123-45183	192.168.1.123:45183	ALIVE	6 (0 Used)	2.0 GB (0.0 B Used)
worker-20160217005720-192.168.1.122-48607	192.168.1.122:48607	ALIVE	6 (0 Used)	2.0 GB (0.0 B Used)

Spark 1.6.0 Spark Master at spark://worker1:7077

URL: spark://worker1:7077
REST URL: spark://worker1:6066 *(cluster mode)*
Alive Workers: 0
Cores in use: 0 Total, 0 Used
Memory in use: 0.0 B Total, 0.0 B Used
Applications: 0 Running, 0 Completed
Drivers: 0 Running, 0 Completed
Status: STANDBY

Workers

Worker Id	Address	State	Cores	Memory

Spark 1.6.0 Spark Master at spark://worker2:7077

URL: spark://worker2:7077
REST URL: spark://worker2:6066 *(cluster mode)*
Alive Workers: 0
Cores in use: 0 Total, 0 Used
Memory in use: 0.0 B Total, 0.0 B Used
Applications: 0 Running, 0 Completed
Drivers: 0 Running, 0 Completed
Status: STANDBY

Workers

Worker Id	Address	State	Cores	Memory

② 关闭后：原有的 Active 节点无法访问。

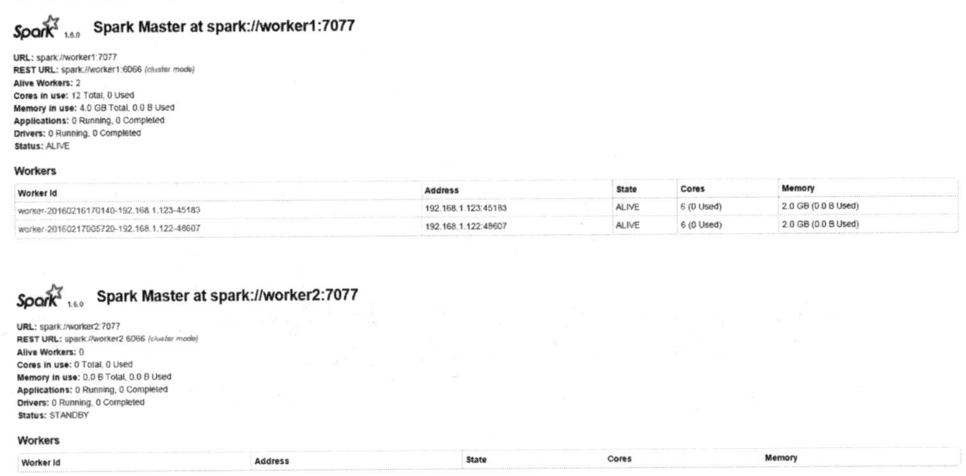

测试结果显示，ZOOKEEPER 集群在 leader 节点出现故障时，集群会选举出新的 leader 为 Spark 的集群提供 Had 的保障。另外，如果 Spark 集群中的 Master 节点遇到故障时，ZOOKEEPER 会从其他状态为 STANDBY 状态的 Master 中选出新的 Master 作为 Active，并保存了集群的元数据来恢复整个集群运行。

15.5 小结

本章节初步介绍了 Spark 集群的几种恢复策略，着重阐释 ZOOKEEPER 的工作机制。通过本章节读者可以初步了解其内部的工作原理。了解 ZOOKEEPER 是如何管理 Spark 集群并保证 Spark 的 HA。同时了解 ZOOKEEPER 和 Spark 整合的相关配置。

第16章 Spark内核架构解密

Spark 内核（Core）是 Spark 的核心，其他所有 Spark 核心组件如 Spark SQL、Spark Streaming、Spark MLib、Spark Graphx 和 Spark R 都是基于 SparkCore 的扩展子框架实现。所以，Spark Core 才是体现 Spark 核心设计思想的地方。本章通过对 Spark 的运行过程涉及的几个关键的 Spark Core 对象如 SparkContext、Driver、Master 和 Worker 等原理的解析来解密 SparkCore 的架构。

16.1 Spark 的运行过程

16.1.1 SparkContext 的创建过程

SparkContext 是整个 Spark 集群的入口。Spark 程序的编写也是要基于 SparkContext 的。其中 RDD 需要通过 SparkContext 来创建，编写的 Spark 的程序也要通过 SparkContext 发布到集群中，用来获取程序运行的计算资源的 SchedulerBackend，是在 SparkContext 实例化的时候创建的。此外在程序运行过程中都是以 SparkContext 为核心进行资源调度的。

值得注意的是，在每个 JVM 中同时只能有一个活动的 SparkContext，如果需要创建新的 SparkContext，需要首先停掉活动的 SparkContext。当然这个限制以后可能会被去掉。

在创建 SparkContext 的时候，需要依赖相应的配置信息，这些配置信息以 SparkConf（其中描述了程序运行时的配置信息，这里面的配置信息将会覆盖系统默认的配置信息）的形式传递给 SparkContext。在该配置信息中可以指定应用程序的名称，程序运行模式、运行时所需要的 CPU 核数以及集群的 Master 路径等信息。

SparkContext 在创建的时候会通过 createTaskScheduler 方法初始化其内部的几大核心组件，其中包括 DAGScheduler、TaskScheduler、SchedulerBackend（SchedulerBackend 在 local 模式下的具体实现是 LocalBackend，在 standalone 模式下的具体实现是 SparkDeploySchedulerBackend）等。createTaskScheduler 方法首先会根据不同的运行模式来创建 TaskScheduler 和 SchedulerBackend。需要说明的是如果是本地运行模式（local），程序一旦失败，不会重新计算。TaskScheduler 根据 SchedulerBackend 进行初始化操作，根据不同的调度算法（目前支持 FIFO 和 FAIR，默认情况下采用 FIFO）创建调度池。最后返回类型为

Tuple2 类型的返回值，其中包含 SchedulerBackend 和 TaskSchedulerscheduler。具体的创建过程如图 16-1 所示。

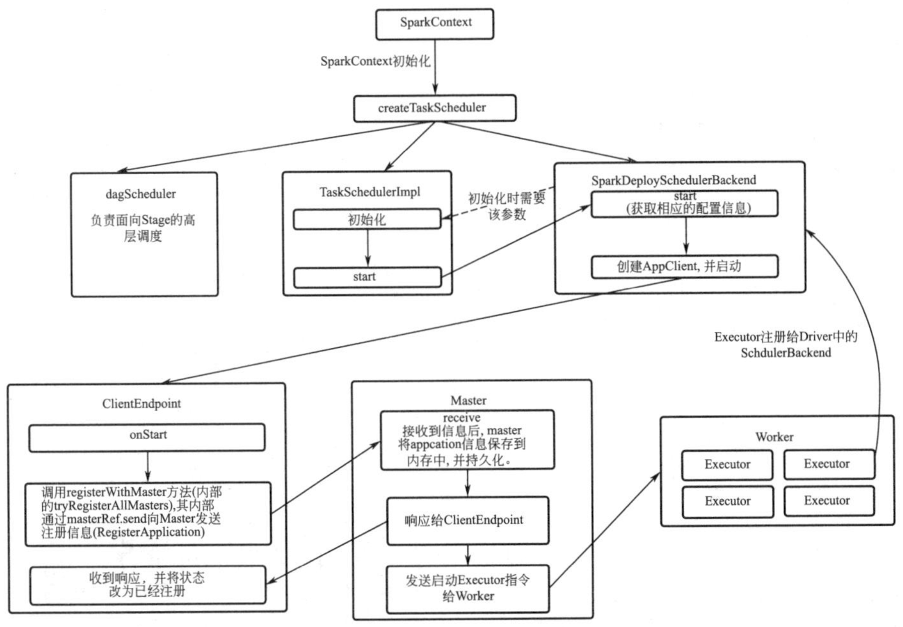

图 16-1　SparkContext 的创建过程

创建 SparkContext 的某个示例代码如下所示。

```
//创建SparkConf对象，用于存储运行时参数
val conf = new SparkConf()
// 设置应用程序名称，在运行程序的监控界面可以看到
    conf.setAppName("Application Name")
// 指定程序的运行模式
conf.setMaster("local") //本地运行模式
//conf.setMaster("spark://Master:7077") //集群裕兴模式
// 通过传入的SparkConf创建SparkContext对象
val sc = new SparkContext(conf)
/**
* 具体的业务代码
*/
// 程序结束后需要关闭SparkContext
sc.stop()
```

16.1.2　Driver 的注册过程

所谓的注册就是进行信息的登记，同样 Driver 的注册也是如此。Driver 的注册过程是主动的。

如图 16-2 所示，DriverProgram 是提交的应用程序，整个程序的核心是 SparkContext。SparkContext 在实例化的时候会创建 SchedulerBackend（其位置于 TaskScheduler 内部），在 standalone 的模式下，其具体的实现是 SparkDeploySchedulerBackend。在 SparkDeploySchedulerBackend 的 start 方法中会创建 AppClient。如下所示。

```
client=newAppClient(sc. env. rpcEnv,masters,appDesc,this,conf)
client. start()
```

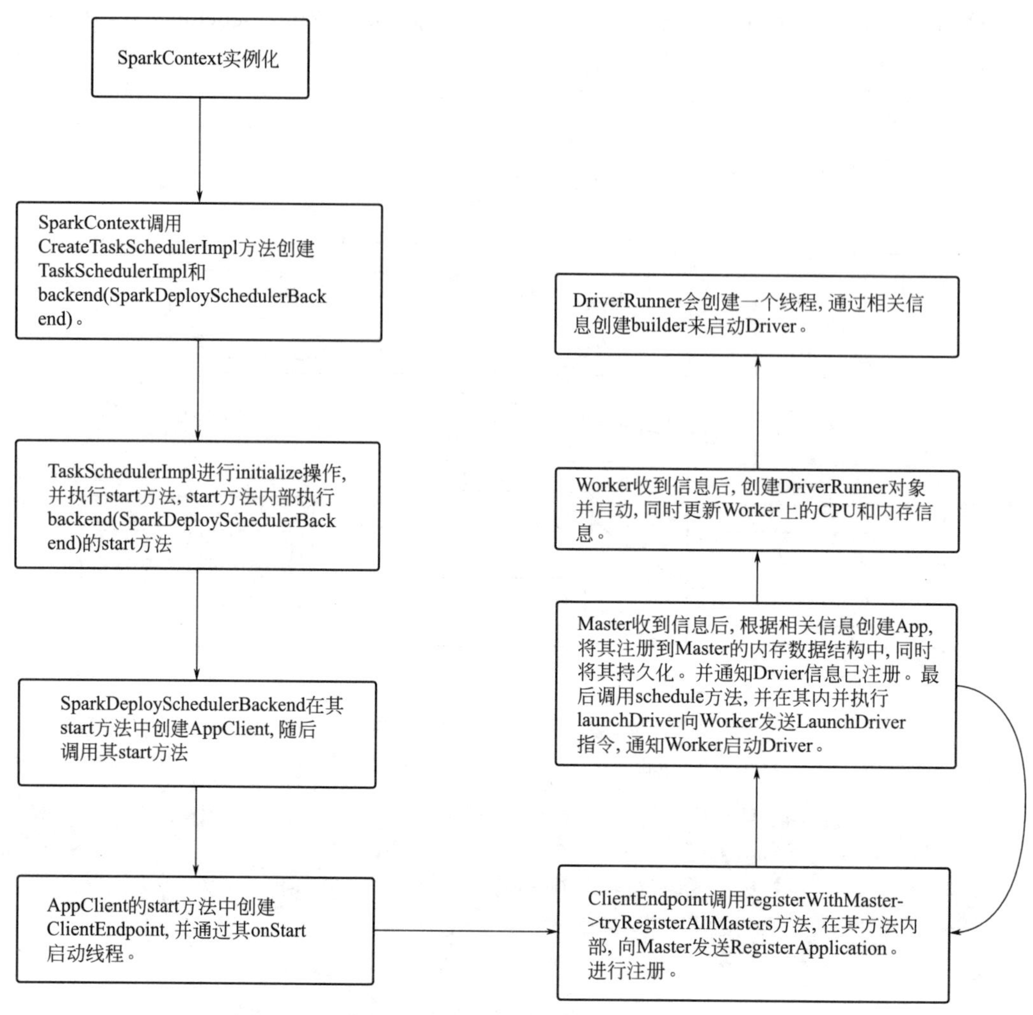

图 16-2　Driver 的注册过程

而在 client. start () 的过程中会通过 registerWithMaster 方法向 Master 进行注册，注册通过 RegisterApplication 携带相关的信息（主要包括程序的描述信息和自身的引用）给 Master。

masterRef. send(RegisterApplication(appDescription, self))

Master 收到信息后进行判断，如果 Master 是 RecoveryState. STANDBY 模式则忽略不进行相应程序。否则会调用 registerApplication 方法将 application 的相关信息（包括 application 本身、application 的 ID、对应的 Driver，地址信息）保存在 Master 中，同时也会进行持久化操作。上述操作之后，Master 会向 Drvier 发送应答信息，同样是通过 RegisteredApplication 传递消息。Master 最后会通过 schedule 方法调度当前可用资源进行计算，相关源码如下所示。

```
case RegisterApplication(description, driver) => {
// TODO Prevent repeated registrations from some driver
    if (state == RecoveryState.STANDBY) {
// ignore, don't send response
    } else {
```

```
    logInfo("Registering app " + description.name)
    val app = createApplication(description, driver)
    registerApplication(app)
    logInfo("Registered app " + description.name + " with ID " + app.id)
    persistenceEngine.addApplication(app)
    driver.send(RegisteredApplication(app.id, self))
    schedule()
  }
}
```

在 schedule 方法内部，会判断当前的状态是否是 ACTIVE，如果是，会在满足 Driver 内存和 CPU 核数的 Worker 上启动 Driver，并将该 Driver 从等待计算的队列中移除。在启动过程中 Worker 和 Driver 会相互记录彼此的信息，同时 Master 会向 Worker 发送启动指令（LaunchDriver），Worker 接收到指令后，会实例化 DriverRunner 对象，并通过其 start 方法启动 Driver，同时更新已使用的资源（内存和 CPU）。最后在每个 Worker 上执行计算任务。

值得注意的是，DriverRunner 会重新启动一个线程来启动 Driver，并将 Drvier 的状态信息（DriverStateChanged）发送给 Worker，Worker 接收到信息之后，会将 Driver 的信息从运行队列中移到完成队列中，更新 Worker 上的资源信息，Worker 也会将 Drvier 的状态信息发送给 Master，Master 接收到信息之后同样会对存储在自身的 Driver 和 Worker 信息进行更新，包括从运行队列转移到完成队列、从持久化引擎中移除、更新 Driver 状态、记录 Drvier 异常、移除 Driver 中存储的 Worker 信息等。最后再次执行 schedule 方法调度当前可用资源进行计算其他任务。

Spark 中各个对象之间的通信是通过 RPC 机制，如果两个对象之间想要进行通信就需要继承 ThreadSafeRpcEndpoint。ThreadSafeRpcEndpoint 继承自 RpcEndpoin，保证了线程之间的通信是安全的。对于 RpcEndpoint 而言，其生命周期包括以下部分：构造（constructor）、启动（onStart）、消息接收（receive *）、停止（onStop）。在 RpcEndpoint 接口中提供了 create 方法来创建 RpcEnv。在 Spark 的新版本中提供了 RpcEnv 的两种默认实现：AkkaRpcEnv 和 NettyRpcEnv。其中 NettyRpcEnv 是默认采用的方式。

Master 在整个集群中负责资源的调度，同时会记录整个集群中各个节点信息机器节点上的的资源信息等，同时会发送相关的指令给 Worker，通知其分配资源进行任务的计算。因此在整个集群运行过程中，Master 和 Worker 不可避免地要进行各种信息的通信。

在 Spark 的启动和运行过程中，Master 和 Worker 都存在通信过程。其中包括在集群启动是 Worker 向 Master 注册；提交程序是 Diver 向 Master 注册；Worker 中 Executor 的状态发生变化；Driver 的状态发生变化以及 Master 和 Worker 之间以一定的频率进行信息传输等，如图 16-3 所示。

Worker 在启动时（通过 onStart 启动）除了会给自身创建工作目录，绑定相关的 webui 外，还会通过 registerWithMaster 方法向 Master 进行主动注册，当然这里的 Master 既包含 Active 状态的也包括 STANDBY 状态的。Worker 向 Master 进行注册的时候会通过 RegisterWorker 携带自身的相关信息，这些信息主要包括 id、workerHost、workerPort、workerRef、cores、memory、workerUiPort、publicAddress，同时 Worker 也会要求 Master 以 RegisterWorkerResponse 方式给予响应。相关源码如下所示。

```
private def registerWithMaster(masterEndpoint: RpcEndpointRef): Unit = {
  masterEndpoint.ask[RegisterWorkerResponse](RegisterWorker(
    workerId, host, port, self, cores, memory, webUi.boundPort, publicAddress))
.onComplete {
    // This is a very fast action so we can use "ThreadUtils.sameThread"
    case Success(msg) =>
Utils.tryLogNonFatalError {
      handleRegisterResponse(msg)
    }
    case Failure(e) =>
     logError(s"Cannot register with master: ${masterEndpoint.address}", e)
     System.exit(1)
   }(ThreadUtils.sameThread)
}
```

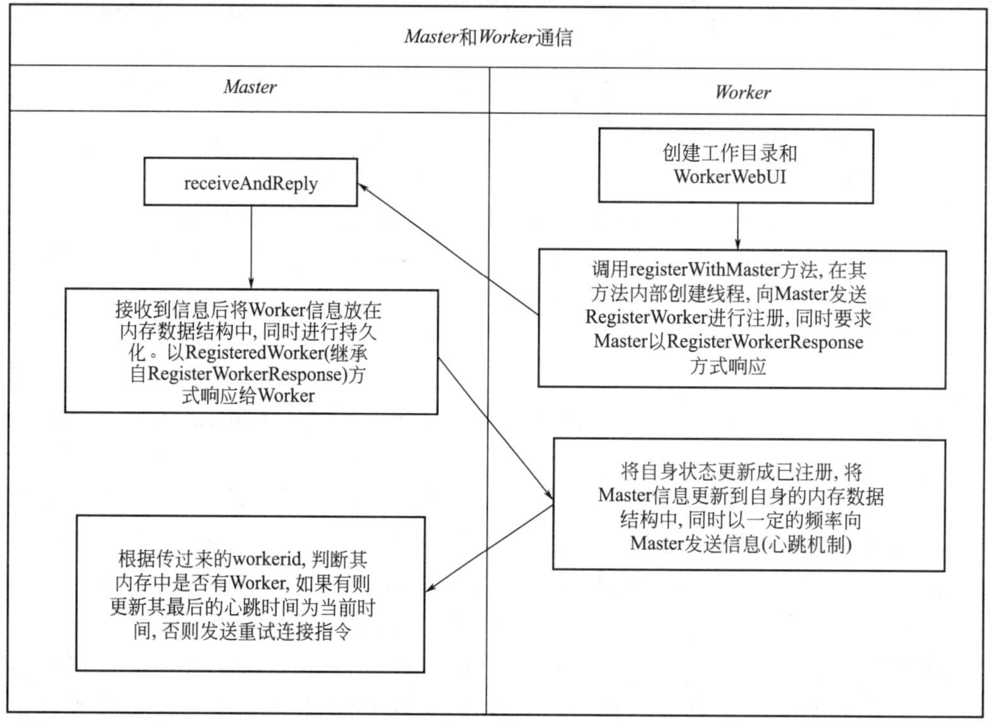

图 16-3　Master 和 Worker 的通信

Master 接收到 Worker 的注册信息（RegisterWorker）后，如果当前的状态是 RecoveryState. STANDBY，那么会以 MasterInStandby（其继承自 RegisterWorkerResponse）的方式回应给注册的 Worker；如果该 Master 已经包含了 Worker 的 id，那么会以 RegisterWorkerFailed（其继承自 RegisterWorkerResponse）的方式回应给注册的 Worker，并通知 Worker 注册的 id 重复。如果非上述两种情况，Master 会根据 Worker 传递过来的信息创建 WorkerInfo 对象，然后调用本身的 registerWorker 方法。相关源码如下所示。

```
override def receiveAndReply(context: RpcCallContext): PartialFunction[Any, Unit] = {
  case RegisterWorker(
    id, workerHost, workerPort, workerRef, cores, memory, workerUiPort, publicAddress) => {
    logInfo("Registering worker %s:%d with %d cores, %s RAM".format(
```

```
        workerHost, workerPort, cores, Utils.megabytesToString(memory)))
      if (state == RecoveryState.STANDBY) {
        context.reply(MasterInStandby)
    } else if (idToWorker.contains(id)) {
        context.reply(RegisterWorkerFailed("Duplicate worker ID"))
    } else {
        val worker = new WorkerInfo(id, workerHost, workerPort, cores, memory,
          workerRef, workerUiPort, publicAddress)
        if (registerWorker(worker)) {
          persistenceEngine.addWorker(worker)
          context.reply(RegisteredWorker(self, masterWebUiUrl))
          schedule()
        } else {
          val workerAddress = worker.endpoint.address
          logWarning("Worker registration failed. Attempted to re-register worker at same " +
            "address: " + workerAddress)
          context.reply(RegisterWorkerFailed("Attempted to re-register worker at same address: "
            + workerAddress))
        }
      }
  }
```

在该方法内部，Master 会对本身存储的 Worker 信息进行遍历，将与进行注册的 Worker 的主机（host）和端口（port）相同的，并且状态为 WorkerState. DEAD 的 Worker 从自己的内存缓存中移除掉。再将要注册的 Worker 的地址放到自身的内存缓存 addressToWorker 中，如果 addressToWorker 中有地址和当前注册的 Worker 相同，Master 会将其对应的 Worker 状态更新为 WorkerState. UNKNOWN，同时将其相关的信息移除。包括 Worker 下的 Excutors 和 Drivers。除了地址信息外，诸如 Worker 本身及其对应的 id，Master 都会进行相应的记录，同时会对其进行持久化操作。

持久化完成后，Master 会对注册的 Worker 进行响应，其同样是通过 RegisteredWorker 方式来通知 Worker。相关源码如下所示。

```
case RegisteredWorker(masterRef, masterWebUiUrl) =>
      logInfo("Successfully registered with master " + masterRef.address.toSparkURL)
      registered = true
      changeMaster(masterRef, masterWebUiUrl)
      forwordMessageScheduler.scheduleAtFixedRate(new Runnable {
        override def run(): Unit = Utils.tryLogNonFatalError {
          self.send(SendHeartbeat)
        }
      }, 0, HEARTBEAT_MILLIS, TimeUnit.MILLISECONDS)
      if (CLEANUP_ENABLED) {
        logInfo (
          s"Worker cleanup enabled; old application directories will be deleted in: $workDir")
        forwordMessageScheduler.scheduleAtFixedRate(new Runnable {
          override def run(): Unit = Utils.tryLogNonFatalError {
            self.se nd(WorkDirCleanup)
          }
        }, CLEANUP_INTERVAL_MILLIS, CLEANUP_INTERVAL_MILLIS, TimeUnit.MILLISECONDS)
      }
```

Worker 接收到 Master 的响应之后，会将自身标记为已注册状态，同时更新自身关于 Master 的相关信息，并取消剩余的重试（已表明注册成功，不需要重试）。然后 Worker 会启动一个线程以一定的频率向 Master 汇报当前 worker 的情况，这就是所谓的心跳机制。相关源码如下所示。

```
case Heartbeat(workerId, worker) => {
   idToWorker.get(workerId) match {
    case Some(workerInfo) =>
      workerInfo.lastHeartbeat = System.currentTimeMillis()
    case None =>
      if (workers.map(_.id).contains(workerId)) {
        logWarning(s"Got heartbeat from unregistered worker $workerId." +
          " Asking it to re-register.")
        worker.send(ReconnectWorker(masterUrl))
} else {
        logWarning(s"Got heartbeat from unregistered worker $workerId." +
          " This worker was never registered, so ignoring the heartbeat.")
      }
   }
}
```

汇报的时候是通过 Heartbeat 传递相关的参数，Master 收到心跳后，会判断当前的队列中是否都有该 Worker，如果有，更新其最后的心跳时间即可，否则会让该 Worker 尝试重新连接 Master。

最后，Master 会通过 schedule 方法对集群中的资源进行重新调度并分配给等待计算的其他程序执行计算任务。

16.1.3 Worker 中任务的执行

在 Spark 集群中，作业的运行实际上是在每个 Worker 所在的节点上进行的。在 Master 调用 schedule 方法的时候，首先会通过 Random. shuffle 将 Worker 的集合打乱（目的是为了进行负载均衡），然后从 Driver 的等待队列中取出 Driver，如果 Worker 的计算资源（内存和 CPU）满足 Driver 的要求，那么会在该 Worker 上启动 Driver。最后执行 startExecutorsOnWorkers 方法。

在 startExecutorsOnWorkers 方法中会根据集群资源的状况，为当前的 Executors 在相应的 Worker 上申请计算资源，然后 Master 会将相应的信息通过 LaunchExecutor 的方式发送到指定的 Worker 上，这些信息中包括执行应用程序的 id、Executor 的 id、应用程序的描述信息以及申请的 core 和内存信息。同时 Master 以 ExecutorAdded 的方式发送给应用程序对应的 Driver 上，而这些信息则包括 Executor 的 id、worker 的 id、worker 的 hostPort、以及申请的 core 和内存信息。相关源码如下所示。

```
private[worker] def handleExecutorStateChanged(executorStateChanged: ExecutorStateChanged):
  Unit = {
  sendToMaster(executorStateChanged)
  val state = executorStateChanged.state
  if (ExecutorState.isFinished(state)) {
    val appId = executorStateChanged.appId
    val fullId = appId + "/" + executorStateChanged.execId
```

```
      val message = executorStateChanged.message
      val exitStatus = executorStateChanged.exitStatus
      executors.get(fullId) match {
        case Some(executor) =>
          logInfo("Executor " + fullId + " finished with state " + state +
            message.map(" message " + _).getOrElse("") +
            exitStatus.map(" exitStatus " + _).getOrElse(""))
          executors-= fullId
          finishedExecutors(fullId) = executor
          trimFinishedExecutorsIfNecessary()
          coresUsed-= executor.cores
          memoryUsed-= executor.memory
        case None =>
          logInfo("Unknown Executor " + fullId + " finished with state " + state +
            message.map(" message " + _).getOrElse("") +
            exitStatus.map(" exitStatus " + _).getOrElse(""))
      }
      maybeCleanupApplication(appId)
    }
  }
}
```

当 Worker 接收到来自 Master 的 LaunchExecutor 指令后，会判断如果该指令来自于 Active 级别的 Master 时，首先会创建 Executor 在本地文件系统的工作目录，然后会根据相关的信息创建 ExecutorRunner 对象，其内部会启动一个线程，在该线程内部会通过 CommandUtils 获取相应的指令来启动一个进程运行 Executor，该进程就是 CoarseGrainedSchedulerBackend，它和 Executor 是一一对应的。而 Executor 才是真正负责 Task 计算的，Executor 在实例化的时候会创建一个线程池，以多线程的方式对任务进行高效的执行。

当 Executor 执行完成后，会通过 StateChanged（其中包括应用程序 id，Excutor 的 id，执行状态、异常信息等）的方式发送给 Worker。Worker 接收到 Executor 的状态信息后，如果该 Excutor 是完成状态，会将该 Executor 及其占用的资源信息从 Worker 的内存数据结构中移除。实际上在 Worker 接收到信息后，也会将状态信息发送给 Master。同样的，在 Master 接收到状态变更信息后也会对自身内存数据结构中的数据进行更新。相比 Worker 而言，Master 多了对 Application 相关信息的处理。

16.1.4 任务的调度过程

Spark 的任务调度是在资源调度的基础上进行的，资源的管理和调度是由 Master 负责的。在 Spark 集群中，对于 Application 的注册、Worker 的注册、Executor 状态的改变，以及 Driver 状态的改变等相关资源的变化都会导致 Master 调用 schedule 方法，对集群中的资源重新调度。而任务调度是之所以需要依赖于资源调度，是因为任务的执行需要相应的计算资源。

Spark 中任务调度是通过 DAGScheduler、TaskScheduler、SchedulerBackend 等进行的作业调度。如图 16-4 所示。其中 DAGScheduler 负责高层调度。Job 提交的时候，DAGScheduler 会对 stage 进行划分，划分的依据主要是根据 RDD 是否进行 Shuffle，即以 RDD 整个过程为主导，从后往前回溯，如果遇到宽依赖就进行一次 Shuffle，并且产生一个 stage，DAGScheduler 会将产生的 Stage 以 TaskSet 的形式交给底层的调度器 TaskScheduler，

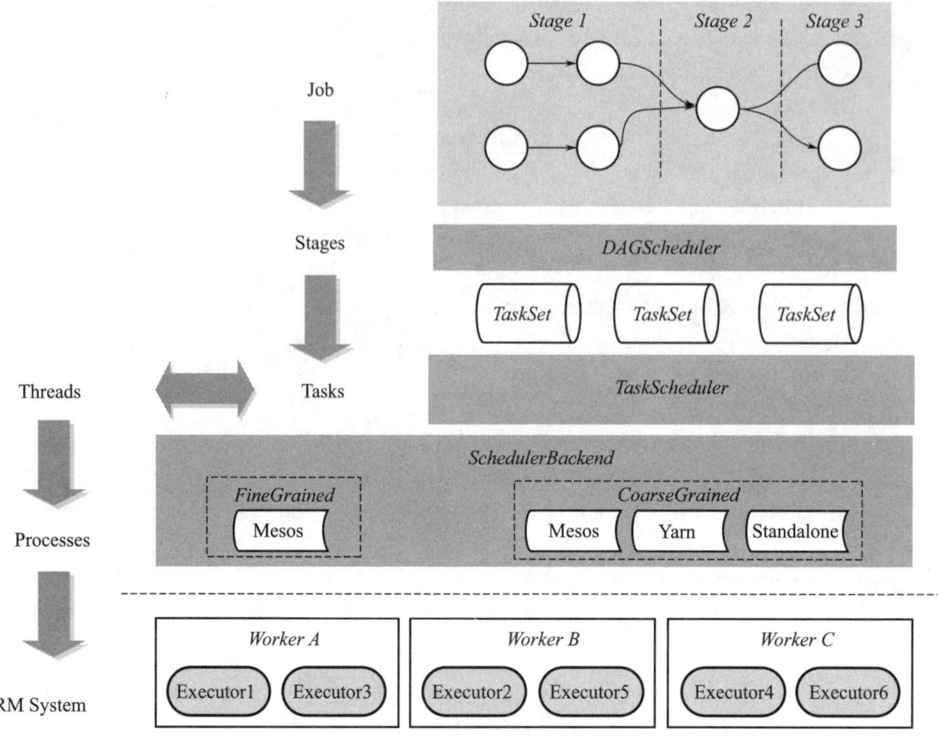

图 16-4　Spark 任务调度

TaskScheduler 接收到 TaskSet 后，会通过其内部的 submitTasks 方法在进行作业的提交。相关源码如下所示。

```
override def submitTasks(taskSet: TaskSet) {
  val tasks = taskSet.tasks
  logInfo("Adding task set " + taskSet.id + " with " + tasks.length + " tasks")
  this.synchronized {
    val manager = createTaskSetManager(taskSet, maxTaskFailures)
    val stage = taskSet.stageId
    val stageTaskSets =
      taskSetsByStageIdAndAttempt.getOrElseUpdate(stage, new HashMap[Int, TaskSetManager])
    stageTaskSets(taskSet.stageAttemptId) = manager
    val conflictingTaskSet =stageTaskSets.exists { case (_, ts) =>
      ts.taskSet != taskSet && !ts.isZombie
    }
    if (conflictingTaskSet) {
      throw new IllegalStateException(s"more than one active taskSet for stage $stage:" +
        s" ${stageTaskSets.toSeq.map{_._2.taskSet.id}.mkString(",")}")
    }
    schedulableBuilder.addTaskSetManager(manager, manager.taskSet.properties)
    if (!isLocal && !hasReceivedTask) {
      starvationTimer.scheduleAtFixedRate(new TimerTask() {
        override def run() {
          if (!hasLaunchedTask) {
            logWarning("Initial job has not accepted any resources; " +
"check your cluster UI to ensure that workers are registered " +
```

```
            "and have sufficient resources")
} else {
          this.cancel()
        }
      }
    }, STARVATION_TIMEOUT_MS, STARVATION_TIMEOUT_MS)
  }
  hasReceivedTask = true
}
backend.reviveOffers()
}
```

TaskScheduler 在提交作业的过程中，会创建 TaskSetManager 用来管理 TaskSet。创建好的 TaskSetManager 会被添加到 schedulableBuilder 中，由 schedulableBuilder 根据不同的调度算法（目前支持 FIFO 和 FAIR 两种策略，默认为 FIFO）来决定 Task 的执行顺序。随后 TaskScheduler 会向 SchedulerBackend（TaskScheduler 初始化时传入的）发送 ReviveOffers 指令，通知其为 Task 准备相应的计算资源，但此时并未分配，只是准备好相应的计算资源。准备好资源后，SchedulerBackend 又会通知 TaskScheduler，此时 TaskScheduler 才会为每个任务分配相应的计算资源，分配的时候会以轮询的方式将任务分配到每个节点上，同时 taskSet 内部也会进行重新洗牌，这样轮询的目的是为了保证负载均衡。SchedulerBackend 在指导 TaskScheduler 分配完资源后，会通过 LaunchTask 向 executor 发送启动 Task 的指令。

16.1.5 Job 执行结果的产生

在 Spark 中，作业的执行首先是由 Driver 会通过 CoarseGrainedSchedulerBackend 向 CoarseGrainedExecutorBackend 发送 LaunchTask 的指令，CoarseGrainedExecutorBackend 接收到指令后，会首先对 TaskDescription 进行反序列化，然后调用 executor 的 launchTask 方法将任务的相关信息（包括 CoarseGrainedSchedulerBackend 本身、Task 的 id、name、attemptNumber、以及序列化之后的任务信息等）传递给 Executor。相关源码如下所示。

```
case LaunchTask(data) =>
    if (executor == null) {
      logError("Received LaunchTask command but executor was null")
      System.exit(1)
} else {
      val taskDesc = ser.deserialize[TaskDescription](data.value)
      logInfo("Got assigned task " + taskDesc.taskId)
      executor.launchTask(this, taskId = taskDesc.taskId, attemptNumber =
taskDesc.attemptNumber,
        taskDesc.name, taskDesc.serializedTask)
    }
```

Executor 接收到信息之后会创建 TaskRunner 对象，并将该对象添加到正在运行的任务队列中，同时调用 Executor 实例化时候创建的线程池来启动任务的执行。相关源码如下所示。

```
def launchTask(
    context: ExecutorBackend,
    taskId: Long,
```

```
    attemptNumber: Int,
    taskName: String,
    serializedTask: ByteBuffer): Unit = {
  val tr = new TaskRunner(context, taskId = taskId, attemptNumber = attemptNumber, taskName,
    serializedTask)
  runningTasks.put(taskId, tr)
  threadPool.execute(tr)
}
```

在线程池启动 TaskRunner 的过程中，首先会创建 Task 的内存管理器 TaskMemoryManager。然后通知 CoarseGrainedExecutorBackend 当前任务的执行状态（当前状态为 TaskState. RUNNING），CoarseGrainedExecutorBackend 也会将收到的状态信息通知 driver。随后 TaskRunner 通过 SparkHadoopUtil 获取相关的配置信息，并根据配置信息会从 HDFS 等文件系统上下载需要运行 Jar 等相关文件，将下载的 Jar 放到当前的内存数据结构中。此时 TaskRunner 会将序列化的 Task 进行反序列化形成要运行的 Task，并绑定 TaskRunner 启动时创建的 TaskMemoryManager。相关源码如下所示。

```
val (taskFiles, taskJars, taskBytes) = Task.deserializeWithDependencies(serializedTask)
updateDependencies(taskFiles, taskJars)
task = ser.deserialize[Task[Any]](taskBytes,Thread.currentThread.getContextClassLoader)
task.setTaskMemoryManager(taskMemoryManager)
```

Task 的 run 方法才是任务真正执行的开始，该方法会返回运行的结果，并更新计数器。在 run 方法内部，ShuffleMapTask 执行 runTask 方法将 RDD 的依赖进行反序列化得到相应的 RDD，再调用 RDD. iterator（）进行遍历计算，遍历的时候会首先判断 RDD 的存储级别是否是 StorageLevel. NONE，如果是（表示没有持久化）则直接进行计算（具体计算的时候是执行 RDD 的 compute 方法），否则将尝试读取 CheckPoint 的数据，如果读不到数据再进行计算。计算的结果会通 shuffleManager 的 Writer 写入到相应的文件。相关源码如下所示。

```
override def runTask(context: TaskContext): MapStatus = {
  // Deserialize the RDD using the broadcast variable.
  val deserializeStartTime = System.currentTimeMillis()
  val ser = SparkEnv.get.closureSerializer.newInstance()
  val (rdd, dep) = ser.deserialize[(RDD[_], ShuffleDependency[_, _, _])](
    ByteBuffer.wrap(taskBinary.value), Thread.currentThread.getContextClassLoader)
_executorDeserializeTime = System.currentTimeMillis()- deserializeStartTime
  metrics = Some(context.taskMetrics)
  var writer: ShuffleWriter[Any, Any] = null
  try {
    val manager = SparkEnv.get.shuffleManager
     writer = manager.getWriter[Any, Any](dep.shuffleHandle, partitionId, context)
    writer.write(rdd.iterator(partition, context).asInstanceOf[Iterator[_ <: Product2[Any, Any]]])
    writer.stop(success = true).get
} catch {
    case e: Exception =>
      try {
        if (writer != null) {
          writer.stop(success = false)
        }
```

```
    } catch {
      case e: Exception =>
        log.debug("Could not stop writer", e)
    }
    throw e
  }
}
```

在 Executor 得到 Task 的执行结果后会将其序列化，通过 CoarseGrainedExecutorBackend 执行 statusUpdate 方法将 Task 的 id、执行完成的状态以及执行的结果等状态信息（StatusUpdate）返回给 DriverEndpoint，并将该任务从正在运行的队列中移除掉。DriverEndpoint（是 CoarseGrainedSchedulerBackend 的内部类）接收到信息后，会对自己存储的计算资源进行更新（因为 Task 计算完成，相应的资源要释放），然后启动其他等待执行的 Task。与此同时，DriverEndpoint 会将状态信息传递给 TaskSchedulerImpl。

16.2 小结

本章主要围绕 Spark 程序的运行过程，对主要的环节结合源码进行分析。通过本章节可以初步了解集群在启动时，Master 和 Worker 的启动以及注册过程、程序执行入口 SparkContext 的创建过程、Driver 的注册过程以及任务如何在 Worker 上执行并产生结果。

第17章 Spark运行原理实战解析

Spark 诞生于 AMPLab，其设计满足了对多数计算场景提供一个通用的计算引擎，面对大数据的数据体量大、快速、类型多样化等特点，Spark 都能够很好地满足需求。从 Spark 诞生以来，越来越受到业界的关注和推崇，尤其在 Hadoop MapReduce 力不从心的时候，Spark 成为一颗耀眼的明星出现在公众视野。

要学好 Spark，就需要对 Spark 内核运行架构有较好地掌握，本章将简要让读者了解 Spark 内核的 Driver 端架构、SparkContext 上下文、DAGScheduler 对 Stage 的划分、任务的提交、资源的调度、Spark 运行逻辑以及不同模式下的 Spark 运行架构实战解析。掌握这些内容才能更好地了解作业运行的内部密码，才能更好地基于集群运行出现的性能问题进行调优，准确定位到错误所在及时作出优化。

17.1 用户提交程序 Driver 端解析

Driver 部分代码包含了 SparkConf、SparkContext，一切应用程序代码由 Driver 端的代码和分布在集群其他机器上的 Executor 代码组成（例如 textFile、flatMap、map 等算子操作），Executor（Executor 是运行在 Worker 节点上的进程里的对象）中实现了线程池并发执行和线程的覆用，且每个线程处理一个 task 任务，task 从 Disk 或 Mem 上读取数据。

Spark Application 的运行不依赖于资源管理器 ClusterManager，也就是说运行时不需要 ClusterManager 的参与（粗粒度分配资源即一次性分配完成），也就是说在程序运行前就以粗粒度分配资源的方式分配好了资源。Driver 运行程序的时候在 main 方法中创建了 SparkContext 对象，SparkContext 本身是程序调度器（包含了高层调度器 DAGScheduler 和底层调度器 TaskScheduler）。

Driver 端是主要用来提交 Spark 程序以及负责资源调度，Driver 端所在的机器一般和 SparkCluster 集群在相同的网络环境下，因为要保证 Driver 和 Executor 进行频繁的通信，并且 Driver 端所在的机器配置基本和 Worker 节点相同，Driver 端所在的机器同样安装了 Spark，但其不属于 Spark 集群的范围。当 Application 提交的时候使用 spark-submit 提交命令（可配置运行时的参数 Mem、CPU 等），一般在生产环境下使用自动化 shell 脚本提交。

Spark 程序的开始后由 SparkConf 设置运行的一些参数，例如使用 setMaster（“Local”）设置本地运行的模式，使用 setAppName 设置运行 Spark 程序的名称，之后触发 Spark 应用程序的入口 SparkContext 的创建来初始化 Spark 应用程序所需要的核心组件 DAGScheduler、TaskScheduler、SchedulerBackend 等。而 DAGScheduler 实现了 Job 中不同的 RDD 分布式数据集构成的 DAG 划分成不同的 Stage。

在 Driver 端由 SparkContext 创建 TaskScheduler 和 SchedulerBackend 是对应的关系，即一个 TaskScheduler 对应一个 SchedulerBackend，当 DAGScheduler 将数据集交由 TaskScheduler 的时候，TaskScheduler 开始负责 Application 中的 Job 调度。Driver 端架构如图 17-1 所示。

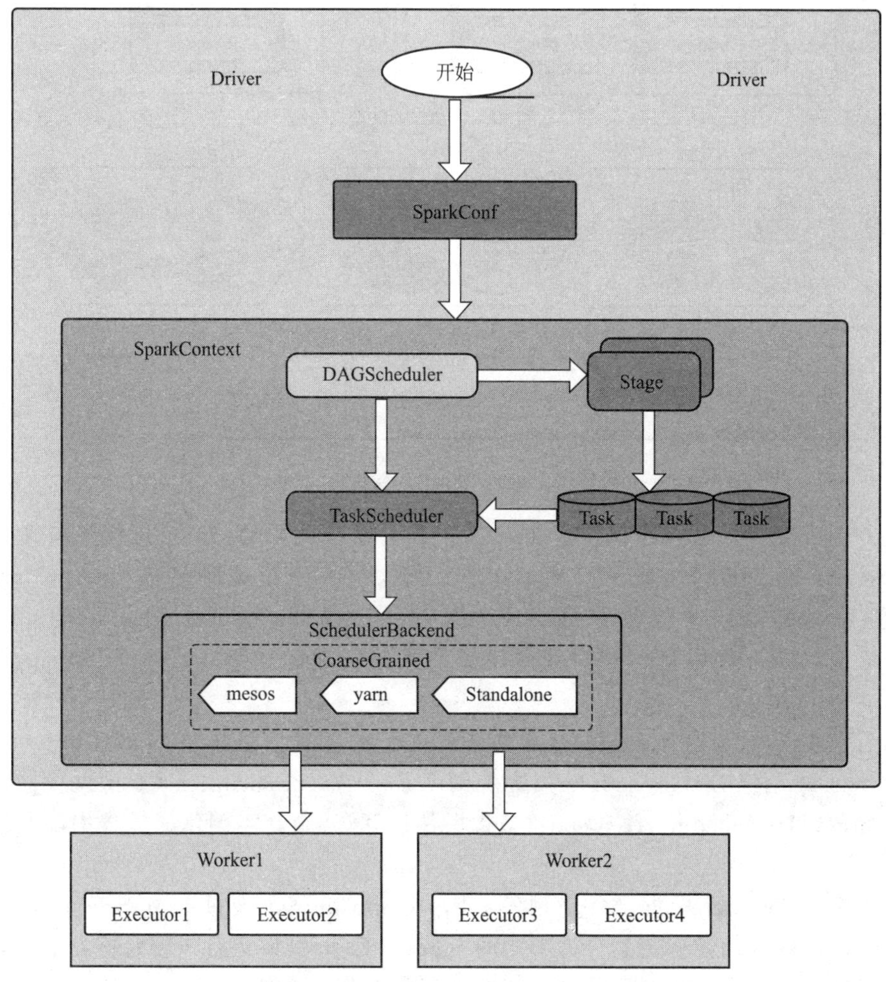

图 17-1　Driver 端架构图

DAGScheduler 划分完 Stage 后将 TaskSet 提交给 TaskScheduler，而 SchedulerBackend 负责将 Application 取得的资源传给 TaskScheduler 完成资源的分配，TaskScheduler 通过 ClusterManager 在集群中启动某个 Worker 上的 Executor，默认情况下为每一个 Executor 开辟一个线程池，由此 TaskSet 在线程池中被不同的线程并发执行。

DAGScheduler 实现了 Job 中不同的 RDD 分布式数据集构成的 DAG 的不同的 Stage 划分，在此过程中形成的 TaskSet 交由 TaskScheduler，需要注意的是 RDD 中的每个 Partition

对应一个 Task，且这些 Task 计算逻辑完全相同，只是数据不同。图 17-2 为 Stage 的划分过程。

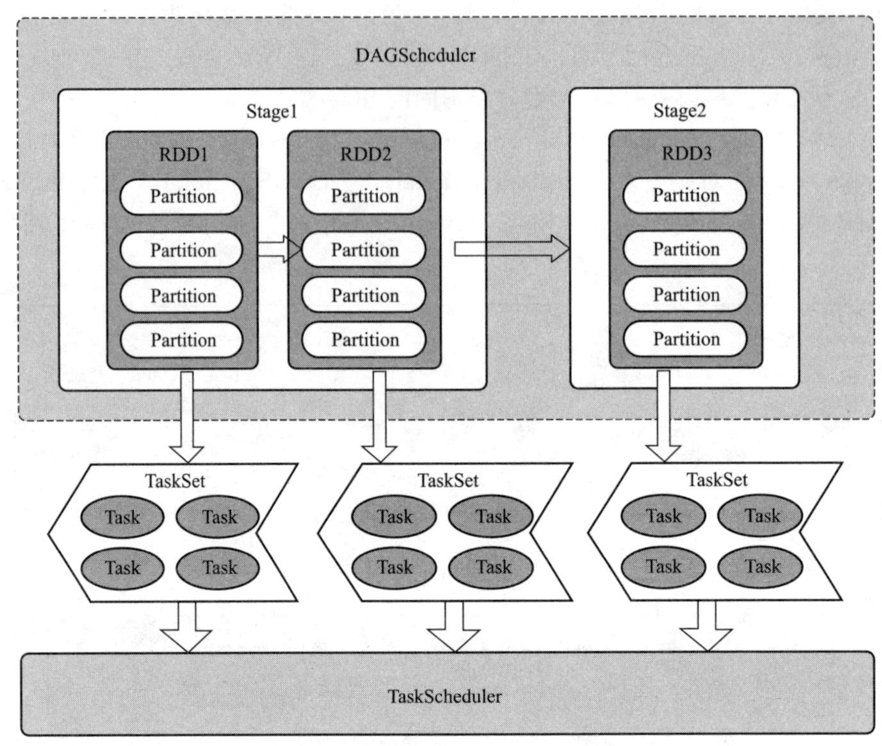

图 17-2 Stage 划分与 TaskSet 生成

首先 Driver 端的应用程序包含了 Executor 和 Driver 代码部分。应用程序本身 Driver 代码（SparkConf 和 SparkContext），SparkConf 中包含了设置程序名称 setAppName、setMaster 等，SparkContext 包含了 DAGScheduler、TaskScheduler、SchedulerBackend 以及 SparkEnv，Executor 代码包含了对业务逻辑的具体实现的代码（map、flatmap、ReduceByKey 等）。

可以看出 Driver 端由 SparkConf 中将程序运行的配置信息传给 SparkContext 上下文，由 SparkContext 创建 DAGScheduler（高层调度器）、TaskScheduler（底层调度器：负责一个作业内部运行）、SchedulerBackend（负责握住计算资源），实例化的过程中注册当前程序给 Master。

然后 DAGScheduler 根据 Actor 触发 job，SparkContext 通过 DAGScheduler 将 job 中 RDD 形成的 DAG 有向无环图划分为不同的 Stage（Task 具体运行于哪台机器上？就是在划分 Stage 的时候确定，这里就是根据数据本地行定位发送到 Executor 的具体位置），TaskScheduler 会将不同 Stage 中的一系列的 Task 发送到对应的 Executor 去执行，具体划分的的过程为不同的（map、flatmap、ReduceByKey）函数 RDDAPI 产生的 RDD 形成的 DAG 进行划分，得到不同的 Stage，而每个 Stage 中的 TaskSet 有相同业务逻辑，只是处理的数据不同，Task 又一一对应 RDD 中的每个 Partition。将划分 Stages 得到的 TaskSet 提交到 TaskScheduler，进而提交给 executor 执行。

Driver 端划分好 Stage 后，提交到 Spark 集群，即提交到 Master，并向 Master 进行注册，由 Master 接收后分配资源的 AppId 并发送指令给 Worker，然后 Worker 分配

Executor，最后由 Executor（并发处理数据分片）中的线程池中的线程并发执行。

17.1.1 SparkConf 解析

SparkConf 是为一个 Spark 应用程序设置相关属性的 key-value 键值对，大多数情况下需要创建一个 SparkConf 的对象然后导入 org. apache. spark. sparkConf。SparkConf 中包含了设置程序名称 setAppName、setMaster 等。如下所示。

```
/**
 * 设置Spark运行模式
 * "local"表示运行在本地模式
 * "local[3]"表示运行本地模式使用3个Core核心
 * "spark://master:7077"表示以Standalone模式运行Spark
 */
def setMaster(master: String): SparkConf = {
  set("spark.master", master)
}

/**
 * 设置Spark应用程序的名称，在Web UI中能够显示名称
 */
def setAppName(name: String): SparkConf = {
  set("spark.app.name", name)
}

/**
 * 设置JAR 文件Set
 */
def setJars(jars: Seq[String]): SparkConf = {
  for (jar <- jars if (jar == null)) logWarning("null jar passed to SparkContext constructor")
  set("spark.jars", jars.filter(_ != null).mkString(","))
}

/**
 * Set JAR files to distribute to the cluster. (Java-friendly version.)
 */
def setJars(jars: Array[String]): SparkConf = {
  setJars(jars.toSeq)
}
```

实际开发 Spark 应用程序的时候，SparkConf 的创建和使用如下所示。

```
/**
 *创建Spark的配置对象SparkConf，设置Spark程序的运行时
 * 的配置信息，例如通过SetMaster来设置程序的连接Spark集群的Master的
 * URL，若设置为local，则代表Spark程序在本地运行，特别适合于
 * 机器配置特别差的初学者。
 */
val conf = new SparkConf()
conf.setAppName("Wow,My First Spark App!")
conf.setMaster("local") //若将改行注释，可以运行于集群环境
```

```
//conf.setMaster ("spark://Master:7077")
//绑定了程序必须运行于集群上
//此时，程序运行于本地模式，无需Spark集群
```

一般情况下，推荐使用函数式编程的思想，链式创建 SparkConf 对象并设置 Spark 程序运行的相关属性，例如“new SparkConf（）.setMaster（"local"）.setAppName（"My-app"）”。

17.1.2 SparkContext 解析

SparkContext 是 Spark 应用程序的上下文和入口，就像 Java 应用程序的入口函数“main”一样，另外 SparkContext 会创建 RDD、累加器和广播。从 Spark 内核的角度将 SparkContext 主要创建了三个核心对象 DAGScheduler、TaskScheduler、SchedulerBackend。

在实际 Spark 应用程序的开发中在 main 方法中创建 SparkContext 对象，作为 Spark 应用程序的入口，如下所示。

```
/**
 * 建SparkContext对象
 * SparkContext是Spark程序所有功能的唯一入口，无论采用Scala
 * Java、Pytho等不同语言；
 * SparkContext：初始化Spark应用程序运行所需要的核心组件，包括
 * DAGSchedular、 taskSchedular、SchedularBacked，
 * 同时还负责Spark应用程序中最为重要的一个对象。
 */
val sc = new SparkContext(conf)
//创建SparkConte对象，通过传入SparkContext实例来定制Spark运行的具体参数和配置信息。
```

并且在 Spark 应用程序结束的时候，关闭 SparkContext 对象。如下所示。

```
//关闭SparkContext对象
sc.stop()
```

可以认为 SparkContext 就是整个 Spark 应用程序的关键点，即 Spark 应用程序是通过 SparkContext 发布到了 Spark 集群中，由 SparkContext 开启整个 Spark 作业运行的大门，然后 SparkContext 创建的高层调度器 DAGScheduler 和底层可插拔调度器 TaskScheduler 指挥 Spark 应用程序的具体运行，如果将 Spark 程序在集群中的运行比作电影放映，毫无疑问 SparkContext 就是电影拍摄的总导演，程序运行结束后 SparkContext 负责关闭 Spark 程序的运行，如果程序运行过程中关闭了 SparkContext 对象，Spark 作业就会被终止。

SparkContext 还负责创建 RDD，如果您对 RDD 的创建和产生熟悉，您可能知道 RDD 的创建有至少 7 种方式，例如其他 RDD 转化、Scala 集合生成、其他类型数据转化等，但是需要注意的是 Spark 程序的第一个 RDD 一定是由 SparkContext 创建的。如下所示。

```
/**
 * 第二步：创建SparkContext对象
 * SparkContext是Spark程序所有功能的唯一入口，无论采用Scala
 * Java、Pytho等不同语言；
 *SparkCon核心作用：初始化Spark应用程序运行所需要的核心组件，包括
 * DAGSchedular、 taskSchedular、SchedularBacked，
 *同时还负责Spark应用程序中最为重要的一个对象。
 */
```

```
val sc = new SparkContext(conf)
//创建SparkConte对象，通过传入SparkContext实例来定制Spark运行的具体参数和
配置等*
/**
 *据具体的数据来源（HDFS、Hbase、Local FS、DB、S3等）通过SparkContext创建
RDD
  * RDD的创建基本有三种方式：
  * 根据外部的数据来源（HDFS）
  * Scala集合
  * 有其他的RDD操作
  *数据会被RDD划分为一系列的Partition、分配到每个Partition的数据属于一个Task
 *处理范围
  */
val lines = sc.textFile("C://Hello.txt",1)  //并行度为1 (partition) ,读取本地文件
//根据类型推断，这边的lines是RDD[String]
```

下面查看 SparkContext 源码定义部分。

```
/**
 * Main entry point for Spark functionality. A SparkContext represents the connection to a
Park
 * cluster, and can be used to create RDDs, accumulators and broadcast variables on that
luster.
 *
 * Only one SparkContext may be active per JVM.  You must `stop()` the active
 *SparkContext before
creating a new one.  This limitation may eventually be removed; see SPARK-2243 for more details.
 *
 * @param config a Spark Config object describing the application configuration. Any
Settings in
*   this config overrides the default configs as well as system properties.
 */

class SparkContext(config: SparkConf) extends Logging with ExecutorAllocationClient {
```

Spark 程序在执行的时候分为 Driver 和 Executor 两部分，Spark 程序的编写核心基础是 RDD，具体包含两部分。

① Spark 应用程序中第一个 RDD，一定是由 SparkContext 来创建的；

② Spark 程序的调度优化也是基于 SparkContext。

Spark 程序的注册要通过 SparkContext 实例化时候产生的对象来完成（其实是由 SchedulerBackend 注册程序，申请计算资源）的。

Spark 程序运行的时候要通过 ClusterManager 获得具体的资源，计算资源的获取也是由 SparkContext 产生的对象来申请的（实际是由 SchedulerBackend 来获取资源的，SchedulerBackend 是由 SparkContext 实例化的时候产生的，也就是说在构造 SparkContext 的时候产生）。

从调度的层面讲调度优化也是基于 SparkContext 的，从程序注册的角度讲也是基于 SparkContext，从程序获取计算资源的角度讲也是基于 SparkContext 获取计算资源，只要 SparkContext 关闭，程序也就结束运行，也就是说 SparkContext 崩溃或者结束的整个 Spark 程序也意味着结束。

SparkContext 构建的顶级三大核心对象：DAGScheduler、TaskScheduler、SchedulerBackend，其功能有以下几个方面。

① SchedulerBackend 是面向 Stage 的调度器；

② TaskScheduler 是一个接口，根据 ClusterManager 的不同会有不同的实现，Standalone 模式下具体的实现是？TaskSchedulerlmpl；

③ SchedulerBackend 是一个接口，根据 ClusterManager 的不同会有不同的实现，Standalone 模式下具体的实现是 SparkDeploySchedulerBackend。

从整个程序运行的角度来讲，SparkContext 包含四大核心对象：DAGScheduler、TaskScheduler、SchedulerBackend、MapOutputTrackerMaster。其中 MapOutputTrackerMaster 在 DAGScheduler 通过 action 算子的触发进行 Shuffle 的时候进而被划分为不同的 Stage，在此过程中 MapOutputTrackerMaster 会获取 shuffle 相关的 map 结果信息。

SparkContext 默认构造器在 SparkContext 实例化的时候需要构造器中的参数赋值见下面所示。

```
// Create and start the scheduler
val (sched, ts) = SparkContext.createTaskScheduler(this, master)
_schedulerBackend = sched
_taskScheduler = ts                  //task
_dagScheduler = new DAGScheduler(this)
_heartbeatReceiver.ask[Boolean](TaskSchedulerIsSet)

// start TaskScheduler after taskScheduler sets DAGScheduler reference in DAGScheduler's
// constructor
_taskScheduler.start() //调用TaskScheduler的start方法
```

createTaskScheduler 在 SparkContext 的默认构造器中，当实例化 SparkContext 的时候需要被调用，调用的时候返回了 SchedulerBackend 和 _ taskScheduler 的实例，基于该内容有创建了 DAGScheduler，DAGScheduler 管理 TaskScheduler。

createTaskScheduler 中采用了 Local 模式 LOCAL _ N _ REGEX、Local 多线程模式、Local 多线程模式重试、Cluster 模式、Yarn 模式、Mesos 模式，默认情况下作业失败了就失败了，不会去重试。

```
/**
 * Create a task scheduler based on a given master URL.
 * Return a 2-tuple of the scheduler backend and the task scheduler.
 *在SparkContext实例化的时候被调用
 */
private def createTaskScheduler(
    sc: SparkContext,
    master: String): (SchedulerBackend, TaskScheduler)= {
  import SparkMasterRegex._

  //本地模式下，一般不会重新尝试执行失败的task
  val MAX_LOCAL_TASK_FAILURES = 1
```

下面为 Standalone 模式。

```
case SPARK_REGEX(sparkUrl) =>
//创建TaskScheduler对象
```

```
val scheduler = new TaskSchedulerImpl(sc)
val masterUrls = sparkUrl.split(",").map("spark://" + _)
//创建SchedulerBackend的对象
val backend = new SparkDeploySchedulerBackend(scheduler, sc, masterUrls)
scheduler.initialize(backend)
(backend, scheduler)
```

从上述源码可以看出传进的 sparkURL 创建 TaskSchedulerImpl 对象，TaskSchedulerImpl 是底层调度器的核心，在创建 TaskSchedulerImpl 后，以 TaskSchedulerImpl 为参数创建了一个 SchedulerBackend 对象，而 SchedulerBackend 进行 initialize 的时候将 SchedulerBackend 传入，具体传入的是 SparkDeploySchedulerBackend，这里 Scheduler 的 initialize 方法如下所示。

```
//先进先出FIFO
def initialize(backend: SchedulerBackend) {
  this.backend = backend
  // temporarily set rootPool name to empty
  rootPool = new Pool("", schedulingMode, 0, 0)
  schedulableBuilder = {
    schedulingMode match {
      case SchedulingMode.FIFO =>
//先进先出的调度模式
        new FIFOSchedulableBuilder(rootPool)
      case SchedulingMode.FAIR =>
        new FairSchedulableBuilder(rootPool, conf)
    }
  }
  schedulableBuilder.buildPools()
}
```

在 Scheduler. initialize 调用的时候会创建 SchedulerPool 调度池，一个任务有两种方式。

SparkDeploySchedulerBackend 有三大核心功能（需要说明的是把 Task 发送给 Executor 通过 SchedulerBackend 来完成的，SparkDeploySchedulerBackend 是被 TaskSchedulerImpl 来管理的）：

① 负责与 Master 链接注册当前程序；

② 接收集群中为当前应用程序而分配计算资源 Executor 的注册并管理 Executor；

③ 负责发送 Task 到具体的 Executor 执行。

TaskSchedulerImpl. start（）后，然后导致 SparkDeploySchedulerBackend. start（）。如下所示。

```
override def start() {
  backend.start()
//这里Backend.start实质是SparkDeploySchedulerBackend.start
  if (!isLocal && conf.getBoolean("spark.speculation", false)) {
    logInfo("Starting speculative execution thread")
    speculationScheduler.scheduleAtFixedRate(new Runnable {
      override def run(): Unit = Utils.tryOrStopSparkContext(sc) {
        checkSpeculatableTasks()
      }
```

```
    }, SPECULATION_INTERVAL_MS, SPECULATION_INTERVAL_MS, TimeUnit.MILLISECONDS)
  }
}
```

当通过 SparkDeploySchedulerBackend 注册程序给 Master 的时候，会把上述 command 提交给 Master，Master 发指令给 Worker 去启动 Executor 所在的进程的时候加载 main（Executor 进程所在的入口类），就是 command 中的 CoarseGrainedExecutorBackend，当然这里可以实现自己的 ExecutorBackend，在 CoarseGrainedExecutorBackend 中启动 Executor（Executor 先注册），Executor 通过线程池并发执行的方式执行 Task。

```
val command = Command("org.apache.spark.executor.CoarseGrainedExecutorBackend",
    args, sc.executorEnvs, classPathEntries ++ testingClassPath, libraryPathEntries, javaOpts)
```

下面是 CoaseGrainedExecutorBackend 中的 run（）方法，其中有 driverUrl、executorId、hostname、cores、APPId 等参数。

```
//首先创建SparkConf对象
val executorConf = new SparkConf
    val port = executorConf.getInt("spark.executor.port", 0)
 //获得executor的port
    val fetcher = RpcEnv.create(
      "driverPropsFetcher",
      hostname,
      port,
      executorConf,
      new SecurityManager(executorConf),
//传入SparkConf的实例
      clientMode = true)
    val driver = fetcher.setupEndpointRefByURI(driverUrl)

    val props = driver.askWithRetry[Seq[(String, String)]](RetrieveSparkProps) ++
      Seq[(String, String)](("spark.app.id", appId))
    fetcher.shutdown()
```

下面代码中使用从 driver 上发送的属性创建了 SparkEnv，然后以键值对的方式循环遍历从 driver 传入的参数，通过 SparkConf 设置该属性（Standalone 模式），然后通过 SparkEnv 调用 createExecutorEnv，主要实现了对 executor 的 port 设置，创建了 CoaseGrainedExecutorBackend（new 出该 main 方法的对象实例）的对象进行 RPC 通信。

```
val env = SparkEnv.createExecutorEnv(
driverConf, executorId, hostname, port, cores, isLocal = false)

// SparkEnv will set spark.executor.port if the rpc env is listening for incoming
// connections (e.g., if it's using akka). Otherwise, the executor is running in
// client mode only, and does not accept incoming connections.
val sparkHostPort = env.conf.getOption("spark.executor.port").map { port =>
    hostname + ":" + port
  }.orNull
env.rpcEnv.setupEndpoint("Executor", new CoarseGrainedExecutorBackend(
env.rpcEnv, driverUrl, executorId, sparkHostPort, cores, userClassPath, env))
```

启动 CoaseGrainedExecutorBackend 的时候需要注册 Executor，当收到 driver 级别的

Executor 注册信息后，才会实例化 Executor 的对象，只有当 Executor 注册成功的时候才会实例化。

```
override def receive: PartialFunction[Any, Unit] = {
  case RegisteredExecutor(hostname) =>
    logInfo("Successfully registered with driver")
    executor = new Executor(executorId, hostname, env, userClassPath, isLocal = false)
```

当 TaskSchedulerImpl 调用 start 方法后导致 SparkDeploySchedulerBackend 调用 start 方法，而 Executor 要向 SparkDeploySchedulerBackend 注册，在 SparkDeploySchedulerBackend 的 start 方法中实例化了 APPClient 对象。

```
val coresPerExecutor = conf.getOption("spark.executor.cores").map(_.toInt)
//每个Executor中使用多少个cores可以在Spark.env中配置
  val appDesc = new ApplicationDescription(sc.appName, maxCores, sc.executorMemory,
    command, appUIAddress, sc.eventLogDir, sc.eventLogCodec, coresPerExecutor)
  client = new AppClient(sc.env.rpcEnv, masters, appDesc, this, conf)
  client.start()
  launcherBackend.setState(SparkAppHandle.State.SUBMITTED)
  waitForRegistration()
  launcherBackend.setState(SparkAppHandle.State.RUNNING)
```

上面 new 除了 AppClient 的实例对象，AppClient 是一个接口，可以和集群进行通信，其参数包含了 masterURL 和应用程序信息（因为要注册应用程序给集群），并且是集群时间的监听器。

```
/**
 * Interface allowing applications to speak with a Spark deploy cluster. Takes a master URL,
 * an app description, and a listener for cluster events, and calls back the listener when various
 * events occur.
 *
 * @param masterUrls Each url should look like spark://host:port.
 *监控集群的正常运行
 */
private[spark] class AppClient(
    rpcEnv: RpcEnv,
    masterUrls: Array[String],
    appDescription: ApplicationDescription,
    listener: AppClientListener,
    conf: SparkConf)
  extends Logging {
```

在 AppClient 中有很重要的类 ClientEndpoint，在 ClientEndpoint 实例化启动的 start 中主要是 registerWithMaster。

```
override def onStart(): Unit = {
    try {
      registerWithMaster(1)
    } catch {
      case e: Exception =>
        logWarning("Failed to connect to master", e)
        markDisconnected()
        stop()
```

```
    }
}
```

在 registerWithMaster 中对 SparkApplication 进行注册。

```
private def registerWithMaster(nthRetry: Int) {
  registerMasterFutures.set(tryRegisterAllMasters())
  registrationRetryTimer.set(registrationRetryThread.scheduleAtFixedRate(new Runnable {
```

registerWithMaster 内部使用 tryRegisterAllMasters 进行注册 SparkApplication。

```
/**
   * Register with all masters asynchronously and returns an array `Future`s for cancellation.
   */
  private def tryRegisterAllMasters(): Array[JFuture[_]] = {
    for (masterAddress <- masterRpcAddresses) yield {
      registerMasterThreadPool.submit(new Runnable {
```

tryRegisterAllMasters 注册 Application 的时候是通过 Thread 完成的。继续跟进代码。

```
masterRef.send(RegisterApplication(appDescription, self))
```

继续跟进查看注册 Application 的内容。

```
private[spark] case class ApplicationDescription(
    name: String,
    maxCores: Option[Int],
    memoryPerExecutorMB: Int,
    command: Command,
    appUiUrl: String,  //在web控制台显示的Application信息
    eventLogDir: Option[URI] = None,
    // short name of compression codec used when writing event logs, if any (e.g. lzf)
    eventLogCodec: Option[String] = None,
    coresPerExecutor: Option[Int] = None,
    user: String = System.getProperty("user.name", "<unknown>")) {
```

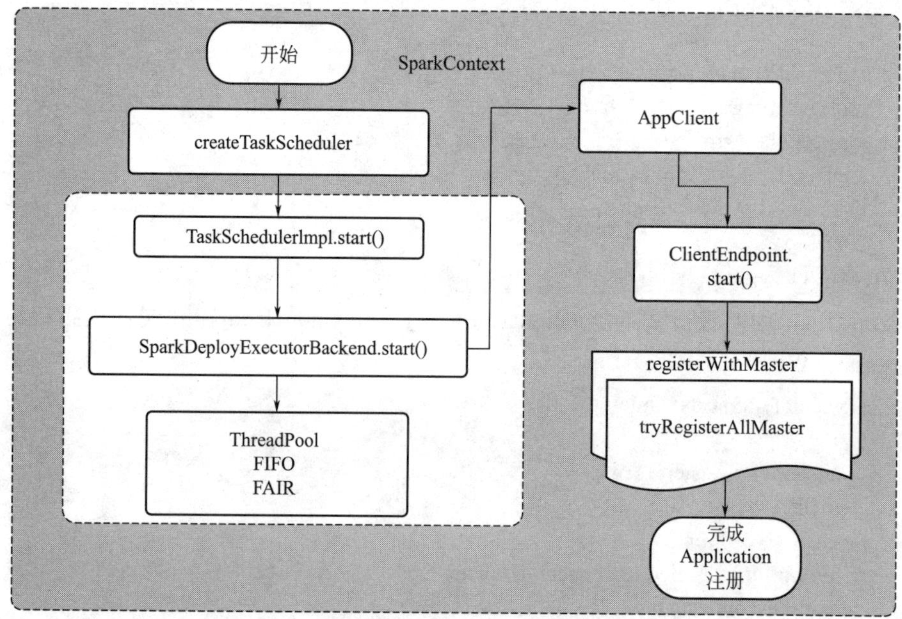

图 17-3 Spark Application 的注册过程

上面看到了 tryRegisterAllMasters 中是通过 Thread 的方式向 Master 注册的，那么在 Master 中可以看到发送的具体注册 Application 的信息。

在 Master 的 receive 方法中看到了通过模式匹配的方式包括了 RegisterApplication 接受注册，分配资源等。图 17-3 表示了 SparkApplication 的注册过程。

上面通过源码的方式详细剖析了提交应用程序后，driver 向 Master 注册 Application 的过程，当注册完成 Application 后 Master 发送指令给 Worker 启动 Executor，并向 driver 中的 SparkDeploySchedulerBackend 注册 Executor 的信息。

SparkContext 中还包含了 TaskScheduler 面向 Stage 的高层调度器，将 action 触发的 job 中的 rdd 划分为若干个 Stage。同时 SparkContext 中也包括 SparkUI，背后是 Jetty 服务，支持通过 web 的方式访问程序的状态。

17.1.3 DAGScheduler 创建

SparkContext 在实例化的时候创建了 DAGScheduler 和 TaskScheduler，TaskScheduler 是通过 org.apache.spark.SparkContext 中的 createTaskScheduler 创建的，而 DAGScheduler 的创建是通过调用其构造函数进行创的，DAGScheduler 保存了对 TaskScheduler 的引用。DAGScheduler 创的源码如下所示。

```
//创建并且开始调度
val (sched, ts) = SparkContext.createTaskScheduler(this, master)
_schedulerBackend= sched
_taskScheduler = ts
 //task
//创建DAGScheduler对象
_dagScheduler = new DAGScheduler(this)
_heartbeatReceiver.ask[Boolean](TaskSchedulerIsSet)
```

由上述创建过程查看构造函数的具体实现，其位置在于 pache.spark.DAGScheduler 的 this 方法，如下所示。

```
def this(sc: SparkContext, taskScheduler: TaskScheduler) = {
 this(
   sc,
   taskScheduler,
   sc.listenerBus,
   sc.env.mapOutputTracker.asInstanceOf[MapOutputTrackerMaster],
   sc.env.blockManager.master,
   sc.env)
}

def this(sc: SparkContext) = this(sc, sc.taskScheduler)
```

跟进 DAGScheduler 的方法，其创建过程实际是通过以下的构造函数完成。

```
private[spark]
class DAGScheduler(
   private[scheduler] val sc: SparkContext,
   private[scheduler] val taskScheduler: TaskScheduler,
   listenerBus: LiveListenerBus,
   mapOutputTracker: MapOutputTrackerMaster,
   blockManagerMaster: BlockManagerMaster,
```

```
    env: SparkEnv,
    clock: Clock = new SystemClock())
  extends Logging {
```

前面提到 DAGScheduler 主要负责将 action 算子触发的 Job 的不同 RDD 划分成若干个 Stage，也就是将 Job 中的 RDD 数据集进行加工包装后，以 TaskSet 的形式传递给 TaskScheduler 进行资源分配处理，然后由 Executor 中的线程并发处理。

DAGScheduler 还会在 SparkContext 中调用 submitJob 或者 runJob 两个方法，以此接收 Job 提交的通道。

DAGScheduler 在划分完 Stage 后，判断不同的 Stage 中的 Task 的类型，形成 TaskSet，而 TaskSet 中是一些 TaskSet［Task］的数组，将 TaskSet 交由 TaskScheduler 后，调用 TaskScheduler 的 submitTask 方法，然后由 TaskScheduler 为其分配计算资源，最后启动 Worker 节点上 Executor 启动 JVM 进程，在线程池中并发执行。

DAGScheduler 还会创建一个 org. apache. spark. scheduler. DAGSchedulerActorSupervisior 对于的 Actor，即 eventActor，主要用来调用 DAGScheduler 中的方法来处理 DAGScheduler 发送给他的各种消息。

17.1.4 TaskScheduler 创建

DAGScheduler 在提交 TaskSet 给底层调度器的时候是面向接口 TaskScheduler 的，这符合面向对象中依赖抽象的原则，带来底层资源调度器的可插拔性，因此使得 Spark 可以运行的众多的资源调度器模式上，例如 Standalone、Yarn、Mesos、Local、EC2，其他自定义的调度器。TaskScheduler 是一个 Trait，如下所示。

```
private[spark] trait TaskScheduler {
  private val appId = "spark-application-" + System.currentTimeMillis
  dcf rootPool: Pool
  def schedulingMode: SchedulingMode
  def start(): Unit
```

在 Standalone 模式下聚焦于 TaskSchedulerImpl，在 SparkContext 实例化的时候通过 createTaskScheduler 创建 TaskScheduler。如下所示。

```
case SIMR_REGEX(simrUrl) =>
    val scheduler = new TaskSchedulerImpl(sc)
    val backend = new SimrSchedulerBackend(scheduler, sc, simrUrl)
    scheduler.initialize(backend)
    (backend, scheduler)
```

在 TaskSchedulerImpl 的 iniitialize 方法中把 SparkDeploySchedulerBackend 传进来从而赋值为 TaskSchedulerImpl 的 Backend；在 TaskSchedulerImpl 调用 Start 方法的时候会调用 backend. start 方法，在 start 方法中会最终注册应用程序。

TaskScheduler 的核心任务是提交 TaskSet 到集群运算并汇报结构。首先为 TaskSet 创建和维护一个 TaskSetManager 并追踪任务到本地性以及错误信息，并且 TaskScheduler 在遇到 Straggle 任务会放到其他的节点进行重试。如下所示。

```
/**
 * A set of tasks submitted together to the low-level TaskScheduler, usually representing
 * missing partitions of a particular stage.
 */
```

```
private[spark] class TaskSet(
//一组Task数组
    val tasks: Array[Task[_]],
    val stageId: Int,
    val stageAttemptId: Int,
    val priority: Int,
    val properties: Properties) {
  val id: String = stageId + "." + stageAttemptId

  override def toString: String = "TaskSet " + id
}
```

向 DAGScheduler 汇报执行情况，包括在 Shuffle 输出 lost 的时候汇报 fetch failed 错误等信息。

SparkContext、DAGScheduler、TaskSchedulerImpl、SparkDeploySchedulerBackend 在应用程序启动的时候只实例化一次，应用程序存在期间始终存在这些对象。

```
    // Create and start the scheduler
    val (sched, ts) = SparkContext.createTaskScheduler(this, master)
    _schedulerBackend = sched
    _taskScheduler = ts
    _dagScheduler = new DAGScheduler(this)
    _heartbeatReceiver.ask[Boolean](TaskSchedulerIsSet)
```

通过 Spark-shell 运行程序来观察 TaskScheduler 内幕。当启动 Spark-shelll 本身的时候命令终端会反馈回来的主要是 ClientEndpoint 和 SparkDeploySchedulerBackend，这是因为此时还没有任何 Job 的触发这是启动 Application 本身而已，因此主要就是实例化 SparkContext 并注册当前应用程序给 Master 且从集群中获得 ExecutorBackend 计算资源。

DAGScheduler 划分好 Stage 后会通过 TaskSchedulerImpl 中的 TaskSetManager 来管理当前要运行的 Stage 中的所有任务 TaskSet，TaskSetManager 会根据 locality aware 来为 Task 分配计算资源、监控 Task 的执行状态（例如重试、慢任务进行推测式执行等）。

每个 TaskScheduler 都对应一个 SchedulerBackend，TaskScheduler 负责用户提交的 SparkApplication 的 Job 的调度工作，当出现 Task 运行过慢时，TaskScheduler 启动备份的 Task 任务来执行，对于运行失败的 Task，会重新尝试运行。SchedulerBackend 负责将领取到的计算资源分配给 TaskScheduler，完成与 Cluster Manager 直接的交互性工作，而 TaskScheduler 将最终为 TaskSet 中的 Task 分配计算资源。

17.1.5 SchedulerBackend 创建

TaskScheduler 内部会握有 SchedulerBackend，SchedulerBackend 从 Standone 的模式来讲具体实现是 SparkDeploySchedulerBackend。如下所示。

```
private[spark] trait SchedulerBackend {
  private val appId = "spark-application-" + System.currentTimeMillis

  def start(): Unit
  def stop(): Unit
  def reviveOffers(): Unit
  def defaultParallelism(): Int
```

在 SparkContext 实例化的时候通过 createTaskScheduler 来创建 TaskSchedulerImpl 和 SparkDeploySchedulerBackend，在 TaskSchedulerImpl 的 initialize 发放中把 SparkDeploySchedulerBackend 传进来从而赋值为 TaskSchedulerImpl 的 backend；在 TaskSchedulerImpl 调用 start 方法的时候会调用 backend. start 方法，在 start 方法中会最终用应用程序。initialize 源码如下所示：

```
def initialize(backend: SchedulerBackend) {
  this.backend = backend
}
```

SparkDeploySchedulerBackend 在启动的时候构建 APPClient 实例并在该实例 start 的时候启动了 ClientEndpoint，ClientEndpoint 会向 Master 注册当前程序；而 SparkDeploySchedulerBackend 的父类 CoarseGrainedSchedulerBackend 在 Start 的时候会实例化为 DriverEndpoint（这就是程序运行的时候的经典的对象 Driver）的消息循环体。如下为 SparkDeploySchedulerBackend 的 start 方法：

```
override def start() {
    super.start()
  // Start executors with a few necessary configs for registering with the scheduler
    val sparkJavaOpts = Utils.sparkJavaOpts(conf, SparkConf.isExecutorStartupConf)
    val javaOpts = sparkJavaOpts ++ extraJavaOpts
    val command = Command("org.apache.spark.executor.CoarseGrainedExecutorBackend",
      args, sc.executorEnvs, classPathEntries ++ testingClassPath, libraryPathEntries, javaOpts)
    val appUIAddress = sc.ui.map(_.appUIAddress).getOrElse("")
    val coresPerExecutor = conf.getOption("spark.executor.cores").map(_.toInt)
    val appDesc = new ApplicationDescription(sc.appName, maxCores, sc.executorMemory,
      command, appUIAddress, sc.eventLogDir, sc.eventLogCodec, coresPerExecutor)
//创建APPClient对象
    client = new AppClient(sc.env.rpcEnv, masters, appDesc, this, conf)
    client.start()
    launcherBackend.setState(SparkAppHandle.State.SUBMITTED)
    waitForRegistration()
    launcherBackend.setState(SparkAppHandle.State.RUNNING)
  }
```

SparkDeploySchedulerBackend 专门负责收集 Worker 上的资源信息，ExecutorBackend 启动的时候会发送 RegisterExecutor 信息向 DriverEndpoint 注册，此时 SparkDeploySchedulerBackend 就掌握了当前应用程序拥有的计算资源，然后启动某个 Worker 上的 Executor 执行 Task。TaskScheduler 就是通过 SparkDeploySchedulerBackend 拥有的计算资源来具体运行 Task。

SchedulerBackend 主要负责申请资源和对 Task 的管理。SparkDeploySchedulerBackend 会启动一个 actor 用来资源的申请，会在 SparkDeploySchedulerBackend 初始化的时候被创建，然后 ClientEndpoint 向 Master 注册 Spark，完成 Worker 上的 ExecutorBackend 创建，同时该 ExecutorBackend 都会被注册到 DriverActor 上。

17.1.6 Stage 划分与 TaskSet 生成

前面已经提到了 Stage 的划分是由 DAGScheduler 来完成的，当用户提交的 Spark 应用程序中遇到 action 操作时，会触发 Job 的生成，也就是遇到 Shuffle 的时候（宽依赖的方式）

将 DAG 划分为不同的 Stage。在实例化 SparkContext 的时候由创建的 DAGScheduler 实现 Stage 的划分同时生成 TaskSet。

Spark Application 中可以由不同的 Action 触发众多的 Job，也就是说一个 Application 中可以有很多的 Job，每个 Job 是由一个或者多个 Stage 构成的，后面的 Stage 依赖前面的 Stage，也就是说只有前面依赖的 Stage 的计算完毕后，后面的 Stage 才会运行。图 17-4 为 Stage 的划分过程。

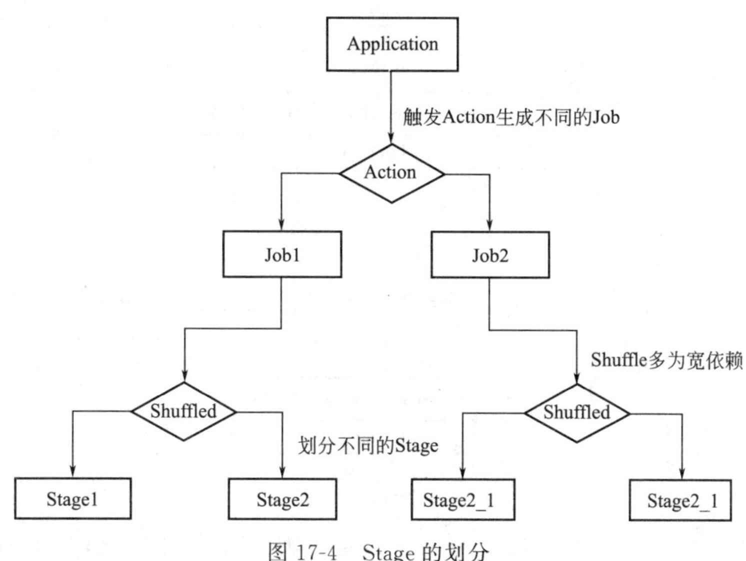

图 17-4 Stage 的划分

Stage 划分依据基本是宽依赖，Shuffle 产生宽依赖呢？例如 reduceByKey，groupByKey 等 action 操作的时候，如图 17-5 所示。

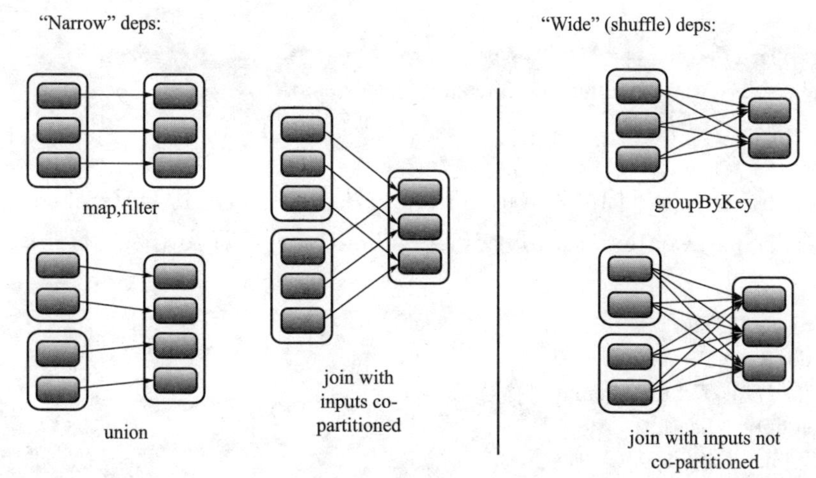

图 17-5 宽依赖和窄依赖

这种划分有两个用处。首先，窄依赖支持在一个结点上管道化执行，例如基于一对一的关系，可以在 filter 之后执行 map；其次，窄依赖支持更高效的故障还原，因为对于窄依赖，只有丢失的父 RDD 的分区需要重新计算。而对于宽依赖，一个结点的故障可能导致来自所有父 RDD 的分区丢失，就需要完全重新执行。因此对于宽依赖，Spark 会在持有各个父分区的结点上，将中间数据持久化来简化故障还原，就像 MapReduce 会持久化 map 的输出一样。图 17-6 为 Stage 划分过程中调用栈。

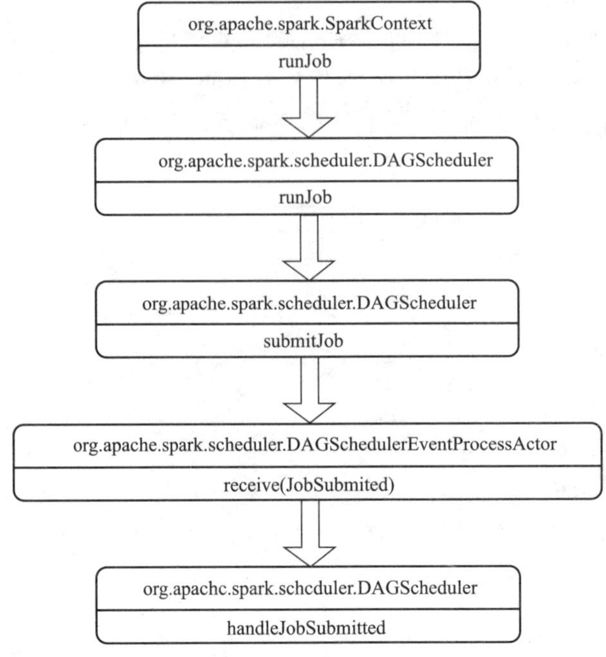

图 17-6　Stage 划分调用栈

从作业调度的层面看，例如 RDD. Scala 中的 collect 方法它是一个 action，一个 action 触发一个 Job，可以看到 collect 方法中具有 runJob。

```
/**
 * Return an array that contains all of the elements in this RDD.
 */
def collect(): Array[T] = withScope {
  val results = sc.runJob(this, (iter: Iterator[T]) => iter.toArray)
Array.concat(results: _*)
}
```

跟踪 runJob，会看到不同的重载的 runJob 方法，一路跟踪到 DAGScheduler 中，从下面源码可以看出 SparkContext. runJob 交给了 submitJob，将提交结果进行匹配成功或失败等情况。

```
def runJob[T, U](
     rdd: RDD[T],
     func: (TaskContext, Iterator[T]) => U,
     partitions: Seq[Int],
     callSite: CallSite,
     resultHandler: (Int, U) => Unit,
     properties: Properties): Unit = {
   val start = System.nanoTime
   val waiter = submitJob(rdd, func, partitions, callSite, resultHandler, properties)
   waiter.awaitResult() match {
    //Job成功
     caseJobSucceeded =>
       logInfo("Job %d finished: %s, took %f s".format
```

```
        (waiter.jobId, callSite.shortForm, (System.nanoTime- start) / 1e9))
    //Job失败
    case JobFailed(exception: Exception) =>
      logInfo("Job %d failed: %s, took %f s".format
        (waiter.jobId, callSite.shortForm, (System.nanoTime- start) / 1e9))
      // SPARK-8644: Include user stack trace in exceptions coming from DAGScheduler.
      val callerStackTrace = Thread.currentThread().getStackTrace.tail
      exception.setStackTrace(exception.getStackTrace ++ callerStackTrace)
      throw exception
  }
}
```

Action（例如 collect）导致了 SparkContext. runJob 的执行，最终导致了 DAGScheduler 中的 submitJob 的执行，其核心是通过发送一个 case class JobSubmitted 的对象给 eventProcessLoop，其中 JobSumitted 的源码如下所示。

```
/**A result-yielding job was submitted on a target RDD */
private[scheduler] case class JobSubmitted(
    jobId: Int,
    finalRDD: RDD[_],
    func: (TaskContext, Iterator[_]) => _,
    partitions: Array[Int],
    callSite: CallSite,
    listener: JobListener,
    properties: Properties = null)
  extends DAGSchedulerEvent
```

从源码可以看出，该提交的作业封装了作业 ID、RDD [_]、对 RDD 操作的函数、有哪些 Partitions 被计算、listener 监听作业状态，并且 JobSubmitted 是一个 case class，而不是 case object 全局唯一的实例。case class 会根据传入的参数产生不同的实例，那么这里 JobSubmitted 会根据一个 Application 中的不同，Job 产生不同实例。

在 DAGScheduler 中的 submitJob 方法中可以看到，eventProcessLoop 通过调用其 post 方法将 JobSubmitted 传入，这里的 post 就是 Java 中 post（post 就是往队列里边放入一个元素），给线程发送一个消息，看下面源码：

```
eventProcessLoop.post(JobSubmitted(
  jobId, rdd, func2, partitions.toArray, callSite, waiter,
  SerializationUtils.clone(properties)))
```

eventProcessLoop 是 DAGSchedulerEventProcessLoop 的具体实例，而 DAGScheduler-EventProcessLoop 是 EventLoop 的子类，具体实现 EventLoop（内部有一个 EventThread 线程）的 onReceive 方法，onReceive 方法转过来回调 doOnReceive

```
private[scheduler] class DAGSchedulerEventProcessLoop(dagScheduler: DAGScheduler)
  extends EventLoop[DAGSchedulerEvent]("dag-scheduler-event-loop") with Logging {
```

而 EventLoop 中是事件队列的形式，队列可以无限期增长，确保子类能够 onReceive，及时处理事件避免 OOM，其内部执行消息循环体，从源码可以看出 onReceive 是一个抽象方法没有具体实现，所以实现的时候应该避免调用在“onReceive”阻断行动，或事件的线程被阻塞以及不能及时处理事件。

```
protected def onReceive(event: E): Unit
```

因为在 EventLoop 中 onReceive 方法并没有实现，所以在 EventLoop 的子类 DAGSchedulerEventProcessLoop 中实现了 onReceive，具体实现是将 DAGSchedulerEvent 作为参数传入，转过来调用 doOnReceive 进行处理。

EventLoop 中开辟一个线程，该线程不断循环一个队列，post 的时候实质是把消息放到队列中，异步处理多个 Job，然后不断循环拿到消息，转过来调用 onReceive 方法，onReceive 具体处理为 doOnReceive 模式匹配收到该消息。

之所以 eventProcessLoop 消息体不直接调用 doOnReceive 方法，要通过 post 往队列里边放一个消息，然后来调用 onReceive 方法，是因为 Spark 异步处理多个 Job，看上去是自己给自己发消息，实则是采用了统一调用规则，不管是主线程自己还是其他线程，使用消息循环器 doOnReceive 非常有利于扩展。doOnReceive 源码如下所示。

```
private def doOnReceive(event: DAGSchedulerEvent): Unit = event match {
    case JobSubmitted(jobId, rdd, func, partitions, callSite, listener, properties) =>
      dagScheduler.handleJobSubmitted(jobId, rdd, func, partitions, callSite, listener, properties)

    case MapStageSubmitted(jobId, dependency, callSite, listener, properties) =>
      dagScheduler.handleMapStageSubmitted(jobId, dependency, callSite, listener, properties)

    case StageCancelled(stageId) =>
      dagScheduler.handleStageCancellation(stageId)

    case JobCancelled(jobId) =>
      dagScheduler.handleJobCancellation(jobId)

    case JobGroupCancelled(groupId) =>
      dagScheduler.handleJobGroupCancelled(groupId)

    case AllJobsCancelled =>
      dagScheduler.doCancelAllJobs()
    //新增Executor
    case ExecutorAdded(execId, host) =>
      dagScheduler.handleExecutorAdded(execId, host)
    //Executor丢失
    case ExecutorLost(execId) =>
      dagScheduler.handleExecutorLost(execId, fetchFailed = false)

    case BeginEvent(task, taskInfo) =>
      dagScheduler.handleBeginEvent(task, taskInfo)

    case GettingResultEvent(taskInfo) =>
      dagScheduler.handleGetTaskResult(taskInfo)

    case completion @ CompletionEvent(task, reason, _, _, taskInfo, taskMetrics) =>
      dagScheduler.handleTaskCompletion(completion)

    case ResubmitFailedStages =>
      dagScheduler.resubmitFailedStages()
}
```

在 doOnReceive 中通过模式匹配然后把执行路由到 JobSubmitted。

在 handleJobSubmitted 中首次创建 finalStage，创建 finalStage 时候会建立父 Stage 依赖链条。源码如下所述。

```
/**
 * Get or create the list of parent stages for a given RDD. The new Stages will be created with
 * the provided firstJobId.
 * 生成rdd的parent Stage。没遇到一个ShuffleDependency，就会生成一个Stage
 */
private def getParentStages(rdd: RDD[_], firstJobId: Int): List[Stage] = {
  val parents = new HashSet[Stage]  //存储父Stage
  val visited = new HashSet[RDD[_]] //存储已经访问到的RDD
  // We are manually maintaining a stack here to prevent StackOverflowError
  // caused by recursively visiting
  val waitingForVisit = new Stack[RDD[_]]
  def visit(r: RDD[_]) { //接收RDD实例
    if (!visited(r)) {
      visited += r
      // Kind of ugly: need to register RDDs with the cache here since
      // we can't do it in its constructor because # of partitions is unknown
      /**
      for (dep <- r.dependencies) {
        dep match {
          case shufDep: ShuffleDependency[_, _, _] =>
            parents += getShuffleMapStage(shufDep, firstJobId) //增加一个Stage
          case _ =>
            waitingForVisit.push(dep.rdd)
        }
      }
    }
  }
  //输入的rdd作为第一个需要处理的RDD。然后从该rdd开始，顺序访问其parent rdd
  waitingForVisit.push(rdd)
  while (waitingForVisit.nonEmpty) {
    visit(waitingForVisit.pop())
  //每次visit如果遇到了ShuffleDependency，那么就会形成一个Stage，否则这些RDD属于同一个Stage
  }
  parents.toList
}
```

此处是 Stage 的划分，若是 ShuffleDependency，则划分一个新的 Stage，若是其他的依赖情况则属于同一个 Stage，不用新建一个 Stage，从后往前回溯的时候，遇到 ShuffleMapStage 就会产生新的 Stage。补充说明：所谓的 Missing 就是说要进行当前的计算了。

Spark 计算速度快的原因之一是数据本地性，因为 Task 对应 RDD 中的各个 Partition，所以数据本地性关乎于 Task 的本地性。下面对具体 Task 任务本地性算法的实现进行了概述。

在 submitMissingTasks 中会通过调用以下代码来获得任务的本地性。

```
val taskIdToLocations: Map[Int, Seq[TaskLocation]] = try {
  stage match {
    case s: ShuffleMapStage =>
      partitionsToCompute.map { id => (id, getPreferredLocs(stage.rdd, id))}.toMap
    case s: ResultStage =>
      val job = s.activeJob.get
      partitionsToCompute.map { id =>
        val p = s.partitions(id)
        (id, getPreferredLocs(stage.rdd, p))
      }.toMap
  }
}
```

具体一个 Partition 中的数据本地性的算法实现为下述代码中。

```
private[spark]
def getPreferredLocs(rdd: RDD[_], partition: Int): Seq[TaskLocation] = {
  getPreferredLocsInternal(rdd, partition, new HashSet)
}
```

在具体算法实现的时候首次查询 DAGScheduler 的内存数据结构中是否存在于当前 Partition 的数据本地性的信息，如果有的话直接返回；如果没有首先回调用 rdd. getPreferredLocations。

假如想让 Spark 运行在 HBase 上面，就需要开发者自定义 RDD，为了保证 Task 计算的数据本地性，最为关键的方式就是实现 RDD 的 getPreferredLocations。

RDD 的本地性：

```
// If the RDD has some placement preferences (as is the case for input RDDs), get those
  val rddPrefs = rdd.preferredLocations(rdd.partitions(partition)).toList
  if (rddPrefs.nonEmpty) {
    return rddPrefs.map(TaskLocation(_))
  }
```

DAGScheduler 计算数据本地性的时候巧妙地借助了 RDD 自身的 getPreferredLocations 中的数据，最大化地优化了数据，因为 getPreferredLocations 中表明了每个 Partition 的数据本地性，虽然当前 Partition 可能被 persist 或者 checkpoint，单数 persist 和 checkpoint 默认情况下肯定是和 getPreferredLocations 中的 Partition 的数据本地性是一致的，所以就极大地简化了 Task 数据本地性算法的实现和效率的优化。

17.1.7 任务提交

在 DAGScheduler 中 handleJobSubmitted 会生成 finalStage，该 finalStage 没有父 Stage 且仅有一个 Partition，然后 handleJobSubmitted 调用 submitStage 来提交该 Stage，Stage 被提交到 Worker 上的 Executor 中执行。

DAGScheduler 中的 submitMissingTaskSet 会完成 DAGScheduler 最后的工作，向 TaskScheduler 提交 Task，具体提交的是逐个将 Stage 中的 Task 提交到 TaskScheduler。源码如下所述。

```
//提交Task并进行资源调度
override def submitTasks(taskSet: TaskSet) {
  val tasks = taskSet.tasks
```

```
  logInfo("Adding task set " + taskSet.id + " with " + tasks.length + " tasks")
  this.synchronized {
    val manager = createTaskSetManager(taskSet, maxTaskFailures)
    val stage = taskSet.stageId
    val stageTaskSets =
      taskSetsByStageIdAndAttempt.getOrElseUpdate(stage, new HashMap[Int,
askSetManager])
    stageTaskSets(taskSet.stageAttemptId) = manager
    val conflictingTaskSet = stageTaskSets.exists { case (_, ts) =>
      ts.taskSet != taskSet && !ts.isZombie
    }
    if (conflictingTaskSet) {
      throw new IllegalStateException(s"more than one active taskSet for stage $stage:" +
        s" ${stageTaskSets.toSeq.map{_._2.taskSet.id}.mkString(",")}")
    }
    //通过addTaskSetManager管理Task
    schedulableBuilder.addTaskSetManager(manager, manager.taskSet.properties)
```

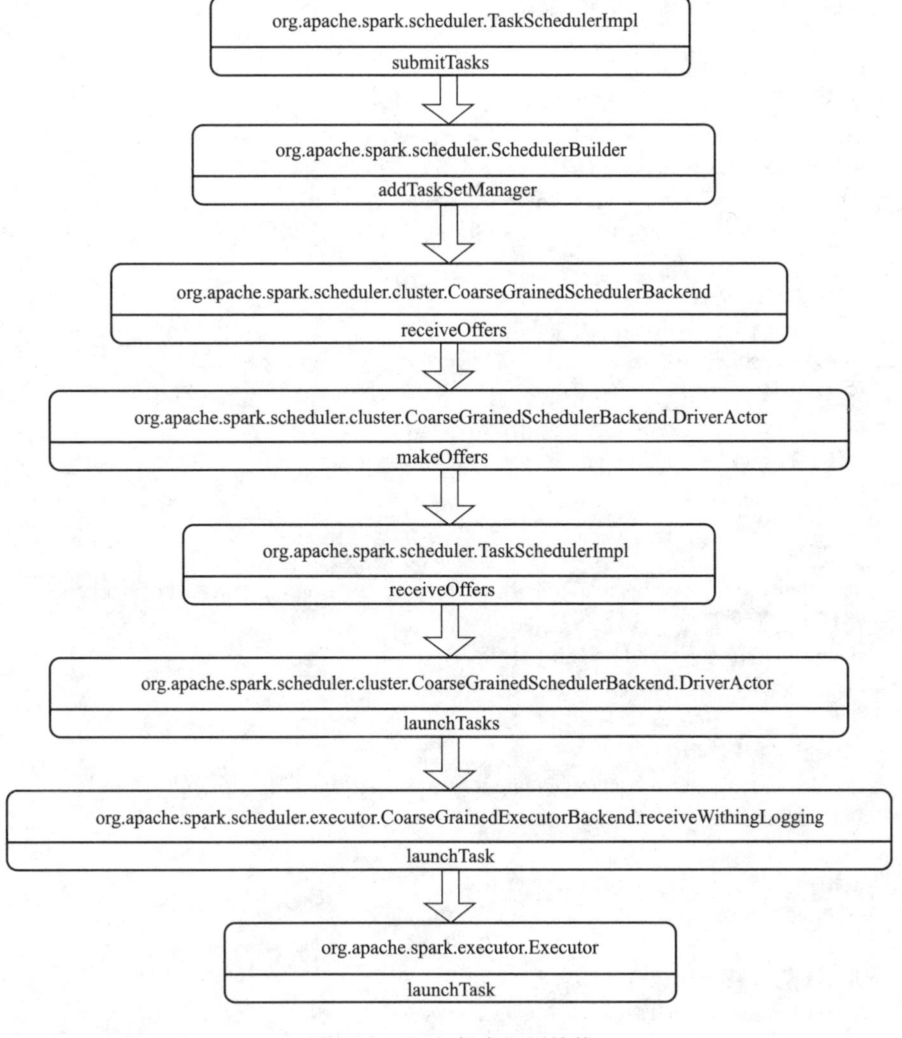

图 17-7 Task 提交调用堆栈

SchedulerBuilder 是一个调度器，支持两种调度模式，分别是先进先出 FIFO 和公平调度 FAIR，一般情况下默认是 FIFO，但也可以根据需要自行设置（设置位置：spark. scheduler. mode），在 SchedulerBuilder 确定了 TaskSetManager 中的 Task 的调度顺序后，有 TaskSetManager 根据本地性的原则确定在某个 Worker 节点上的某个 Executor 中由线程并发执行。

这些 Task 会根据数据本地性被分配到 Worker 上并发执行。图 17-7 为调用堆栈情况。

根据图 17-7 Task 任务提交的过程可以看出，DAGScheduler 提交了 TaskSet 后会交由 TaskSetManager 进行管理，TaskSetManager 在收到 TaskSet 后会根据 Task 的本地性原则为 Task 分配计算资源，并且对 Task 的执行情况进行监控，如果 Task 在执行的过程中出现失败会进行重新（一般默认为 4 次）提交。

CoarseGrainedSchedulerBackend 会进行相应资源调度，并对 Task 分配具体的计算资源，其分配资源的时候将 Task 包装成二维的数组 TaskDescription，包含 TaskID、ExecutorID 和 Task 执行环境的依赖等信息。

```
def resourceOffers(offers: Seq[WorkerOffer]): Seq[Seq[TaskDescription]] = synchronized {
// Mark each slave as alive and remember its hostname
// Also track if new executor is added
//感知集群状态
var newExecAvail = false
for (o <- offers) {
  executorIdToHost(o.executorId) = o.host
  executorIdToTaskCount.getOrElseUpdate(o.executorId, 0)
  if (!executorsByHost.contains(o.host)) {
    executorsByHost(o.host) = new HashSet[String]()

 //启动新的Executor
    executorAdded(o.executorId, o.host)
newExecAvail = true
  }
  for (rack <- getRackForHost(o.host)) {
    hostsByRack.getOrElseUpdate(rack, new HashSet[String]()) += o.host
  }
}

// Randomly shuffle offers to avoid always placing tasks on the same set of workers.
//将可用资源打散
val shuffledOffers = Random.shuffle(offers)
 // Build a list of tasks to assign to each worker.

//将task存储在数组中
val tasks = shuffledOffers.map(o => new ArrayBuffer[TaskDescription](o.cores))
val availableCpus = shuffledOffers.map(o => o.cores).toArray

//获取已经排好序的TaskSetManager
val sortedTaskSets = rootPool.getSortedTaskSetQueue

//遍历Task，重新计算
  for (taskSet <- sortedTaskSets) {
    logDebug("parentName: %s, name: %s, runningTasks: %s".format(
```

```
       taskSet.parent.name, taskSet.name, taskSet.runningTasks))
     if (newExecAvail) {
       taskSet.executorAdded()
     }
}
//Task本地性
// NOTE: the preferredLocality order: PROCESS_LOCAL, NODE_LOCAL, NO_PREF,
RACK_LOCAL, ANY
  var launchedTask = false
  for (taskSet <- sortedTaskSets; maxLocality <- taskSet.myLocalityLevels) {
    do {
      launchedTask = resourceOfferSingleTaskSet(
          taskSet, maxLocality, shuffledOffers, availableCpus, tasks)
    } while (launchedTask)
  }

 if (tasks.size > 0) {
  hasLaunchedTask = true
 }
 return tasks
}
```

之后由 DriverActor 将上面返回的 Tasks 发送到 Executor 上并发执行。

17.2 Spark 运行架构解析

Spark 目前可以部署到不同的资源平台上，包括在 Local 模式下，Standalone、Yarn、Mesos、EC2 等。本章主要以 Standalone 模式解析为主，在本章后面部分会解析 Spark 部署在 Yarn 资源管理器上的 Cluster 和 Client 不同模式架构和简单的实战演示。

Spark 部署在不同的模式下会有不同的效果，一般为了简单的测试可以运行在 Local 模式，当然 Standalone 模式在没有其他计算架构的情况下可以符合绝大多数的数据计算场景，Spark 中部署 Yarn 模式可以同时管理不同的计算框架，例如同时部署 Flink，Hadoop 的 MapReduce 分布式处理架构以及 Spark 等。

Spark 部署在 Mesos 上会包含不同的调度模式粗粒度和细粒度调度模式，Mesos 一般的组成包括 Master、Slaves、FrameWork、Executor 四部分，也可以结合使用 Zookeeper，在实际生产集群中选出 3 个节点做 Master，即使其中 Alive 的 Master 终止，可以选举其中的两个 StandBy 的 Master 接替工作，相当于为集群的正常使用做了热备，实现高可用性，保证集群处理的安全和可靠。

17.2.1 Spark 基本组件介绍

目前 Spark 已经发布到了 1.6.X 版本，预计在 2.0 版本出来会有极大的性能改善，包括 Tungsten 的实现，速度更快，性能更加完善，愈加符合实际业务需要。目前 Spark 已经发展到类似于一站式的满足各种计算场景，例如批处理、实时流处理、图计算以及机器学习库等非常具有高可用性，帮助企业实现最小的产出收获最大的效益。

(1) SparkCore 中相关的术语

Spark 基本术语如表 17-1 所示。

表 17-1 基本术语及概念

Term(术语)	Meaning(解释)
Application	运行于 Spark 上的用户程序，由集群上的一个 driver program(包含 SparkContext 对象)和多个 executor 线程组成
Applicationjar	Jar 包中包含了用户 Spark 应用程序，如果 Jar 包要提交到集群中运行，不需要将其他的 Spark 依赖包打包进行
Driver program	包含 main 方法的程序，负责创建 SparkContext 对象
ClusterManager	集群资源管理器，例如 Mesos，Hadoop Yarn
Deploy mode	部署模式，用于区别 driver program 的运行方式：集群模式(cluter mode)，driver 在集群内部启动；客户端模式(client mode)，driver 进程从集群外部启动
Worker Node	工作节点，集群中可以运行 Spark 应用程序的节点
Executor	Worker node 上的进程，该进程用于执行具体的 Spark 应用程序任务，负责任务间的数据维护(数据在内存中或磁盘上)。不同的 Spark 应用程序有不同的 Executor
Task	运行于 Executor 中的任务单元，Spark 应用程序最终被划分为经过优化后的多个任务的集合(在下一节中将详细阐述)
Job	由多个任务构建的并行计算任务，具体为 Spark 中的 action 操作，如 collect，save 等
Stage	每个 Job 将被拆分为更小的 task 集合，这些任务集合被称为 stage，各 stage 相互独立(类似于 MapReduce 中的 map stage 和 reduce stage)，由于它由多个 task 集合构成，因此也称为 TaskSet

(2) Spark Streaming

Spark Streaming 是将流式的计算分解成一系列的短小的批处理作业，是一个对实时数据进行高容错、高吞吐量的流式处理系统。其数据源也很多样例如 Flume、Kafka、HDFS、S3 等，然后将处理后的数据存储到 HDFS，Database 中，可以进行复杂的操作，例如 map、reduce、join 等。如图 17-8 表示 Spark Streaming 的数据源，图 17-9 表示 Spark Streaming 的处理过程。

图 17-8 Spark Streaming 的数据源

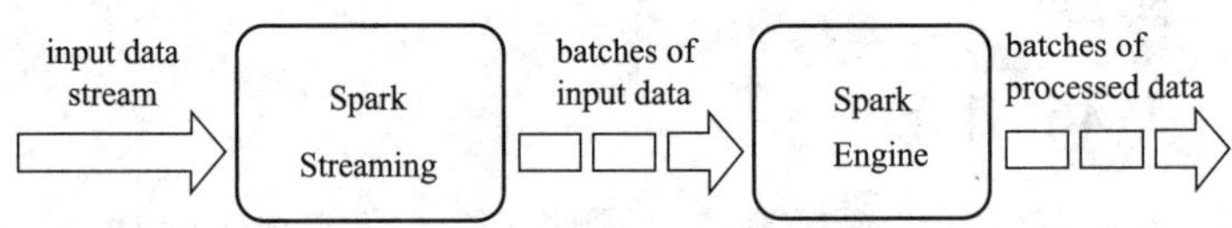

图 17-9 Spark Streaming 的处理过程

Spark Streaming 将批处理作为引擎，将不同数据源输入的数据分成一段段的 Stream，然后将其转换成 RDD，之后再由 Spark Streaming 中对 DStream 的操作转换为对 RDD 的操作，那么这里的操作底层实现还是 Spark 内核的执行流程，将处理的中间结果存储到内存或者磁盘。

(3) Spark GraphX

Spark 图计算是 Spark 关于图的并行计算的 API，其核心是 Resilient Distributed property Graph，他是一个有向多重图，其属性包括点和边，同时包含了 Table 和 Graph 视图，其存储也只需要一份，不同的视图不同的操作符，极具灵活性。

Spark GraphX 集成了 ETL、迭代式的图计算，已经提供了很多的算法，并有 Spark 社区人员不断加入新的算法使其更满足业务和需求。

(4) Spark SQL

Spark SQL 是一种用来处理结构化数据的 Spark 重要组件，SparkSQL 之所以是除了 SparkCore 以外是最大的和最受关注的组件，提供了一个叫做 DataFrames 的可编程抽象数据模型。

数据来源很强大，可以处理一切存储介质和各种格式的数据（同时可以方便地扩展 SparkSQL 的功能表支持更多类型的数据，例如 Kudu）。

SparkSQL 把数据仓库的计算能力推向了新的高度，不仅是很强大的计算速度（尤其是在 Tungsten 成熟后会更加无可匹敌），更为重要的是把数据仓库的计算的复杂度推向了历史上全新的高度（SparkSQL 后续推出的 dataFrame 可以让数据仓库直接让机器学习、图计算等复杂的算法库来对数据仓库进行复杂的深度数据价值的挖掘），SparkSQL 比 Shark 快了至少一个数量级。

Hive 进行多维度的数据分析，但是瓶颈较大，SparkSQL（DataFrame、DataSet）不仅是数据仓库的引擎，也是数据挖掘的引擎，更为重要的是 SparkSQL 是数据科学技术和分析引擎。Hive＋SparkSQL＋DataFrame 将是未来可预测的最核心的大数据技术组合，将在国内得到极大的使用需求。

(5) MLlib

MLlib 是 Spark 框架的一个基本组件即 Spark 的机器学习库，它提供了常用的机器学习算法以及测试数据等，并且由 Spark 开发团队陆续迭代更新。它由常见的学习算法和实用程序，包括分类、回归、聚类、协同过滤、降维、以及低层次的优化原语和更高级别的管道。

17.2.2 Spark 的运行逻辑

Spark 运行逻辑主要是将 Spark 应用程序打包提交到 Spark 运行集群上，对作业运行的处理是从文件系统例如从 HDFS 上读取需要处理的数据，然后 Spark 根据自身的运行逻辑和架构将待处理的数据通过高层调度器 DAGScheduler 依据宽依赖划分为不同的 Stage，整个处理划分过程都是对 action 算子的操作触发的，然后形成一个有向无环图 DAG，在此过程中对 RDD 的划分形成 TaskSet，然后提交 TaskSet 并为其分配计算资源，然后根据数据本地性启动某些 Worker 上的 Executor，在线程池中并发执行，最后将执行的结果放到文件系统 HDFS 或者数据库中。

当 TaskSchedulerImpl 接收到 DAGScheduler 划分的 Stage（TaskSet），接着会通过

TaskSchedulerImpl. submitTasks 将 TaskSet 加入到 TaskSetManager 中进行管理，然后由 SchedulableBuilder 确定 TaskSetManager 的调度顺序，之后按照 TaskSetanager 的 locality aware 来确定每个 Task 运行在哪个 TaskSchedulerBackend 中。

```
schedulableBuilder.addTaskSetManager( manager,manager.taskSet.properties)
```

CoarseGrainedSchedulerBackend. reviveOffers 给 DriverEndpoint 发送 ReviveOffiers，ReviveOffersv 本身是一个空的对象，只是起到触发底层资源调度的作用，在有 Task 提交或者计算资源变动的时候会发送 ReviveOffers 这个消息作为触发器。

DriverEndpoint 接受 ReviveOffers 消息并路由到 makeOffers 具体的方法中：在 makeOffers 中首先准备好所有可以用于计算的 workOffers（代表了所有可用的 ExecutorBackend 中可以使用的 ExecutorBackend 中可以使用的 Cores 等信息）。

```
/**
 * Represents free resources available on an executor.
 */
private[spark]
case class WorkerOffer(executorId: String, host: String, cores: Int)
```

当为每个 Task 具体分配计算资源，输入 ExecutorBackend 及其上可用的 Cores 时，输出 TaskDescription 的二维数组，在其中确定了每个 Task 具体运行在哪个 ExecutorBackend 上：（TaskSchedulerImpl. resourceOffers） ResourceOffers 到底是如何确定 Task 具体运行在哪个 ExecutorBackend 上的呢？算法的实现具体主要有如下几步。

① 通过 Random. Shuffle 方法重新洗牌所有的计算资源以寻求计算的负载均衡；

② 根据每个 ExecutorBackend 的 Cores 的个数申明类型为 TaskDescription 的 ArrayBuffer 数组；

③ 如果有新的 ExecutorBackend 所分配给的 Job，此时会调用 executorAdded 来获取最新的完整的可用计算资源；

④ 要求最高级别的优先级本地性；

⑤ 通过调用 TaskSetManger 的 resourceOffer 最终确定每个 Task 具体运行在哪个 ExecutorBackend 的具体的 Locality Level；

⑥ 通过 launchTask。

在此过程中需要明确以下几点。

① Task 默认的最大重试次数是 4 次（TaskSchedulerImpl 中）。

```
def this(sc: SparkContext) = this(sc, sc.conf.getInt("spark.task.maxFailures", 4))
```

② Spark 应用程序目前支持两种调度器：FIFO、FAIR，可以通过 Spark-env. sh 中 spark. sheculer. mode 进行具体的设置（默认为 FIFO）。

③ TaskSetManager 中要负责为 Task 分配计算资源：此时程序已经具备集群中的计算资源了，根据计算本地性原则确定 Task 具体要运行在哪个 ExecutorBackend 中。

④ TaskDescription 中已经确定好了 Task 具体运行在哪个。在确定 Task 具体运行在哪个 ExecutorBackend 上的算法是由 TaskSetManager 的 ResourceOffer 方法决定的。

```
/**
 * Description of a task that gets passed onto executors to be executed, usually created by
 * [[TaskSetManager.resourceOffer]].
 */
```

```
private[spark] class TaskDescription(
    val taskId: Long,
    val attemptNumber: Int,
    val executorId: String,
    val name: String,
    val index: Int,    // Index within this task's TaskSet
    _serializedTask: ByteBuffer)
  extends Serializable {
```

⑤ 对于数据本地性从高到低依次为：PROCESS _ LOCAL，NODE _ LOCAL，NO _ PREF，RACK _ LOCAL，ANY，其中 NO _ PREF 是指集群本地性（优先级的判断是根据数据本地性）。

⑥ 每个 Task 默认是采用一个线程进行计算的。

⑦ DAGScheduler 是从数据层面考虑 perferedLocation 的，而 TaskSccheduler（高层调度）内存是从具体计算 Task（底层调度）角度考虑计算的本地性。

⑧ Task 进行广播时候的 AkkFrameSize 大小是 128MB，如果任务大于等于 128MB-200K 的话，则 Task 会直接丢弃掉；如果小于 128MB-200K 的话会通过 CoarseGrainedExecutorBackend 去 launchTask 到具体的 ExecutorBackend 上。

在 SparkContext 实例化的时候调用 createTaskScheduler 来创建 TaskSchedulerImpl 和 SparkDeploySchedulerBackend，同时在 SparkContext 实例化的时候会调用 TaskSchedulerImpl 的 start，在 start 方法中会调用 SparkDeploySchedulerBackend 的 start，在该 start 方法中会创建 AppClient 对象并调用 AppClient 对象的 start 方法，在该 start 方法中会创建 ClientEndpoint，在创建 ClientEndpoint 会传入 Command 来指定具体为当前应用程序启动的 Executor 进行的入口类的名称为 CoarseGrainedExecutorBackend，然后 ClientEndpoint 启动并通过 tryRegisterMaster 来注册当前的应用程序到 Master 中，Master 接受到注册信息后如何可以运行程序，则会为该程序生成 Job ID 并通过 schedule 来分配计算资源，具体计算资源的分配是通过应用程序的运行方式、Memory、cores 等配置信息来决定的。

最后 Master 会发送指令给 Worker，Worker 中为当前应用程序分配计算资源时会首先分配 ExecutorRunner，ExecutorRunner 内部会通过 Thread 的方式构建 ProcessBuilder 来启动另外一个 JVM 进程，这个 JVM 进程启动时候加载的 main 方法所在的类，就是在创建 ClientEndpoint 时传入的 Command 来指定具体名称为 CoarseGrainedExecutorBackend 的类，此时 JVM 在通过 ProcessBuilder 启动的时候获得了 CoarseGrainedExecutorBackend 后加载并调用其中的 main 方法，在 main 方法中会实例化 CoarseGrainedExecutorBackend 本身这个消息循环体，而 CoarseGrainedExecutorBackend 在实例化的时候会通过回调 onStart 向 DriverEndpoint 发送 RegisterExecutor 来注册当前的 CoarseGrainedExecutorBackend，此时 DriverEndpoint 收到到该注册信息并保存在了 SparkDeploySchedulerBackend 实例的内存数据结构中，此时 Driver 就获得了计算资源，同时并发送 RegisterExecutor 给 CoarseGrainedExecutorBackend，然后启动 Executor，执行 Task。图 17-10 为 SparkContext 运行机制。

Driver 端提交程序 Tasks 后，由 Master 检测没有问题便进行资源的分配和 AppId 的分配，然后发送指令给 Worker 节点，Worker 节点默认会分配一个 Executor，然后在 Executor 的线程池中进行并发执行。Master 收到提交的程序，Master 根据以下三点为程序分配资源。

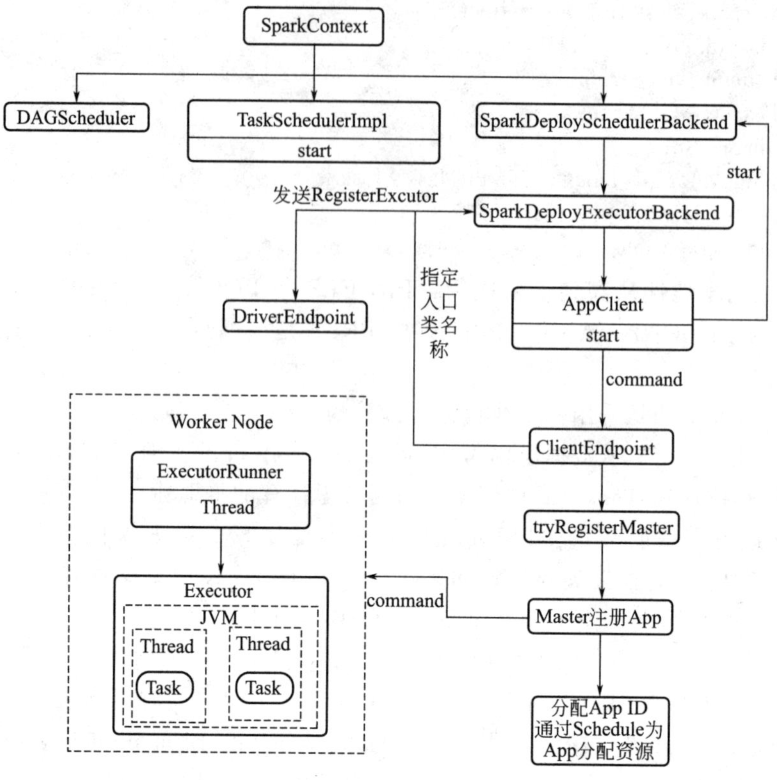

图 17-10　SparkContext 运行机制

① spark-env. sh 和 spark-defaults. sh；

② spark-submit 提供的参数；

③ 程序中 SparkConf 配置的参数。

Master 负责资源的管理和调度，因此主要是 Master 中的 Schedule 方法中实现，每当新程序提交或集群资源发送变化的时候调用该方法。需要说明的是作业的调度只有在 Master 是 Alive 的状况下进行，其内部是将 Master 中保留的 Worker 信息随机打乱，然后随机取一个 Alive 的 Worker 参与资源分配。

在为应用程序分配 Executor 的时候需要判断是否还需要分配 Cores，若不需要则不会为其分配 Executor，只有 Application 对每个 Executor 的内存和 Cores 满足要求的情况下才会被分配，分配完 Executor 后，Master 会通过远程通信的方式发指令给 Worker 具体启动 ExecutorBackend。

CoarseGrainedExecutorBackend 启动时向 Driver 注册 Executor 其实质是注册 ExecutorBackend 实例，由此可知 CoarseGrainedExecutorBackend 和 Executor 是一一对应的关系，它是 Executor 运行所在的进程名称，Executor 才是真正处理 Task 的对象，Executor 内部是通过线程池的方式来完成 Task 的计算工作。

Executor 具体是如何工作的呢？当 Driver 发送过来 Task 的时候，其实是发送给了 CoarseGrainedExecutorBackend 这个 RpcEndpoint，而不是直接发送给了 Executor，ExecutorBackend 在收到 Driver 中发送过来的消息后会提供调用 launchTask 来交给 Executor 去执行，Executor 是真正负责 Task 计算的，其在实例化的时候会实例化一个线程池来准备 Task 的计算的；创建 ThreadPool 中以多线程并发执行和线程复用的方式高效地执行 Spark

发过来的 Task。

Driver 向 ExecutorBackend 发送 LaunchTask，launchTask 执行任务，接收到 Task 执行的命令后，会首先把 Task 封装到 TaskRunner 中，然后交给线程池中的线程处理；TaskRunner 其实是 Java 中的 Runnable 接口的具体实现，在真正工作的时候会交给线程池中的线程去运行的。如图 17-11 所示为 Executor 执行机制。

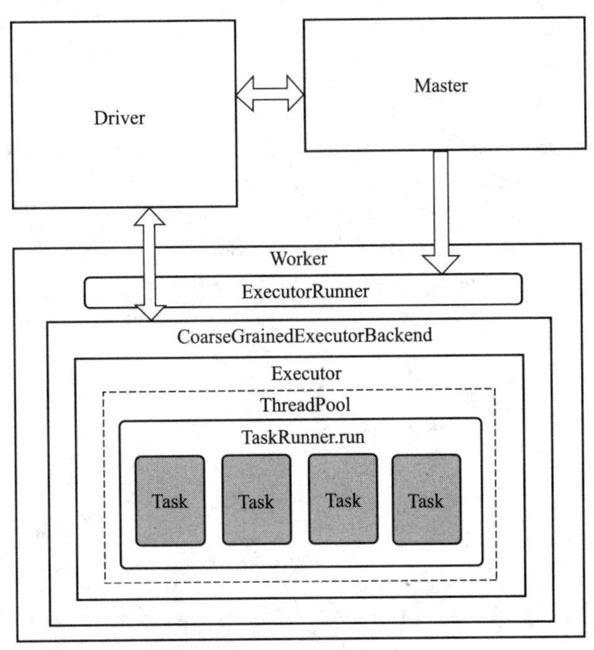

图 17-11　Executor 执行机制

Worker 管理当前 Node 的资源，并接受 Master 的指令来分配具体的计算资源（Executor）。Worker 进程通过一个 Proxy（代理句柄）为 ExecutorRunner 的对象实例远程启动 ExecutorBackend 进程，实际工作的时候会通过 TaskRunner（一个 Runner 的接口，线程一般会用 Runner 的接口封装业务逻辑，有 run 方法所以可以回调）来封装 Executor 接收到的 Task，然后从 ThreadPool 中获取一个线程执行，执行完成后释放并覆用。

划分的 Stage 中，在最后一个 Stage 前的其他 Stage 都进行 shuffleMapTask，这是是对数据进行 shuffle，shuffle 的结果保存在 Executor 所在节点的本地文件系统中，最后一个 Stage 中的 Task 就是 ResultTask，负责结果数据生成。

Driver 会不断发送 Task 给 Executor 进行执行，所以 Task 都正确执行，到程序运行结束；若超过执行次数的限制，或者没有执行时会停止，待正确执行后会进入下一个 stage，若没有执行成功，高层调度器 DAGSchedular 会进行一定次数的重试，若还是执行不成功就意味着整个作业的失败。

每个 Task 具体执行 RDD 中的一个 Partition，（默认情况下为 128M，但最后一个记录跨两个 Block）基于该 Partition 具体执行，定义的一系列内部函数，直到程序执行完成。然后将程序执行结果数据保存到 HDFS、HBase 或数据库中。

下面以 Spark-shell 的运行为例进行说明，启动 Hadoop 集群的 HDFS 以及 Spark 集群。然后运行 Spark-shell。首先在 Hadoop 的 sbin 目录下启动 HDFS 如下所示。

```
1. ./start-dfs.sh
```

然后在 Spark 安装目录下的 sbin 目录下启动 Spark，如下所示。

```
./start-all.sh
```

启动后 jps 查看启动进程，如下所示。

```
root@Master:/usr/local/spark-1.6.0-bin-hadoop2.6/# jps
23432 jps
6543 NameNode
7456 Master
9734 Worker
8398 SecondaryNameNode
```

然后开始运行 Spark-shell，如下所示。

```
root@Master:/usr/local/spark/spark-1.6.0-bin-hadoop2.6/bin# ./spark-shell --master spark://Master:7077
log4j:WARN No appenders could be found for logger
(org.apache.hadoop.metrics2.lib.MutableMetricsFactory).
log4j:WARN Please initialize the log4j system properly.
log4j:WARN See http://logging.apache.org/log4j/1.2/faq.html#noconfig for more info.
Using Spark's repl log4j profile: org/apache/spark/log4j-defaults-repl.properties
To adjust logging level use sc.setLogLevel("INFO")
Welcome to
      ____              __
     / __/__  ___ _____/ /__
    _\ \/ _ \/ _ `/ __/  '_/
   /___/ .__/\_,_/_/ /_/\_\   version 1.6.0
      /_/

Using Scala version 2.10.5 (Java HotSpot(TM) 64-Bit Server VM, Java 1.7.0_71)
Type in expressions to have them evaluated.
Type :help for more information.
```

Spark-shell 开始运行，可以看到 SparkUI 的地址，以及 ClientEndpoint 注册，Spaprk-DeploySchedulerBackend 连接上集群后取得 App，然后 ClientEndpoint 获取 Worker 上的 Executor，同样 SpaprkDeploySchedulerBackend 也获取到 Executor，如下所示。

```
INFO client.AppClient$ClientEndpoint: Connecting to master spark://Master:7077···
INFO cluster.SpaprkDeploySchedulerBackend:Connected to Spark cluster with app ID-
20160220202236-0000

INFO client.AppClient$ClientEndpoint: Executor added: app-20160220202236-0000/0 on
worker-20160220200130-192.168.1.11 with 2 cores

INFO cluster.SparkDeployScheudulerBackend: Grained Executor ID  app-20160220202236-
0000/0 on hostPort-192.168.1.11 with 2 cores,1GB RAM

 INFO client.AppClient$ClientEndpoint: Executor added: app-20160220202236-0000/1 on
workere 20160220200130-192.168.1.12 with 2 cores

INFO cluster.SparkDeployScheudulerBackend: Grained Executor ID  app-20160220202236-
0000/1 on hostPort-192.168.1.12 with 2 cores,1GB RAM
```

之后可以看到数据块的存储 BlockManager，以及 Executor 的执行，如下所示。

```
INFO storage.BlockManagerMaster：Registerd Blockmanger
INFO client.APPClient$ClientEndpoint：Executor updated：20160220202236-0000/0 is now
```

```
RUNNING
INFO client.APPClient$ClientEndpoint：Executor updated：20160220202236-0000/1 is now
RUNNING
INFO Cluster.SparkDeploySchedulerBackend：SchedulerBackend is ready for Scheduling
beginning after reached minRegisterResoucesTatio:0.0
INFO repl.SparkILoop:Created spark context..
Spark context available as sc.
```

Spark-shell 运行完成后，运行 WordCount 程序，图 17-12 为运行 WordCount 部分过程，如下所示。

```
scala> sc.TextFile("/library/wordCount/input/Data").flatMap(_.split(" ")).map(word
=>(word,1)).reduceByKey(_+_,1).saveAsTextFile("/myResoult/")
```

```
INFO scheduler.DAGScheduler: ShuffleMapStage 0 (map at <console>:28) finished in 11.817 s
INFO scheduler.TaskSchedulerImpl: Removed TaskSet 0.0, whose tasks have all completed, from pool
INFO scheduler.DAGScheduler: looking for newly runnable stages
INFO scheduler.DAGScheduler: running: Set()
INFO scheduler.DAGScheduler: waiting: Set(ResultStage 1)
INFO scheduler.DAGScheduler: failed: Set()
INFO scheduler.DAGScheduler: Submitting ResultStage 1 (MapPartitionsRDD[5] at saveAsTextFile at
```

图 17-12　WordCount 运行部分过程

从图 17-12 可以看出 ShuffleMapStage 运行结束后，然后将该 Stage 对应的 TaskSet 从调度池中移除（前面已经提到过调度的两种方式 FIFO 以及 FAIL），然后开始寻找新的 Stage，恰恰剩下最后一个 ResultStage，开始被执行。

```
TaskSetManager: Starting task 2.0 in stage 0.0 (TID 0, Worker1, partition 2,NODE_LOCAL, 2150 bytes)
TaskSetManager: Starting task 0.0 in stage 0.0 (TID 1, Worker3, partition 0,NODE_LOCAL, 2150 bytes)
TaskSetManager: Starting task 3.0 in stage 0.0 (TID 2, Worker2, partition 3,NODE_LOCAL, 2149 bytes)
```

图 17-13　Task 的执行

图 17-13 是 Task 执行的过程，具体执行通过日志可以看出，包括 Task 运行在哪个 Worker 节点上，对应的是哪个 Partition，Task 任务是由 TaskSetManager 进行管理，以及数据本地性（NODE _ LOCAL 是当前节点）等。由此可以更好地对 Spark 运行逻辑有一个更加清晰的理解和认识。

17.3　Spark 在不同集群上的运行架构

Spark 一般可以部署不同的模式，例如 Yarn 资源管理器、Mesos 等，会包含不同的调度模式粗粒度和细粒度调度模式，Mesos 一般的组成包括 Master、Slaves、FrameWork、Executor 四部分，也可以结合使用 Zookeeper，在实际生产集群中选出 3 个节点做 Master，即使其中 Alive 的 Master 终止，可以选举其中的两个 StandBy 的 Master 接替工作，相当于为集群的正常使用做了准备，实现高可用性，保证集群处理的安全和可靠。

17.3.1　Spark 在 Standalone 模式下的运行架构

通常情况下使用 Spark Standalone 模式可以满足大多数的计算场景，在同一个集群上需要运行不同的计算框架，例如同时需要 Hadoop 的 MapReduce 计算框架或者 Flink 等，此时需要 Yarn 或者 Mesos 等资源管理器管理，管理集群在不同的计算框架下协同处理。

Standalone 模式下，用户首先提交 Spark 应用程序，之后创建 SparkContext，新创建的 SparkContext 会根据编程时设置的参数或系统默认的配置连接到 ClusterManager 上，ClusterManager 会根据用户提交时的设置（如：占用 CPU、MEN 等资源情况），来为用户程序分配计算资源，启动相应的 Executor；而 Driver 根据用户程序调度的 Stage 的划分，即高层调度器（RDD 的依赖关系），若有宽依赖，会划分成不同的 Stage，每个 Stage 由一组完全相同的任务组成（业务逻辑相同，处理数据不同的 Tasks 组成），该 Stage 分别作用于待处理的分区，待 Stage 划分完成和 TaskSet 创建完成后，Driver 端会向 Executor 发送具体的 task，当 Executor 收到 task 后，会自动下载运行需要的库、包等，准备好运行环境后由线程池中的线程开始执行，因此 Spark 执行是线程级别的。

Hadoop 运行比 Spark 代价大很多，因 Hadoop 中的 MapReduce 运行的 JVM 虚拟机不可以复用，而 Spark 运行的线程池中的线程可以进行复用。

执行 Task 的过程中，Executor 会将执行的 Task 汇报给 Driver，Driver 收到 Task 的运行状态情况后，会根据具体状况进行更新等。图 17-14 为 Standalone 模式的整体架构。

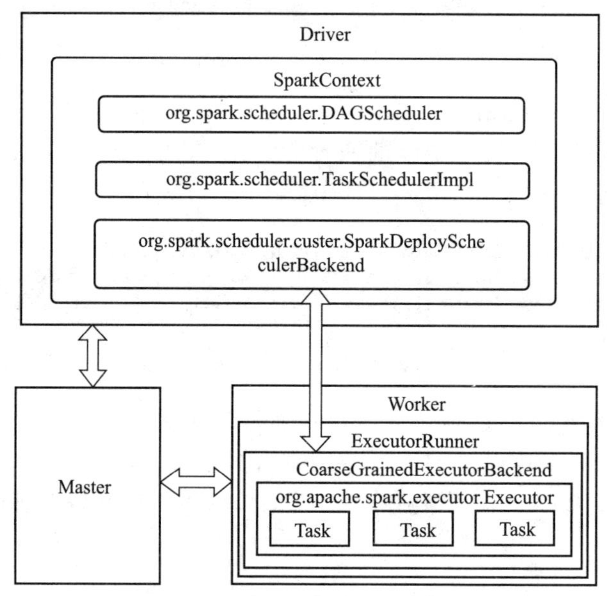

图 17-14　Standalone 模式整体架构

Task 根据不同的 Stage 的划分，会被划分为两种类型：

① shuffleMapTask，在最后一个 Stage 前的其他 Stage 都进行 shuffleMapTask，此时是对数据进行 shuffle，shuffle 的结果保存在 Executor 所在节点的本地文件系统中；

② ResultTask，负责生成结果数据。

Driver 会不断发送 Task 给 Executor 进行执行，所以 Task 都正确执行或者超过执行次数的限制，或者没有执行时会停止，待正确执行后会进入下一个 stage，若没有正确执行成功，高层调度器 DAGSchedular 会进行一定次数的重试，若还是执行不成功就意味着整个作业的失败。

Worker 是集群运行节点，Worker 上不会运行 Application 的程序，因为 Worker 管理当前 Node 的资源资源，并接受 Master 的指令来分配具体的计算资源 Executor（在新的进程中分配），Worker 不会向 Master 发送 Worker 的资源占用情况（MEN、CPU），Worker

向 Master 发送心跳，只有 Worker ID，因为 Master 在程序注册的时候已经分配好了资源，同时只有故障的时候才会发送资源的情况。图 17-15 为 Spark 任务调度过程：

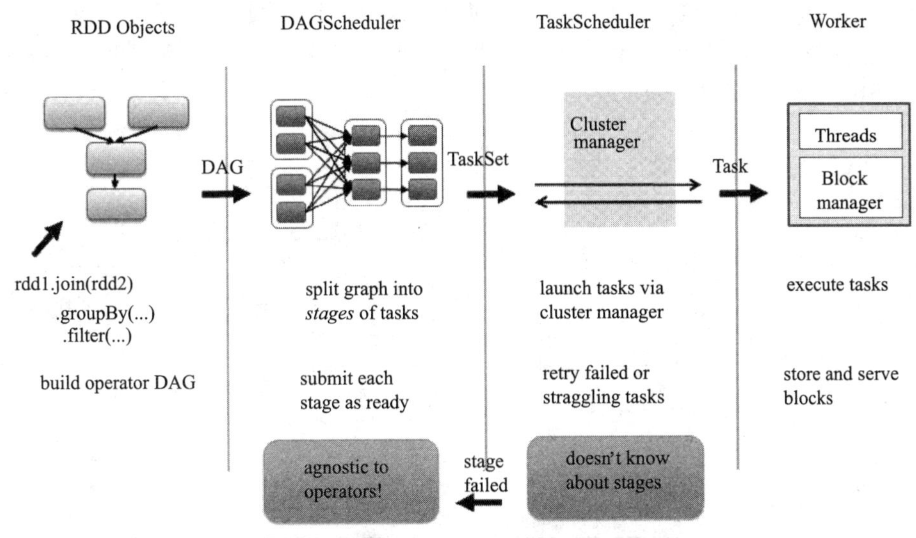

图 17-15　Spark 任务调度

DAGScheduler 负责将 Job 中 Rdd 构成的 DAG 划分为不同的 Stage。

Job 中包含了一系列 Task 的并行计算，Job 由 action（如 SaveAsTextFile）触发的，其前面是若干 RDD，而 RDD 是 Transformation 级别的，transformation 是 Lazy 的，因为是 Lazy 的所以不计算往前推（如：WordCount 中 collect 触发了 job，执行的时候由 mapPartitionRDD 往前推到 hadoopRDD，之后开始一步步执行），若两个 RDD 之间回数的时候是窄依赖的话就在内存中进行迭代，也是 Spark 之所以快的原因，不是因为基于内存，而是因为其调度和容错及其他内容。

DAGScheduler 根据宽依赖触发 Stage 的划分，在 RDD 中将依赖划分成了两种类型，窄依赖（narrow dependencies）和宽依赖（wide dependencies）。

窄依赖是指父 RDD 的每个分区都只被子 RDD 的一个分区所使用（一对一）。

宽依赖就是指父 RDD 的分区被多个子 RDD 的分区所依赖（一对多）。

首先在 Spark 集群上通过 spark-submit 的方式提交 Spark 应用程序，在用户提交的 driver 端会实例化 SparkContext，同时会创建 DAGScheduler、TaskScheduler、SchedulerBackend 三大核心对象，在此过程中 DAGScheduler 将 Job 划分成 Stage，Stage 内部是由 Task 组成的 TaskSet，然后将 TaskSet 依次提交给 TaskSheduler，TaskScheduler 和 SchedulerBackend 负责执行 TaskSet 并为其分配计算资源，然后将 Spark Application 注册给 Master，Master 接收到注册信息后，分配 appId 和计算资源，同时 Master 将用户提交的程序发送指令给 Worker 分配计算资源，Worker 默认启动一个 Executor 给一个 Task，Worker 进程通过 proxy 句柄为 ExecutorRunner 对象实例远程启动 ExecutorBackend，即 CoaseGrainedExecutorBackend，ExecutorBackend 里面 Executor 通过 TaskRunner 封装 Task，Task 从 Executor 中的 ThreadPool 线程池获取一条线程执行 Task，每个 Task 计算 RDD 中的一个 Partition，对于执行完成后线程回收复用，下一个 Task 会循环直到整个程序运行完成知道最后一个 Stage 中的 Task 称为 ResultTask（前面 Stage 中的 Task 都是 ShuffleMapTask，为下一个 Stage 做数据准备），生成 job 的结果。

17.3.2 Spark on yarn 的运行架构

Yarn 是 Hadoop 推出整个分布式（大数据）集群的资源管理器，负责资源的管理和分配，基于 Yarn 我们可以在同一个大数据集群上同时运行多个计算框架，例如 Spark，MapReduce、Storm 等。如图 17-16 所示。

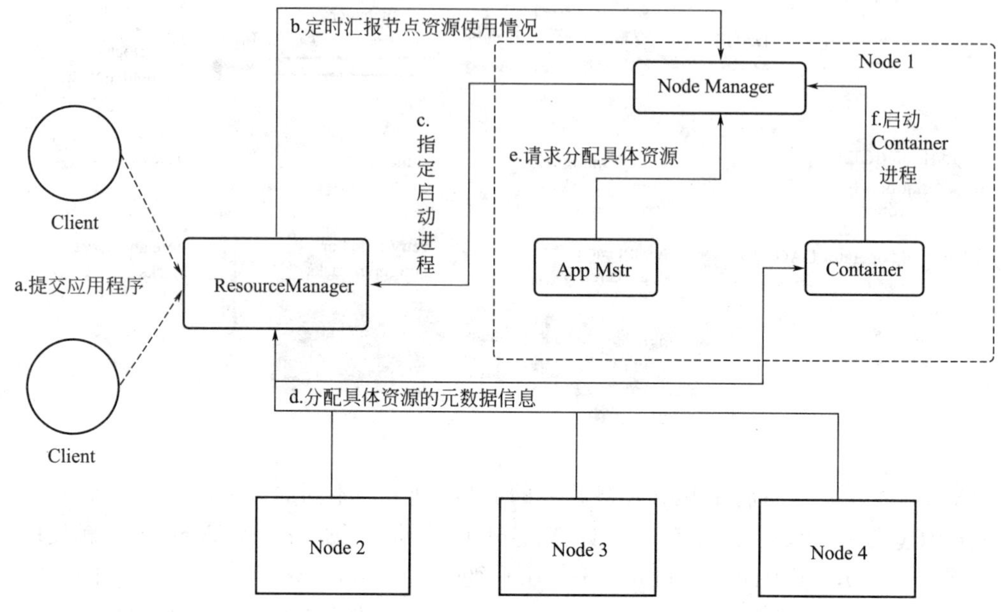

图 17-16 Spark On YarnCluster 运行机制

Yarn 有两个主体对象：ResourceManager 及 NodeManager。

ResourceManager，负责集群的资源管理（CPU、内存等，有可能包含磁盘 IO 和网络 IO，具体跟集群资源管理器的能力有关），例如 ResourceManager 管理 10000 台机器的资源，那么每台机器的资源是由 Node Manager 来管理的，并且 Node Manager 会定期向 ResourceManager 汇报具体资源管理的使用情况，所以基于每台 Node Manager 向 ResourceManager 汇报当前节点的资源使用情况，ResourceManager 就会知道整个集群的资源使用情况，从而达到管理集群资源的目的。

NodeManager 管理节点的资源，并向 ResourceManager 汇报每台机器的资源使用情况。

当 Client 客户端提交应用程序的时候，首先提交给 ResourceManager，ResourceManager 会根据自己对整个集群资源的掌握情况，选择在某一台机器上启动一个具体的进程，此时 ResourceManager 就是整个集群的 Master，负责应用程序运行的具体资源的管理，也就是说 ResourceManager 在具体某台机器上启动进程来运行由 Client 提交的应用程序，当然作为 Master 的 ResourceManager 不会主动到某台具体的机器上启动具体的进程，而是指定 NodeManager 来启动具体的节点上的进程来运行应用程序任务。

而 App Mstr 是应用程序的 Driver，一旦应用程序提交后会向 ResourceManager 进行注册和申请资源，ResourceManager 接收到注册请求会根据实时掌握的集群资源使用情况，判断具体哪台机器节点有资源可以满足应用程序的运行，那么 ResourceManager 会将具体可以运行应用程序的机器节点和某些机器节点可以分配多少计算资源的元数据分配给 App Mstr，然后 App Mstr 拿到 ResourceManager 提供的资源的元数据向 NodeManager 分配具体的资源（APP Mastr 就是当前节点的 Master，具体资源就是 MEN 和 CPU 等），ResourceManager 实质上是告诉了 APP Mastr 资源的元数据信息，APP Mastr 向 Node Manager 请求分配据具体的每台节点的计算资源，Container 封装了 MEN 和 CPU，所以 ResourceManager 根据集群资源的使用情况告诉 APP Mastr 能够向 Node Manager 分配多少 Container，然后 Node Manager 会根据元数据的信息启动具体的 Container，Container 封装了 MEN 和 CPU，它是一个个的 JVM 虚拟机，元数据就是机器，App Mstr（负责作业运行，相对于 Driver）。

这里可以用项目人员开发为例进行解释：Container 的关系就是项目经理和工程师之间的关系，ResourceManager 就相当于公司的资源总管，那么在具体地进行应用程序的执行的时候，项目经理（App Mstr）向资源总管（ResourceManager）申请资源，而整个项目（应用程序）的具体执行有项目经理（App Mstr）负责。

如果 AppMstr 无法一次性地分配好资源，或者提交的应用程序需要 5000 个 Container 也就是 JVM 虚拟机去运行应用程序，而 ResourceManager 发现无法一次性申请运行 5000 个 JVM 的资源，此时 AppMstr 依然告诉 ResourceManager 需要能够运行 5000 个 Container 的计算资源，能够分配到基于先进行分配，AppMstr 启动能够运行的 JVM 进程计算已经分配到资源的 Container，没有分配到的 Container 在队列中等待，当有 Container 运行完成，释放资源后队列中的 Container 继续进行复用该怎样进行计算，这个过程中集群中的资源和作业一直进行动态变化。Container 的启动是由 NodeManager 启动，Container 要向 Nodemanage 汇报资源信息，Container 要向 App Mstr 汇报计算信息。

Standalone 模式下 Client 提交应用程序给 Master，Master 管理集群的资源使用情况，如果是 Cluster 的模式，Master 会主动到某台机器上启动 Driver 进程，同时让 Worker 启动 Executor 进程来运行 Task 任务。

在具体的机器节点上，APP Mastr 就是当前节点的 Master，具体资源就是 MEN 和 CPU 等，Container 封装了 MEN 和 CPU，它是一个个的 JVM 虚拟机，元数据就是机器，APPMstr（负责作业运行，相对于 Driver）和 Container 的关系就是项目经理和员工程师的关系。

Client 端向 ResourceManager 提交 Application，ResourceManager 后根据集群资源状况决定在具体某个 Node 上来启动当前提交的 Application 的任务调度器 Driver（ApplicationMaster），然后决定命令具体的某个 Node 上的资源管理器 NodeManager 来启动一个新的 JVM 基础运行程序的 Driver 部分，当 ApplicationMaster 启动的时候会下载当前 Application 相关的 Jar 等各种资源并基于此决定向 ResourceManager 申请资源的具体内容（例如需要多少个 Container，和 Container 的配置），ResourceManager 接受到 ApplicationMaster 的资源分配的请求后会最大化的满足资源分配的请求，并把资源的元数据信息发送给 ApplicationMaster，ApplicationMaster 收到资源的元数据信息后，会根据元数据的信息发送指令给具体机器上的 NodeManager，让 NodeManager 来启动具体的 Container，Container 在启动后必须向 ApplicationMaster 注册，当 ApplicationMaster 获得了用于计算的 Container 后，开始并行任务的调度和计算，直到作业运行完成。

如果 ResourceManager 第一次没有能够完全完成资源分配的请求，后续 ResourceManager 发现集群中有新的可用资源时，会主动向 ApplicationMaster 发送新的可用资源的元数据信息，以提供跟多的资源用于当前程序的运行。

由 client 向 ResourceManager 提交请求，并上传 jar 到 HDFS，包括四个步骤。

① 连接到 ResourceManager。

② 从 RM ASM（ApplicationsManager）中获得 metric、queue 和 resource 等信息。

③ upload app jar and spark-assembly jar。

④ 设置运行环境和 container 上下文（launch-container. sh 等脚本）。

在 yarn-client 模式下，Driver 运行在 Client 上，通过 ApplicationMaster 向 RM 获取资源。本地 Driver 负责与所有的 Executor，container 进行交互，并将最后的结果汇总。结束掉终端，相当于 kill 掉这个 spark 应用。一般来说，如果运行的结果仅仅返回到 terminal 上时需要配置这个。图 17-17 为 Spark on Yarn client 架构。

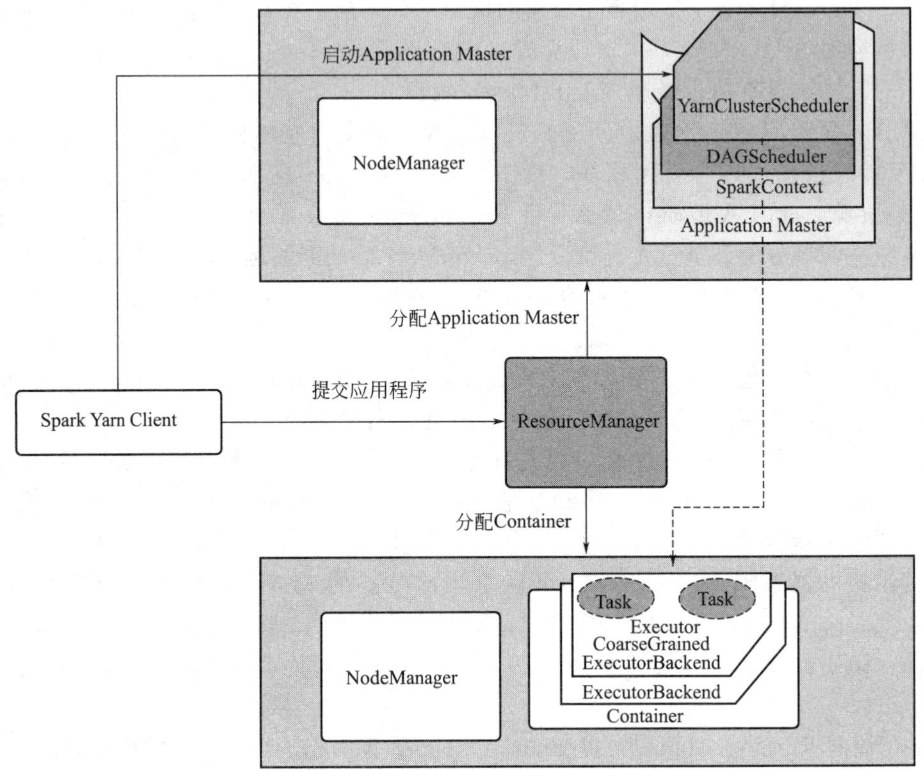

图 17-17　Spark On YarnClient 运行原理

需要说明的是 Standalone 模式下 Client 提交应用程序给 Master，Master 管理集群的资源使用情况，如果是 Cluster 的模式，Master 会主动到某台机器上启动 Driver 进程，同时让 Worker 启动 Executor 进程来运行 Task 任务。

在具体的机器节点上，APPMastr 就是当前节点的 Master，具体资源就是 MEN 和 CPU 等，Container 封装了 MEN 和 CPU，它是一个个的 JVM 虚拟机，元数据就是机器，APPMstr（负责作业运行，相对于 Driver）和 Container 的关系就是项目经理和工程师的关系。

Container 的启动是由 NodeManager 启动，Container 要向 Nodemanage 汇报资源信息，Container 要向 App Mstr 汇报计算信息。

Spark on Yarn 的两种运行模式实战，此时不需要启动 Spark 集群，只需要启动 Yarn 即可，Yarn 的 ResourceManager 就相对于 Spark Standalone 模式下的 Master（启动 spark 集群是要用到 standalone，现在有 yarn 了，就不用 spark 集群了）。

需要说明的是无论什么模式，只要当前机器运行了 Spark，首先需要安装 Spark、Scala、Java，不用启动 Spark 集群，但是 Spark On Yarn 模式下，应用程序提交后资源和集群的管理模式不同，但作业实际是运行在 Spark 集群架构中，Standalone 模式下有 Master、Worker、Spark On Yarn 模式下 Master 相当于 ResourceManager。

17.3.3 Spark 在不同模式下的应用实战

选择 Spark on Yarn 运行模式，此时不需要启动 Spark 集群，只需要启动 Yarn 即可，Yarn 的 ResourceManager 就相对于 Spark Standalone 模式下的 Master。

Spark on Yarn 的两种运行模式唯一的决定因素是当前 Application 从任务调度器 Driver 运行在什么地方。Cluster：如果 Spark 运行在 on Yarn 上，根本就没必要启动 Spark 集群，Master 是 ResourceManager；Client：Driver 运行在当前提交程序的客户机器上。

Standalone 模式下启动 Spark 集群，也就是启动 Master 和 Worker，其实启动的是资源管理器，真正作业计算的时候和集群资源管理器没有任何关系，因此 Spark 的 Job 真正执行作业的时候不是运行在启动的 Spark 集群中的，而是运行在一个个 JVM 中的，只要在 JVM 所在的集群上安装配置了 Spark 即可。

实战运行 Spark on yarn，以下为 Spark on yarn 的 Client 模式，首先启动 Hadoop 的分布式文件系统 HDFS。

```
./start-dfs.sh
```

启动 Yarn：

```
./start-yarn.sh
```

启动 Hadoop 集群的 jobHistoryServer：

```
./mr-jobhistory-daemon.sh start historyserver
```

启动 Spark 集群的 historyserver：

```
./start-history-server.sh
```

JPS 查看进程：

```
root@Master:/usr/local/spark/spark-1.6.0-bin-hadoop2.6/sbin# jps
3675 DataNode
5020 ResourceManager
5743 HistoryServer
3553 NameNode
5170 NodeManager
5773 Jps
3832 SecondaryNameNode
5651 JobHistoryServer
```

然后提交应用程序：

```
./spark-submit --class org.apache.spark.examples.SparkPi --master yarn --deploy-mode client ..
/lib/spark-examples-1.6.0-hadoop2.6.0.jar 1000
```

以上命令提交作业，其中：

① --class org. apache. spark. examples. SparkPi：运行 SparkPi；

② --master yarn：指定作业运行在 yarn 上；

③ --deploy-mode client：使用 yarn 的 Client 模式运行作业；

④ . /lib/spark-examples-1. 6. 0-hadoop2. 6. 0. jar：SparkPi 所在的 jar 的位置；

⑤ 1000：指定并行度 1000（即 50000JVM 虚拟机）将并行任务变成 1000 个（1000 台虚拟机）。

在此可以看看 Spark 实例中的 Pi 的代码，如下所示。

```
package org.apache.spark.examples

import scala.math.random
import org.apache.spark._

/** Computes an approximation to pi */
object SparkPi {
  def main(args: Array[String]) {
    val conf = new SparkConf().setAppName("Spark Pi")
    val spark = new SparkContext(conf)
    val slices = if (args.length > 0) args(0).toIntelse 2
    val n = math.min(100000L * slices, Int.MaxValue).toInt // avoid overflow
    val count = spark.parallelize(1 until n, slices).map { i =>
      val x = random * 2- 1
      val y = random * 2- 1
      if (x*x + y*y < 1) 1 else 0
    }.reduce(_+_)
    println("Pi is roughly " + 4.0 * count / n)
    spark.stop()
  }
}
```

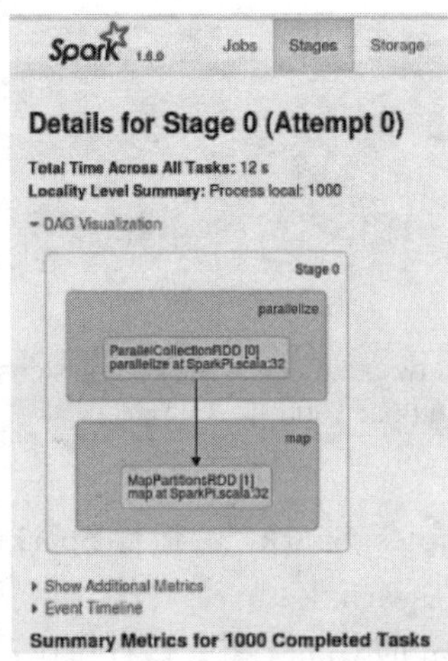

图 17-18　Spark on yarn Client 运行 Stage 划分情况

实际运行的时候在 Master 上多增加了一个 SparkSubmit 进程，是因为使用 SparkSubmit 提交的应用程序，那么在 Worker 节点上增加了 CoaseGrainedExecutorBackend 进程，ResourceManager 为 NodeManager 分配的，和 Standalone 模式下 Master 为 Worker 分配的 CoaseGrainedExecutorBackend 相同，为了作业运行，在 Worker1 上启动了 ApplicationMaster 进程，而在其他的 Worker 节点上启动了 CoaseGrainedExecutorBackend 进程，此时可以通过 http：//Master：4040 查看日志信息。运行 Spark PI 结束后，Spark 集群的 18080 端口查看 History Server 中 Job 的运行情况，如图 17-18 所示。

查看 Hadoop 的 8088 端口，如图 17-19 所示。

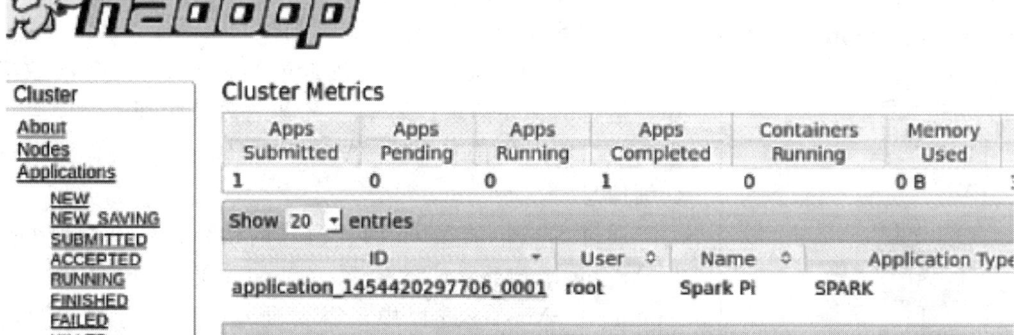

图 17-19　Client 模式下 Hadoop8088 端口情况

Client 模式运行完成后，下面开始 Cluster 模式下运行 Spark Pi，提交程序到 Spark 集群，如下所示。

```
./spark-submit --class org.apache.spark.examples.SparkPi --master yarn --deploy-mode cluster ..
/lib/spark-examples-1.6.0-hadoop2.6.0.jar 1000
```

当运行过程中没有启动 yarn 和 spark-all 的时候运行提交上述作业，会提示找不到 Server，此时集群会一直尝试 retry 连接，如图 17-20 所示。

```
etries=10, sleepTime=1000 MILLISECONDS)
NFO ipc.Client: Retrying connect to server: Master/192.168.112.219:8032. Already tried 4 time(s); retry policy is RetryUpToMaximumCount
etries=10, sleepTime=1000 MILLISECONDS)
NFO ipc.Client: Retrying connect to server: Master/192.168.112.219:8032. Already tried 5 time(s); retry policy is RetryUpToMaximumCount
etries=10, sleepTime=1000 MILLISECONDS)
NFO ipc.Client: Retrying connect to server: Master/192.168.112.219:8032. Already tried 6 time(s); retry policy is RetryUpToMaximumCount
etries=10, sleepTime=1000 MILLISECONDS)
NFO ipc.Client: Retrying connect to server: Master/192.168.112.219:8032. Already tried 7 time(s); retry policy is RetryUpToMaximumCount
etries=10, sleepTime=1000 MILLISECONDS)
NFO ipc.Client: Retrying connect to server: Master/192.168.112.219:8032. Already tried 8 time(s); retry policy is RetryUpToMaximumCount
etries=10, sleepTime=1000 MILLISECONDS)
NFO ipc.Client: Retrying connect to server: Master/192.168.112.219:8032. Already tried 9 time(s); retry policy is RetryUpToMaximumCoun
```

图 17-20　重连集群

当在 retry 的过程中启动了 yarn 后，集群自动连接，此时上传 jar 文件到 HDFS 文件系统，如下代码所示：

```
Uploading resource file:/usr/local/spark/spark-1.6.0-bin-hadoop2.6/lib/spark-assembly-1.6.0-
hadoop2.6.0.jar ->hdfs://Master:9000/user/root/.sparkStaging/application_1454420297706_
0001/spark-assembly-1.6.0-hadoop2.6.0.jar
```

在 HDFS 和 Yarn 均启动的情况下，Spark on yarn 模式下运行过程如图 17-21 所示：

```
INFO yarn.Client: Application report for application_1454420297706_0003 (state: ACCEPTED)
INFO yarn.Client: Application report for application_1454420297706_0003 (state: ACCEPTED)
INFO yarn.Client: Application report for application_1454420297706_0003 (state: ACCEPTED)
INFO yarn.Client: Application report for application_1454420297706_0003 (state: ACCEPTED)
INFO yarn.Client: Application report for application_1454420297706_0003 (state: ACCEPTED)
INFO yarn.Client: Application report for application_1454420297706_0003 (state: ACCEPTED)
INFO yarn.Client: Application report for application_1454420297706_0003 (state: ACCEPTED)
INFO yarn.Client: Application report for application_1454420297706_0003 (state: ACCEPTED)
INFO yarn.Client: Application report for application_1454420297706_0003 (state: ACCEPTED)
INFO yarn.Client: Application report for application_1454420297706_0003 (state: ACCEPTED)
INFO yarn.Client: Application report for application_1454420297706_0003 (state: RUNNING)
```

图 17-21 Spark on yarn Cluster 运行过程

通过 http：//Master：8088 查看作业运行情况，运行 Stage 如图 17-22 所示。

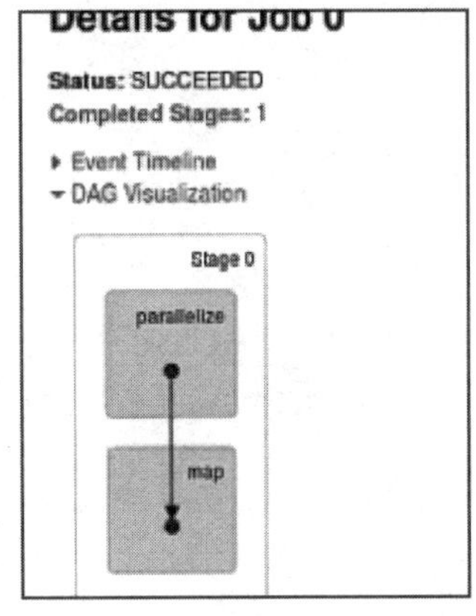

图 17-22 运行 SparkPi 产生的 DAG

因为 SparkPi 中只有 map，所以只有一个 Stage。

cluster 的模式下 driver 在 App Master 中，driver 收集了作业运行的主要信息，而 yarn 资源管理器管理了作业运行信息的端口，因此在 http：//Master：8088 中查看作业运行的详细信息。

通过 Spark on Yarn 两种模式的运行，可以看到 Driver 与 ApplicationMaster 的关系。Cluster 模式下 Driver 位于 ApplicationMaster 进程中，需要通过 Hadoop 默认指定的 8088 端口，通过 Web 控制台查看当前的 Spark 程序运行的信息，例如进度、资源的使用（Cluster 的模式中 Driver 在 App Master 中）。

Cluster 模式下，Driver 为提交代码的机器上，此时 ApplicationMaster 依旧位于集群中且只负责资源的申请和 launchExecutor，此时启动后的 Eexcutor 并不会向 ApplicationMaster 进程注册，而是向 Driver 进行注册。

从实践的角度，需要注意几点。

① 在 Spark on Yarn 的模式下 HadoopYarn 的配置 yarn. nodemanager. local-dirs 会覆盖 Spark 的 Spark. local. dir；

② 在实际生产环境下一般都是采用 Cluster，我们会通过 HistoryServer 来获取最终全部的集群运行的信息；

③ 如果想直接看运行的日志信息，可以使用以下命令：yarn logs-ApplicationId＜app ID＞。

17.4 Spark 运行架构的实战解析

Spark 默认是粗粒度，即 spark 作业提交的时候就会为我们作业分配资源，后续运行的过程中一般使用已分配的资源，除非资源发生异常需要重新分配。

Worker 进程负责当前 worker 节点上的资源的使用。从分布式的节点架构讲 Spark 是 Master/Slave 分布式的，从作业的运行架构来讲主要分为 Driver 和众多的 Worker，而从集群静态部署来讲分为 Master 和 Worker。

下面以 WordCount 为例在 Spark 集群上运行，在 IDEA 或者 Eclipse 中开发 WordCount 程序，代码如下所示。

```
import org.apache.spark.{SparkContext, SparkConf}

object WordCount {

  def main(args: Array[String]) {

    /**
     * 第一步：创建Spark的配置对象SparkConf，设置Spark程序的运行时
     * 的配置信息，例如通过SetMaster来设置程序的连接Spark集群的Master的
     * URL，若设置为local，则代表Spark程序在本地运行，特别适合于
     * 机器配置特别差的初学者。
     *
     */
    val conf = new SparkConf()
    conf.setAppName("Wow,My First Spark App!")
    conf.setMaster("local") //若将改行注释，可以运行于集群环境
    //conf.setMaster("spark://Master:7077") //绑定了程序必须运行于集群上
    //此时，程序运行于本地模式，无需Spark集群
    /**
     *第二步：创建SparkContext对象
     * SparkContext是Spark程序所有功能的唯一入口，无论采用Scala
     * Java、Pytho等不同语言；
     * SparkCo核心作用：初始化Spark应用程序运行所需要的核心组件，包括
     * DAGSchedular、 taskSchedular、SchedularBacked，
     * 同时还负责Spark应用程序中最为重要的一个对象。
     */
    val sc = new SparkContext(conf)
    //创建SparkConte对象，通过传入SparkContext实例来定制Spark运行的具体参数和配置信息。
    /**
     *第三步：根据具体的数据来源（HDFS、Hbase、Local FS、DB、S3等）通过Spark
     Context创建RDD
     * RDD的创建基本有三种方式：
     *  1、根据外部的数据来源（HDFS）
     *  2、Scala集合
     *  3、有其他的RDD操作
     * 数据会被RDD划分为一系列的Partition、分配到每个Partition的数据属于一个Task的处
     理范围
     */

    val lines = sc.textFile("C://Hello.txt",1)  //并行度为1（partition）,读取本地文件
    //根据类型推断，这边的lines是RDD[String]

    /**
     * 第四步：对初始的RDD进行Transformation级别的处理（高阶函数：map、filter、
     flatmap等）
     * 来进行具体的数据计算；
     * 4.1、将每一行的字符串拆分成单个的单词
```

```
   */
  val words = lines.flatMap(line => line.split(" "))
  //对每一行的字符串进行单词拆分并把所以行的拆分结果通过flat合并成一个大的单词集合
  /**
   * 第五步：对初始的RDD进行Transformation级别的处理（高阶函数：map、filter、
   flatmap等）
   * 来进行具体的数据计算；
   *  4.2 在单词拆分的基础上对每个单词实例计数为1，也就是word =>(word,1)
   */

  val pairs = words.map{ word => (word,1)}
  /**
   * 第六步：对初始的RDD进行Transformation级别的处理（高阶函数：map、filter、
   flatmap等）
   * 来进行具体的数据计算；
   *  4.3、在每个单词实例计数为1的基础上统计每个单词在文件中出现的总次数
   */
  val wordCounts=pairs.reduceByKey(_+_)
  //对相同的Key，进行value的累积（包括local和Reduce级别同时进行Reduce）

  // wordCounts.foreach(wordNumberPair => println(wordNumberPair._1 + ":"+
  wordNumberPair._2))
   wordCounts.collect.foreach(wordNumberPair => println(wordNumberPair._1 + ":"
 +wordNumberPair._2))
  //关闭SparkContext对象
  sc.stop()
  /**
  }
}
```

首先启动集群：

```
./start-dfs.sh
```

启动 Spark 集群：

```
./start-all.sh
```

启动 Hadoop 集群的 jobHistoryServer：

```
./mr-jobhistory-daemon.sh start historyserver
```

启动 Spark 集群的 historyserver：

```
./start-history-server.sh
```

JPS 查看进程：

```
root@Master:/usr/local/spark/spark-1.6.0-bin-hadoop2.6/sbin# jps
3675 DataNode
5020 ResourceManager
5743 HistoryServer
3553 NameNode
5170 NodeManager
5773 Jps
3832 SecondaryNameNode
5651 JobHistoryServer
```

将 WordCount 程序以 Jar 文件导出后，上传到集群中的 HDFS 上，然后在 Spark 安装目录下的 bin 目录下使用 Spark-submit 进行提交，如下所示：

```
./spark-aubmit --class com.dt.spark.WordCount_cluster --master spark://Master:7077
/root/Documents/SparkApps/WordCount.jar
```

运行完成后，将结果打印出来，如图 17-23 所示。

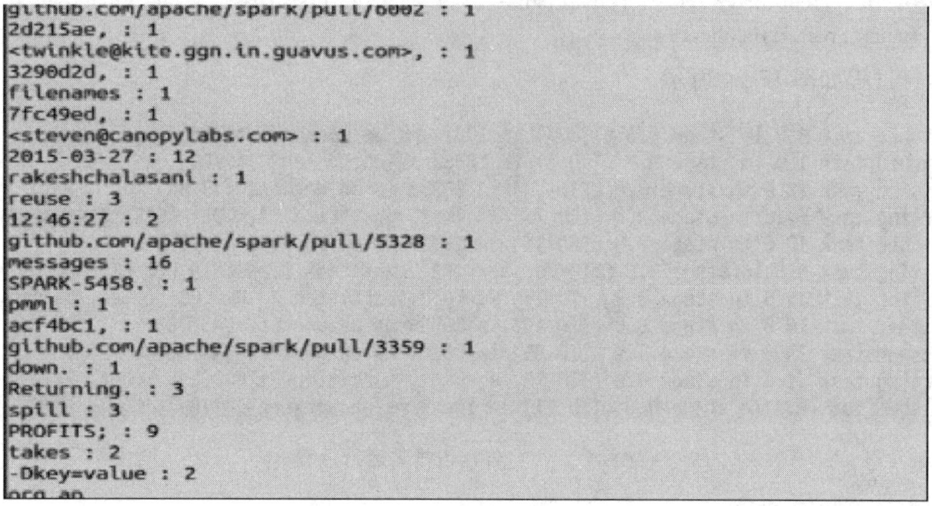

```
github.com/apache/spark/pull/6002 : 1
2d215ae, : 1
<twinkle@kite.ggn.in.guavus.com>, : 1
3290d2d, : 1
filenames : 1
7fc49ed, : 1
<steven@canopylabs.com> : 1
2015-03-27 : 12
rakeshchalasani : 1
reuse : 3
12:46:27 : 2
github.com/apache/spark/pull/5328 : 1
messages : 16
SPARK-5458. : 1
pmml : 1
acf4bc1, : 1
github.com/apache/spark/pull/3359 : 1
down. : 1
Returning. : 3
spill : 3
PROFITS; : 9
takes : 2
-Dkey=value : 2
org.ap
```

图 17-23　WordCount 运行结果

下面使用 Spark-shell 的方式分析了解 Spark 运行架构，运行模式为 Standalone，首先在作业运行前启动 HDFS 以及 Spark 集群，之后运行 Spark-shell，如下所示。

```
./spark-shell --master spark://Master:7077
```

此时 web 控制台 http：//Master：8080，多了一个 Running Applications，该 Application 有 Master 分配的 ApplicationId 和由 Master 的分配的资源（MEM、CPU 的 Cores 等），而分配的 Cores 的个数和 MEM 在 spark. env. sh 中做了配置，Spark-shell 中默认情况下没有任何的 job，但是在 Standalone 模式中默认为粗粒度的资源分配模式，提交应用程序后，Master 已经进行了粗粒度的资源分配。

使用 spark-shell 运行程序会在 Worker 上多了进程 CoarseGrainedExecutorBackend，默认情况下 Worker 节点为程序分配一个 Executor，而 CoarseGrainedExecutorBackend 进程里有 Executor，Executor 会通过并发线程池并发执行的方式执行 Task。

```
root@Worker1:~# jps
3920 Worker
3244 DataNode
2565 jps
6455 CoarseGrainedExecutorBacken
```

应用程序在提交的时候会进行注册并由 Master 分配 ID 和计算资源，无论一个应用程序中有多少的 Action 导致的作业在运行，也不会产生资源冲突的情况，因为作业运行的资源实际上是在粗粒度的方式下在程序注册的时候分配好的，多个 job 可以采取资源复用和排队执行的方式运行完应用程序。

默认的资源分配方式在每个 Worker 上启动一个 ExecutorBackend 进程，且默认情况下会最大地占用 CPU 和 MEM，若不加限制，集群上除了 Spark 还有其他程序的话，Spark 运行就会占用最大的资源，给人一种 Spark 很占内存的感觉，若有多套计算框架，就需要资源管理器 yarn 或者 mesos。

因为 Spark-shell 只是一个程序，没有 Job，所以下面在 Spark-shell 运行的基础上进行广告点击排名的运行，如下所示。

```
scala> sc.TextFile("/library/wordCount/input/Data").flatMap(_.split(" ")).map(word
=>(word,1)).reduceByKey(_+_,1).map(pair =>(pair._2 , pair._1)).sortByKey(false).map(pair
=>(pair._2,pair._1)).collect
```

运行过程如图 17-24 所示。

```
Finished task 3.0 in stage 0.0 (TID 7) in 10377 ms on Worker2 (41/88)
Finished task 1.0 in stage 0.0 (TID 3) in 10382 ms on Worker2 (42/88)
Finished task 28.0 in stage 0.0 (TID 27) in 10332 ms on Worker2 (43/88)
Starting task 84.0 in stage 0.0 (TID 75, Worker2, partition 84,NODE_LOCAL, 2175 bytes)
Finished task 18.0 in stage 0.0 (TID 15) in 10388 ms on Worker2 (44/88)
Starting task 8.0 in stage 0.0 (TID 76, Worker2, partition 8,PROCESS_LOCAL, 2162 bytes)
Starting task 29.0 in stage 0.0 (TID 77, Worker2, partition 29,PROCESS_LOCAL, 2171 bytes
Finished task 20.0 in stage 0.0 (TID 19) in 10390 ms on Worker2 (45/88)
Finished task 33.0 in stage 0.0 (TID 31) in 10367 ms on Worker2 (46/88)
Starting task 31.0 in stage 0.0 (TID 78, Worker2, partition 31,PROCESS_LOCAL, 2170 bytes
Finished task 7.0 in stage 0.0 (TID 11) in 10414 ms on Worker2 (47/88)
```

图 17-24　广告点击排名运行过程

运行结果如图 17-25 所示。

```
27 20:30:23 INFO spark.MapOutputTrackerMaster: Size of output statuses for shuffle 0 is 136 bytes
27 20:30:23 INFO scheduler.TaskSetManager: Finished task 0.0 in stage 2.0 (TID 89) in 429 ms on Worker1 (1/1)
27 20:30:23 INFO scheduler.TaskSchedulerImpl: Removed TaskSet 2.0, whose tasks have all completed, from pool
27 20:30:23 INFO scheduler.DAGScheduler: ResultStage 2 (collect at <console>:28) finished in 0.432 s
27 20:30:23 INFO scheduler.DAGScheduler: Job 0 finished: collect at <console>:28, took 14.438056 s
Array[(String, Int)] = Array(("",49018), (Commit:,3060), (INFO,2707), (to,1834), (-0800,1593), (in,1558), (the
 (0,1036), (2015-08-30,1028), (of,1027), (and,982), (#,974), ("   at",948), (-0700,927), (-,769), (org.apache.
heReplicationMonitor:,641), (at,486), (from,479), (a,466), (=,456), (is,435), (after,409), ([SQL],393), (on,3
, (milliseconds,337), (with,332), (Rescanning,318), (block(s),318), (millisecond(s).,318), (directive(s),318),
ordCount.java:67),314), (Reduce,313), ((SUM(1),,309), (The,286), (by,269), (not,243), (file,241), (Add,241), (
,227), (1,224), (org.apache.flink.runtime.executiongr...
```

图 17-25　广告点击排名的运行结果

因为广告点击排名中使用了 collect 算子的 action 操作，所以会触发 Job，DAGScheduler 会将 RDD 划分成不同的 Stage，划分的依据为宽依赖导致的 Shuffle，关于 Stage 的划分前面已经详细介绍过，划分完 Stage 后将形成的 TaskSet 交由 TaskScheduler 的 TaskSetManager 进行管理，并为其分配计算资源，由 ExecutorBackend 启动 Worker 节点上的 Executor，最后在线程池中并发执行 Task。如图 17-26 所示 Stage 的划分以及 DAG 的形成。

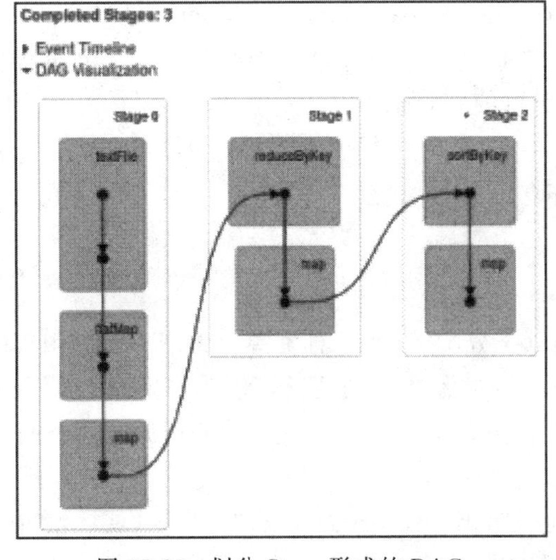

图 17-26　划分 Stage 形成的 DAG

从上面运行的 job 可以看出一个 job 产生了三个 stage（数据在 Stage 内部是 pipeline 流过去的，依次是 HadoopRDD、MapPartitionRDD、MapPartitionRDD、ShuffledRDD...），两个 shuffledRDD，因为 shuffledRDD是宽依赖（每个父RDD都被子RDD使用），因为遇到宽依赖就断开，遇到窄依赖就把当前的RDD加入该Stage中，所以形成三个stage。

▸ Show Additional Metrics
▸ Event Timeline

Summary Metrics for 88 Completed Tasks

Metric	Min	25th percentile	Median
Duration	52 ms	0.2 s	0.5 s
GC Time	0 ms	0 ms	0 ms
Input Size / Records	0.0 B / 0	716.0 B / 12	1335.0 B / 27
Shuffle Write Size / Records	0.0 B / 0	609.0 B / 61	1239.0 B / 117

从Stage中可以看出，一共有88Task，因为这里有88个文件且每个文件大小均小于HDFS默认的文件块BLOCK大小，即128MB，每个文件就是一个Partition，共有88个Partitions，而默认情况下一个Task对应执行一个Partition，所以就有88个Task，而每个Task都运行在Executor中，并发复用执行。每个Worker运行多少个Task，主要由于数据本地行，数据在哪尽量在哪儿排队并发运行，以此减少网络传输IO，减少通信频率。

一次性最多在Executor中能运行多少个并发的task，取决于当前节点Executor中Cores数量，在实际运行的时候哪个task执行完成，就会将资源回收到线程池中进行复用，对于一次没法全部运行的任务，就会形成task排队的情况，为了应对这种情况，优化的方法（避免oom），指定多个Executor线程池，增加分片数量，每个分片中的数据就小，所以减少OOM情况，获取更多的MEM和资源，但前提是Spark运行在拥有其他大数据框架的集群中。

下面以Cache的情况运行，Cache过后会新生成一个Job，如下所示。

```
1. scala> val cached = sc.TextFile("/library/wordCount/input/Data").flatMap(_.split(" ")).map(word
   =>(word,1)).reduceByKey(_+_,1).map(pair =>(pair._2 , pair._1)).cache
```

Storage

RDDs

RDD Name	Storage Level	Cached Partitions
ShuffledRDD	Memory Deserialized 1x Replicated	1

由于进行了Cache操作，所以讲ShuffledRDD Cache到了内存中，且只有一个副本，因为数据不足128MB，所以只有一个Partition，对应一个Task，这里Executor在Worker2上，那么作业运行在Worker2上，充分说明了运行任务的时候基于数据本地性。下面测试在ReduceByKey的时候，不传入第二个参数，即不指定作业运行的并行度，分别运行以下代码。

```
2. scala> val cached = sc.TextFile("/library/wordCount/input/Data").flatMap(_.split(" ")).map(word =>
   (word,1)).reduceByKey(_+_).map(pair => (pair._2 , pair._1)).cache
3. scala> cached.map(pair = > (pair._2,pair._1)).collect
```

划分Stage形成的DAG如图17-27所示。

可以看出DAG中Stage8已经kipped了，是Application内部进行的优化，查看job得知本次作业运行产生了88个并行度，即88个Partition，因为不指定并行度的话，作业会继承并行度。

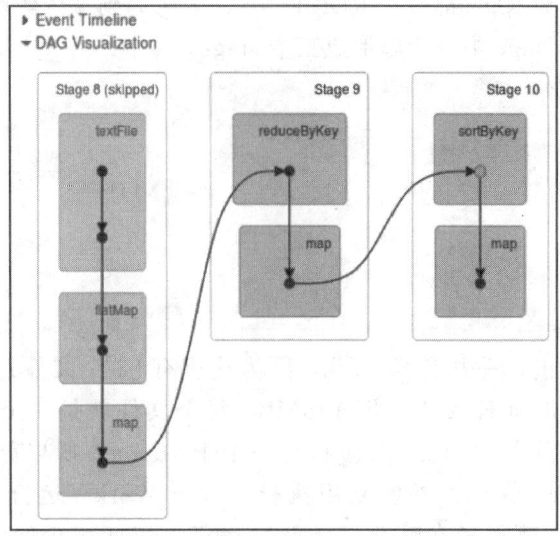

图 17-27　划分 Stage 形成的 DAG

下面将开发好的广告点击排名程序发布到集群中运行，代码如下。

```
import org.apache.spark.SparkConf
import org.apache.spark.SparkContext

/**
  * @author css-kxr
  * HomeWord：广告点击排序
  */
object HitCount {
  def main(args:String){
    /**
      * first:create new object of SparkConf,There is none Parameters
      */
    val conf = new SparkConf()
    conf.setAppName("That is so cool Rank the Hit rates advertisements")

    /**
      *Second:create new object of SparkContext that is a only program Entrance to run
      */
    val sc = new SparkContext(conf)

    /**
      * third：According to the data resource HDFS,S3& DB to create RDD
      * textFile("Path of files",1)  number 1 = one Partition
      */
    val lines = textFile("C://spark.kong.dt.job//spark-1.6.0-bin-hadoop-2.6.0//README.MD",
)

    /**
      * fourth:Ececute the lines like mapmr filter and so o
      */
```

```
    val rank = lines.flatMap(line => line.split(" ")).map(word =>(word,1)).reduceByKey(_+_).map
  (w =>(w._2,w._1))
      .sortByKey(false).collect()
    val sort = rank.reverse.map(words =>(words._2,words._1)).foreach(println)
    sc.stop()
  }
}
```

在实际的开发中，一般都会在 Windows 系统中开发后以 Jar 文件的形式导出后，同时上传到集群 HDFS 文件系统中，然后在集群中运行。

从 CorseGrainedExecutorBackend 的角度来看是 Worker Process 来管理当前节点上的 MEM 和 CPU，但是真正管理资源的是 Master，Worker Process 只是走个形式，因为 CorseGrainedExecutorBackend 进程是被 Worker Process 分配的，实质上是通过 Master 来管理 Worker 节点的资源。

每个 Worker 上包含一个或者多个 ExecutorBackend 进程，而每个 ExecutorBackend 中包含一个 Executor 对象，该对象拥有一个线程池，而每个线程又可以使用多个 Task 任务。

从 DAG 逻辑角度来看数据在 Stage 内部是 pipeline 流过去的，因为有两次 ShuffledRDD，所以 job 被划分成三个 Stage。

从 Hadoop 的 MapReduce 角度看 stage0 是 stage1 的 mapper；而 stage1 是 stage0 的 reducer；stage1 是 stage2 的 mapper；而 stage2 是 stage1 的 reducer，可以将 Spark 看做一个 MapReduce 的更加具体的实现。

17.5 小结

总体而言，spark 在集群启动的时候，有个全局的资源管理器 Master，负责整个集群资源的管理以及接受程序提交并为程序分配资源，而每个 Worker 节点上都一个 Worker Process 来管理当前机器上的计算资源，当应用程序提交的时候，Master 就会为所提交的应用程序在每个节点上默认分配一个 CorseGrainedExecutorBackend 进程，然后会最大化地应用当前机器的 MEM 和 CPU，当 Driver 实例化没有问题的时候，Driver 本身会进行作业的调度来驱动 CorseGrainedExecutorBackend 中的 Executor 中的线程，来具体并发执行 task，这就是 Spark 并发执行的过程。

Spark 对批处理、实时流处理、交互式查询、图计算有几乎完美的集成，因而 Spark 避免了多种运算场景下需要部署不同集群带来的资源损耗，它为分布式数据集的处理提供了一个有效的框架，基于此框架使用者可以完成多种体量巨大、复杂数据的计算和处理。由此也奠定了 Spark 在大数据计算领域的霸主地位，Spark 也因此逐渐成为未来大数据领域最有价值的技术。